Lecture Notes in Mathematics

Edited by A. Dold and B. Eckmann

1380

H. P. Schlickewei E. Wirsing (Eds.)

Number Theory

Proceedings of the Journées Arithmétiques
held in Ulm, FRG, September 14–18, 1987

Springer-Verlag
New York Berlin Heidelberg London Paris Tokyo Hong Kong

Editors

Hans Peter Schlickewei
Eduard Wirsing
Abteilung für Mathematik, Universität Ulm
Oberer Eselsberg, 7900 Ulm, Federal Republic of Germany

Mathematics Subject Classification (1980): 11-XX, 11-06

ISBN 3-540-51397-3 Springer-Verlag Berlin Heidelberg New York
ISBN 0-387-51397-3 Springer-Verlag New York Berlin Heidelberg

Printed in Germany

Printing and binding: Druckhaus Beltz, Hemsbach/Bergstr.
2146/3140-543210

Résumé
and
Acknowledgement

The 15th Journées Arithmétiques were held at the University of Ulm in 1987. 148 number theorists had joined here to exchange the results of their work. The mornings were devoted to the 10 main lectures delivered to the plenum, while the afternoon lectures had to run in three parallel sections; there were 68 of them. As an organizer I tried to secure as much space and time as possible for any type of personal contact, and I like to think that this part of the conference may not have been the least fruitful one.

A proceedings volume that is refereed like any scientific journal may not duplicate work that is being published elsewhere. Thus many interesting lectures given at the conference are not found in here. But the volume gives a good cross-section of the Journées Arithmétiques 1987 and thereby of present activity in number theory.

The conference could happen only because of various contributions from many sides:

The basic financial support came from

Deutsche Forschungsgemeinschaft,
Centre National de la Recherche Scientifique,
Ministerium für Wissenschaft und Kunst Baden-Württemberg,
Ambassade de France en République Fédérale d'Allemagne,
Ulmer Universitätsgesellschaft.

Further generous funding and material help was received from

Bausparkasse Schwäbisch Hall AG,
Commerzbank AG, Ulm
Robert Bosch GmbH, Stuttgart,
Daimler Benz AG, Stuttgart,
Dornier GmbH, Stuttgart,
I B M Stuttgart,
Jacobs Suchard, Bremen,
Mineralbrunnen Überkingen-Tainach AG, Bad Überkingen,
Sparkasse Ulm,
Springer-Verlag, Berlin, Heidelberg, New York, Tokyo,
Ulmer Volksbank,
Vereinigte Eos-Isar Lebensversicherung AG, München,
Vereinigte Krankenversicherung AG, München,
Voith GmbH, Heidenheim,
Weinwerbezentrale Badischer Winzergenossenschaften, Karlsruhe,
Württembergische Metallwarenfabrik AG, Geislingen/Steige.

The University of Ulm provided rooms and the assistance of its administration.

We cordially thank all institutions, firms and persons that made the Journées Arithmétiques 1987 in Ulm possible.

Personally I wish to thank Professor G. Henniart who successfully battled the authorities and generally organized everything on the French side and Dr. J.H. Goguel and Frau W. Boremski, from the Mathematics Department II in Ulm. Without their constant effort I could not possibly have managed the conference.

E. Wirsing

CONTENTS

APPLICATIONS OF CAYLEY-CHOW FORMS

W. Dale Brownawell*
Department of Mathematics, Penn State University
University Park, PA 16802, USA

I. Introduction.

During the past ten years, Cayley-Chow forms have become a powerful tool for effective elimination. The purpose of this article is to make these recent developments available to non-specialists and to survey the applications which have already been attained. The Cayley-Chow form possesses a venerable geometric history and has appealingly intuitive properties. Indeed, the coordinates of any zero of the Cayley-Chow form of a variety of dimension d are simply the coefficients of $d+1$ hyperplanes having a common point on the variety.

Since potential users of Cayley-Chow forms sometimes seem inhibited by foundational uncertainties, we will be rather complete in our exposition of the basic concepts in Section II below. We state clearly the basic tools and give an indication of the most straightforward applications in Sections III and V. The scheme of proof of Philippon's criterion is more intricate and, besides, has already been given in the terminology adopted here in [B6].

Yu.V. Nesterenko has shown in a series of papers [N2]-[N4] that the Cayley-Chow form responds to many questions of effective elimination which could formerly be approached only through the classical resultant in low(est) dimension. In particular he realized that one could define the valuation of a Cayley-Chow form at a point of an ambient space, and he proved that any smallness would be preserved during elimination corresponding to hypersurface intersection.

Nesterenko initially applied these ideas in [N2] to bound the order of zero of a polynomial in a solution of a system of linear first-order differential equations. This important bound partially effectivized the crucial Shidlovsky lemma in the Siegel-Shidlovsky theory for algebraic independence.

In his thesis [P2], P. Philippon used the Cayley-Chow form to perfom elimination for the elliptic Lindemann theorem. Then in [P3],[P4] he adopted and adapted Nesterenko's constructions and proved two more very important properties of Cayley-Chow forms. These properties were used to obtain his deep generalization of Gelfond's criterion and ensuing results on algebraic independence.

Later the present author was able to employ the ideas of Nesterenko and Philippon to obtain a local generalization of Liouville's inequality [B5] giving a somewhat more direct approach [B6],[BT] to the applications to independence of Philippon. A variant of this inequality then provided the basis for the (analytic!) proof [B2] of the existence of coefficients in the Nullstellensatz satisfying essentially optimal bounds on their degrees, as was kindly pointed out to the author by C. Berenstein and A. Yger.

*Research supported in part by an NSF grant.

II. Definition and Properties of Cayley-Chow Forms

A. Planes through a Point

Because the Cayley-Chow form is used to express the condition that hyperplanes have a point in common on a given variety, we make a preliminary elementary remark on the representation of an arbitrary hyperplane in projective space $\mathbb{P}_n(k)$ through a given point $\underline{x} = [x_0:\ldots:x_n]$. First of all, note that all linear relations $\underline{u}\cdot\underline{x} = \Sigma_i\ u_i x_i = 0$ on the coordinates of $\underline{x}$ are generated by the relations between the various pairs of coordinates:

$$\underline{u}_{jk} = (\ldots 0 \ldots \underset{j}{x_k} \ldots 0 \ldots \underset{k}{-x_j} \ldots 0) = \begin{array}{c} \vdots \\ j \\ \vdots \\ k \\ \vdots \end{array} \underbrace{\underset{\ldots j \ldots k \ldots}{\begin{bmatrix} * & & 1 \\ & & \\ -1 & & * \end{bmatrix}}}_{\sigma_{jk}} \begin{bmatrix} x_0 \\ \vdots \\ x_n \end{bmatrix},$$

$0 \leq j < k \leq n$. Since a generic skew symmetric matrix S has the form

$$S = \sum_{j<k} s_{jk}\sigma_{jk}$$

for indeterminants s_{jk}, the coefficients $\underline{u}$ of a generic hyperplane passing through $\underline{x}$ are given by $\underline{u} = S\underline{x}$.

B. Cayley-Chow Forms

Let R be a unique factorization domain with field of quotients k of characteristic zero. Let V be a variety (irreducible) in $\mathbb{P}_n$ of dimension d corresponding to a homogeneous prime ideal $\mathfrak{P}$ in $k[\underline{x}] = k[x_0,\ldots,x_n]$. For $j = 0,\ldots,d$ let $\underline{u}_j = (u_{j0},\ldots,u_{jn})$ be an n+1-tuple of new variables and denote by H_j the hyperplane

$$H_j: \quad u_{j0}x_0 + \cdots + u_{jn}x_n = 0.$$

Then the hyperplanes $H_0,\ldots,H_{d-1}$ intersect V in $\deg \mathfrak{P}$ points (this is one standard definition of $\deg \mathfrak{P}$), say $\underline{\alpha}_k = (\alpha_{k0}:\ldots:\alpha_{kn})$, $k = 1,\ldots,\deg \mathfrak{P}$. One can always normalize the choice of projective coordinates by demanding that a certain non-zero one of them be equal to 1. These points can be considered to be "generic zeros" of $\mathfrak{P}$, and their coordinates are algebraic over $\mathfrak{R} = R[\underline{u}_0,\ldots,\underline{u}_{d-1}]$. Any automorphism of the algebraic closure of the field of quotients $\mathfrak{F}$ of $\mathfrak{R}$ leaves the equations for $V, H_0,\ldots,H_{d-1}$ invariant and therefore permutes the $\underline{\alpha}_k$. In fact it can be shown (Lemma 2, [N3]) that the $\underline{\alpha}_k$ are conjugate over $\mathfrak{F}$. Consequently the product $\Pi_k(\alpha_{k0}u_{d0}+\ldots+\alpha_{kn}u_{dn})$ is an irreducible element of $\mathfrak{F}[\underline{u}_d]$. We can therefore choose $a \in k(\underline{u}_0,\ldots,\underline{u}_{d-1})$ (and even in $R[\underline{u}_0,\ldots,\underline{u}_{d-1}]$) so that the polynomial

$$F(\underline{u}_0,\ldots,\underline{u}_d) = a\ \Pi_k\ (\alpha_{k0}u_{d0}+\ldots+\alpha_{kn}u_{dn}) \tag{2.1}$$

has coefficients in R, but no (non-unit) factors in $\mathfrak{R}$. We shall call it an (R-*integral*) *Cayley-Chow form of* V or $\mathfrak{P}$.

The condition

$$F(\underline{u}_0,\ldots,\underline{u}_d) = 0$$

is precisely the condition that H_d pass through a point of $V \cap H_0 \cap \ldots \cap H_{d-1}$, i.e. that $H_0,\ldots,H_d$ intersect in at least one point of V. By the symmetry of this condition, the irreducibility of F as a polynomial in $\underline{u}_d$, and the lack of factors from $\mathfrak{R}$, we see that F is invariant up to a factor from R under permutations of $\underline{u}_0,\ldots,\underline{u}_d$. In particular

$$(\text{total}) \deg F = (d+1)\deg \mathfrak{P}.$$

More generally we extend the notion of R-integral Cayley-Chow form to any product

$$F = F_1^{e_1}\ldots F_r^{e_r}$$

of R-integral Cayley-Chow forms F_i of homogeneous prime ideals of $k[\underline{x}]$ (of the same dimension for our purposes).

There is a way of assigning such a Cayley-Chow form to any unmixed homogeneous ideal of $k[\underline{x}]$ (see [N2]), but we shall not need that here. We remark that the more usual definitions took R to be a field [VdW] or a PID [N2],[N3],[P4], but that we plan to use the more general situation in a later paper. Another contrast to [P4] is that Philippon takes hypersurfaces of arbitrary degree for his H_i.

C. Degrees and Valuations of Cayley-Chow Forms

One can associate some invariants with such an F. For example there is always the partial degree $\delta(F) = \deg_{\underline{u}_i} F$, $i = 0,\ldots,d$. We will be interested in the cases $R = \mathbb{Z}$ and $R = \mathbb{C}[z]$ where R has a valuation. Then we have the height of F defined as $H(F) = \max |\text{coeff of } F|$ and coefficient degree $\deg_z F$, respectively.

If there were a canonical way of chosing a basis for ideals, we could also define the valuation of a homogeneous ideal $\mathfrak{J}$ at a projective point $\underline{z}$ (with coordinates in a field extension K of k with valuation extending one on k) simply by

$$\|\mathfrak{J}\|_{\underline{z}} = \max |B(\underline{z})|/\|\underline{z}\|^{\deg B},$$

where B runs over the canonical basis of $\mathfrak{J}$ and the denominator is introduced to ensure definition on projective space. Here we mean that $\|\underline{z}\| = \max |z_i|$ under the valuation on K (in particular in $\mathbb{C}\langle\langle z\rangle\rangle$, $|f| = \exp(-\text{ord } f)$, where ord denotes the "order of zero" of the power series).

Therefore before we proceed, we want to consider how closely the Cayley-Chow form F of a prime homogeneous ideal $\mathfrak{P}$ (still of dimension d). determines $\mathfrak{P}$. Let $S^{(0)},\ldots,S^{(d)}$ be d+1 generic skew symmetric matrices, $S^{(i)} = (s_{jk}^{(i)})$, and write

$$F(S^{(0)}\underline{x},\ldots,S^{(d)}\underline{x}) = \Sigma\, p_\sigma(x)\sigma, \qquad (2.2)$$

where now σ runs through the monomials in the $s_{jk}^{(i)}$, $0 \le j < k \le n$, which are homogeneous of degree $\delta(F)$ for each $i = 0,\ldots,d$. Let $z \in \mathbb{P}_n(K)$, where K is an extension field of k. Then since $S^{(0)}\underline{z},\ldots,S^{(d)}\underline{z}$ are $d+1$ generic hyperplanes through $\underline{z}$,

$$F(S^{(0)}\underline{z},\ldots,S^{(d)}\underline{z}) = 0 \iff \underline{z} \text{ is a zero of } \mathfrak{P}.$$

But clearly also in terms of (2.2),

$$F(S^{(0)}\underline{z},\ldots,S^{(d)}\underline{z}) = 0 \iff \underline{z} \text{ is a zero of all } p_\sigma(x);$$

in other words, $\mathfrak{P} = \text{radical}\langle..p_\sigma(x)..\rangle$. In fact it is a theorem of Krull (Lemma 11 of [N2]) that

$$\langle..p_\sigma(\underline{x})..\rangle = \mathfrak{P}\cap\mathfrak{E},$$

where $\mathfrak{E} = k[\underline{x}]$ or else $\mathfrak{E}$ is embedded in $\mathfrak{P}$. Thus the polynomials p_σ are tantamount to a canonical basis of $\mathfrak{P}$. Now we can apply a slight modification of the original strategy for defining the absolute value of an ideal.

For any Cayley-Chow form F we make the sustitution of equation (2.2) and set

$$\|F\|_{\underline{z}} = \frac{\max |p_\sigma(\underline{z})|}{\|\underline{z}\|^{\deg F}}.$$

Philippon [P4] takes the Mahler measure of the analog of $F(S^{(0)}\underline{x},\ldots,S^{(d)}\underline{x})$ when working in $\mathbb{C}$ or $\mathbb{C}_p$ to establish counterparts of all the remaining results of this section.

D. Principal Ideals

1. Chow Forms of Principal Ideals

Let $P(\underline{x}) \in R[\underline{x}]$ be homogeneous and irreducible of degree d and consider the n generic hyperplanes

$$\begin{array}{ccc} u_{00}x_0 + \cdots + u_{0n}x_n & = & 0 \\ \vdots & & \\ u_{n-1,0}x_0 + \cdots + u_{n-1,n}x_n & = & 0, \end{array}$$

i.e. hyperplanes with undetermined coefficients u_{ij}. According to Cramer's rule, they meet in a unique point whose projective coordinates can be taken as the cofactors Δ_i of the x_i in

$$\begin{bmatrix} x_0, & \cdot & \cdot & \cdot & \cdot & , x_n \\ u_{00}, & \cdot & \cdot & \cdot & \cdot & , u_{0n} \\ \vdots & & & & & \\ u_{n-1,0}, & \cdot & \cdot & \cdot & & , u_{n-1,n} \end{bmatrix}.$$

Therefore the condition that this point lies on the hypersurface $P = 0$ is that $P(\Delta_0,\ldots,\Delta_n) = 0$. Thus the form

$$F(\underline{u}_0,\ldots,\underline{u}_{n-1}) = P(\Delta_0,\ldots,\Delta_{n-1})$$

is a Cayley-Chow form of the principal ideal (P) multiplied by a GCD of the coefficients of P itself. Clearly $\delta(F) = \deg_{x_i} P$, $i = 0,\ldots,n$ and

$$H(F) \le H(P)((n+1)!)^{\delta(F)} \text{ when } R = \mathbb{Z}, \quad \deg_z F \le \deg_z F \text{ when } R = \mathbb{C}[z].$$

2. Valuation of a Cayley-Chow Form of a Principal Ideal

Lemma. $[x_0:\ldots:x_n] = [\Delta_0':\ldots:\Delta_n']$, *where* Δ_i' *is the cofactor of* x_i *in*

$$\begin{bmatrix} \underline{x} \\ S^{(0)}\underline{x} \\ \vdots \\ S^{(n-1)}\underline{x} \end{bmatrix} = \begin{bmatrix} x_0 & \ldots & x_n \\ \Sigma_j s_{0j}^{(0)} x_j & \ldots & \Sigma_j s_{nj}^{(0)} x_j \\ \cdot & \cdot\ \cdot\ \cdot & \cdot \\ \Sigma_j s_{0j}^{(n-1)} x_j & \ldots & \Sigma_j s_{nj}^{(n-1)} x_j \end{bmatrix}.$$

Proof. By the remarks in the first section, the vectors $S^{(0)}\underline{x},\ldots,S^{(n-1)}\underline{x}$ form a basis for the dual space of $\mathbb{C}^{(n+1)}\cdot\underline{x}$. Thus

$$\det \begin{bmatrix} 0\ldots0 & \overset{i}{x_j} & 0\ldots0 & \overset{j}{-x_i} & 0\ldots0 \\ & & S^{(0)}\underline{x} & & \\ & & \vdots & & \\ & & S^{(n-1)}\underline{x} & & \end{bmatrix} = 0,$$

and therefore $x_j\Delta_i' - x_i\Delta_j' = 0$. Consequently

$$F(S^{(0)}\underline{x},\ldots,S^{(n-1)}\underline{x}) = Q(\Delta_0',\ldots,\Delta_n') = \left(\frac{\Delta_i'}{x_i}\right)^{\deg Q} Q(x_0,\ldots,x_n).$$

Thus when $R = \mathbb{Z}$, we see that

$$\|F\|_{\underline{z}} \le \|Q\|_{\underline{z}} n^{\deg Q},$$

where $\|Q\|_{\underline{z}} = |Q(\underline{z})|/(\max|z_i|)^{\deg Q}$. In the non-archimedean case, $\|F\|_{\underline{z}} \le \|Q\|_{\underline{z}}$.

E. Resultants

Let $Q \in R[\underline{x}]$ be homogeneous and F the R-integral Cayley-Chow form appearing in (2.1). Then we define the *resultant* $R(F,Q)$ to be the product

$$R(F,Q) = a^{\deg Q}\prod_k Q(\alpha_{k0},\ldots,\alpha_{kn}).$$

Nesterenko established the following properties of the resultant:

1. *If* $d > 0$, *then* $R(F,Q)$ *is a product of Cayley-Chow forms associated to the minimal prime components of* $(\mathfrak{P},Q)$.
2. *If* $d > 0$, *then* $\delta(R(F,Q)) = \delta(F)\cdot\deg Q$.
3. *If* $R = \mathbb{C}[z]$, *then*

$$\deg_z R(F,Q) \le \deg_z F\cdot\deg Q + \delta(F)\deg_z Q,$$

$$\operatorname{ord} R(F,Q) \ge \min\{\operatorname{ord} F,\ \operatorname{ord} Q\}.$$

4. *If* $R = \mathbb{Z}$, *then for an explicit constant* c, *depending on* n,

$$H(R(F,Q)) \le H(F,Q) = H(F)^{\deg Q}H(Q)^{\delta(F)}\exp\{\delta(F)\deg Q\},$$

$$\|R(F,Q))\|_{\underline{z}} \le H(F,Q)\max\{\|F\|_{\underline{z}},\|Q\|_{\underline{z}}\}.$$

The proofs are given in Lemma 4 [N3], Lemmas 5 and 6 and equation (28) of [N4]. There only an inequality is asserted for (2), but, as remarked in [B2], the proof gives the equality which had been noticed earlier in [P6].

This result is the principal tool for our applications. It allows one to reduce the dimension while controlling size. Philippon established two further powerful inequalities, which we will come to later.

III. Zero Estimates for Linear Differential Equations

A. The Siegel-Shidlovsky Setting

In 1928 C.L. Siegel began the investigation of the algebraic independence of values of E-function solutions of systems of linear first order differential equations (see [Sh] for a complete exposition of the results mentioned in this section)

$$\begin{aligned} y_1' &= a_{11}y_1 + \cdots + a_{1n}y_n \\ &\cdots\cdots \\ y_n' &= a_{n1}y_1 + \cdots + a_{nn}y_n, \end{aligned} \tag{3.1}$$

$a_{ij} \in L(z)$, $[L:\mathbb{Q}] < \infty$. A prototypical E-function is represented by the generalized hypergeometric function

$${}_pF_{q-1}(z^{q-p}) = \Sigma\ \frac{(\mu_1)_n\cdots(\mu_p)_n}{(\nu_1)_n\cdots(\nu_{q-1})_n}\ \frac{\underline{z}^{(q-p)n}}{n!}$$

$\mu_i, \nu_j \in \mathbb{Q}$, $-\nu_j \notin \mathbb{Z}_{\geq 0}$.

Although Siegel could completely treat the cases $n = 1,2$ of (3.1) and systems which decomposed into such subsystems, the general case was settled in a certain sense by A.B. Shidlovsky in 1955 [Sh]. Shidlovsky showed that the values of an E-function solution $(f_1,\dots,f_n)$ of (3.1) at an algebraic non-singular point α of the system would be algebraically independent exactly when the functions themselves were algebraically independent over $L(z)$. The crucial advance established a general lower bound on the rank of a certain matrix in terms of the order of zero of an auxiliary function; this step has become known as Shidlovsky's Lemma.

B. Zero Estimates

Nesterenko realized that to obtain effective dependence on degree in a quantitative analogue of Shidlovsky's independence results, it would suffice to be able to bound the order of zero at the origin of an arbitrary polynomial expression $P(z,f_1,\dots,f_n)$. To this end, he introduced the operator

$$D = t\Sigma_i(a_{i1}x_1+\dots+a_{in}x_n)\frac{\partial}{\partial x_i} + t\frac{\partial}{\partial z},$$

where t is the least common denominator for the a_{ij}, on the ring $T = \mathbb{C}[z,x_1,\dots,x_n]$ to mirror differentiation in $\mathbb{C}[z,f_1,\dots,f_n]$:

$$DP|_{\underline{x}=\underline{f}} = t\frac{d}{dz} P(z,f_1,\dots,f_n).$$

He defined $\text{ord } P = \text{ord}_{z=0}P(z,f_1,\dots,f_n)$ and for a prime ideal $\mathfrak{P}$ in T, we have the definition from Section II above of $\text{ord } \mathfrak{P}$ via the Chow form of its homogenization in $\mathbb{C}[z,x_0,\dots,x_n]$. The following is a major result of [N1]:

THEOREM. (Nesterenko) *If* $f_1,\dots,f_n$ *are algebraically independent* (*over* $\mathbb{C}(z)$) *solutions of* (3.1), *then there is a constant* c^* *depending on* (3.1), *and* $f_1,\dots,f_n$ *such that*

$$D\mathfrak{P} \subset \mathfrak{P} \Rightarrow \text{ord } \mathfrak{P} \leq c^*.$$

The zero estimate now attainable using results in the literature is the following:

THEOREM. *There is a constant* c, *depending effectively on* c^* *and on* (3.1) *such that if* $P \in \mathbb{C}[z,x_1,\dots,x_n]$ *is non-zero, then*

$$\text{ord}_{z=0}P(z,f_1,\dots,f_n) \leq c\delta_0\delta^n + c^*\delta^{2n},$$

where $\delta_0 = \deg_z P$, $\delta = \deg_{\underline{x}} P$.

Note that this result is sharp except for the δ^{2n} term. Nesterenko's original result [N2] had $c\delta_0\delta^{(n+1)^{(n+1)}}$ on the right, which however for transcendence applications is practically as good. In the meantime somewhat more general results have been established by C.F. Osgood [O] and by D. Bertrand and F. Beukers [BB] based on an approach of G.V. Chudnovsky. Very recently Nesterenko has announced an essentially optimal result [N9] even in the very general case that the functions satisfy only a system of differential equations in which the polynomials on the right hand side of (3.1) can be polynomials of higher degree.

Outline of Proof: We prove inductively that if ord P is large enough, then there exist polynomials $L_1(=P), L_2, \ldots, L_n$ such that each $L_i \in \mathbb{Z}P+\mathbb{Z}DP+\ldots+\mathbb{Z}D^{e_i}P$, $e_i = \delta+\ldots+\delta^{i-1}$ and a Cayley-Chow form F_i with the following properties:

Each F_i is a power product of the Cayley-Chow forms of the the isolated prime components (of dimension $n-i$ over $\mathbb{C}(z)$) of $I_i = ({}^hL_1, \ldots, {}^hL_i)$ having ord > c* (assumed > 0 for simplicity) and

i) $\operatorname{ord} F_i \geq \operatorname{ord} P - c^*(\delta+\delta^2+\ldots+\delta^i)$,

ii) $\deg_z F_i \leq i\delta_0\delta^{i-1} + \tau((i-1)\delta^i+(i-2)\delta^{i+1}+\ldots+\delta^{2(i-1)})$, where $\tau = \max\{\deg t, \deg ta_{ij}\}$,

iii) $\delta(F) \leq \delta^i$ if $i \leq n$.

Case i = 1.

The assertions all follow from the remarks on the Cayley-Chow form of a principal ideal once we delete all factors Q of P with $\operatorname{ord} Q \leq c^*\deg_{\underline{x}} Q$.

Induction Step.

The induction relies on two basic facts, the first of which follows from putting together ideas from [N1],[N2],[BM]:

a) If $\mathfrak{Q}$ is an isolated $\mathfrak{P}$-primary component of the $\underline{x}$-homogeneous ideal $\mathfrak{A}$ of T, then either $D\mathfrak{P} \subset \mathfrak{P}$ or else $D^e\mathfrak{A} \not\subset \mathfrak{P}$ for some $e \leq \exp(\mathfrak{Q})$. The second result that we need is a consequence of the development of bounds on the exponent of primary components whose generators have degree satisfying known bounds [BM],[MW],[P5],[B7]:

b) If $\mathfrak{Q}$ is an isolated $\mathfrak{P}$-primary component of I_i with $z \notin \mathfrak{P}$, then $\exp(\mathfrak{Q}) \leq \delta^i$. Furthermore there are at most δ^i such components.

Let us assume now that the claim has been established for some $i \leq n$. By a) if $\mathfrak{Q}$ is an isolated $\mathfrak{P}$-primary component of I_i, then some $D^jD^{e_k}P \notin \mathfrak{P}$ for some $1 \leq j \leq \delta^i$ and $0 \leq k \leq i$, i.e. $D^\ell P \notin \mathfrak{P}$ for $\ell \leq \delta+\ldots+\delta^i = e_{i+1}$. By the technique of taking sufficiently general $\mathbb{Z}$-linear combinations (see [BM], [MW]), we find $L_{i+1} \in \mathbb{Z}P+\mathbb{Z}DP+\ldots+\mathbb{Z}D^{e_{i+1}}P$ not lying in any underlying prime ideal of F_i.

Now when $i < n$, produce F_{i+1} by removing all irreducible factors from the resultant $R(F, L_{i+1})$ having ord $\le c^*$. Using the fundamental inequalities for resultants, we see that properties i)-iii) follow by induction. On the other hand, when $i = n$, we see that $F_{i+1} = F_{n+1} \in \mathbb{C}[z]$ and so

$$\text{ord } F_{n+1} \le \deg F_{n+1} \le (n+1)\delta_0\delta^n + t(n\delta^{n+1} + \dots + 2\delta^{2n-1} + \delta^{2n}).$$

Combining this inequality with part i) gives the inequality claimed.□

The original result of Nesterenko was extended to the case that the coordinates of $\underline{f}$ are not necessarily algebraically independent over $\mathbb{C}(z)$ by N.G. Tai and by the author (unpublished). A.B. Shidlovsky implicitly and S. Lang explicitly gave measures of algebraic independence in the general Siegel-Shidlovsky setting in terms of the height of the polynomials. Nesterenko's work [N1],[N2] established the dependence on the degree as well.

Very recent work of various authors promises to completely effectivize these results by different approaches.

IV. The Gelfond-Philippon Criterion

A. Gelfond's Criterion

The other classical method for algebraic independence was developed by A.O. Gelfond [G]. It used the upper bound for the number of zeros of exponential polynomials which reached its classical formulation in the theorem of R. Tijdeman [Tij]. That zero estimate was used to satisfy the hypothesis of Gelfond's criterion, which we give in the formulation of [B1].

Gelfond's Criterion. *Let* $\alpha \in \mathbb{C}$. *Suppose that* $a > 1$ *and that* $\{\delta_n\}$ *and* $\{\sigma_n\}$ *are two positive, strictly increasing unbounded sequences satisfying* $\delta_{n+1} \le a\delta_n$ *and* $\sigma_{n+1} \le a\sigma_n$. *If there is a sequence of non-zero polynomials* $P_n \in \mathbb{Z}[x]$ *with* $\deg P_n \le \delta_n$, $\deg P_n + \log \text{ht } P_n \le \sigma_n$ *and* $\log |P_n(\alpha)| \le -(2a+1)\delta_n\sigma_n$, *then each* $P_n(\alpha) = 0$.

To show transcendence degree ≥ 2, the Thue-Siegel Lemma (= Dirichlet Box Principle) is used to construct, for every large enough parameter N, an auxiliary function (exponential polynomial) whose values on one sector of a lattice will be polynomials in the numbers under consideration. If these numbers generated a field of transcendence degree one, say all algebraic over $\mathbb{Q}(\theta)$, then we would use Tijdeman's theorem to obtain a non-zero value of our function and take the norm down to $\mathbb{Q}(\theta)$ to obtain $P_N(\theta)$. We obtain a contradiction from Gelfond's criterion, and thus at least two of the numbers must be algebraically independent.

B. Philippon's Generalization

Philippon's generalization [P4] of Gelfond's criterion to higher dimensions was a major breakthrough for algebraic independence. In addition to furnishing the first sharp tools for independence of more than two numbers in the Gelfond-Schneider setting, it was a technical tour de force.

Nesterenko had already shown [N4] that his development of the Cayley-Chow form was sufficient for a systematic proof of the results attacked by Chudnovsky in 1974 [C]. However by incorporating in an ingenious way two additional features, Philippon established a criterion which is sharp and which demonstrates lower bounds for transcendence degrees which are in general exponentially better that the previous ones (optimal to within a factor of 2).

We state the criterion for affine polynomials. So let us remark that for a polynomial $P \in R_a = \mathbb{Z}[x_1,\dots,x_n]$, *size* $P = \deg P + \log H(P)$. For an affine prime ideal $\mathfrak{P}$ of R_a, we mean by *size* $\mathfrak{P}$ the sum of the degree and log height of its homogenization ${}^h\mathfrak{P}$, i.e. size $\mathfrak{P}$ = size of Cayley-Chow form of ${}^h\mathfrak{P}$.

Theorem (Philippon). *Let* $\omega \in \mathbb{C}^n$. *Suppose that* $\mathfrak{P} < R_a$ *is a prime ideal of dimension* d *and size at most* $\sigma_d \geq 1$ *vanishing at* ω. *For* $a > 1$ *and* $N \geq N_0$, *let* $\{D_N\},\{S_N\}$ *denote monotonically increasing, unbounded sequences of positive integers such that* $D_{N+1} \leq aD_N$, $S_{N+1} \leq aS_N$. *Assume that* $C > 0$ *is sufficiently large and that for each* $N \geq N_0$ *there is an ideal* J_N *generated by homogeneous polynomials* $P_k \in R_a$ *such that*

i) J_N *has only finitely many zeros within the ball* $B_{\rho_N}(\omega)$ *of radius* $\rho_N = \exp(-CD_N^{d}S_N\sigma_d)$,

ii) $\deg P_k \leq D_N$, size $P_k \leq S_N$,

iii) $\log|P_k(\omega)| \leq C^d \log \rho_N$.

Then for all $N \geq N_1$, *the point* ω *is a zero of* J_N.

Note that we have chosen the formulation as in [B6] to more closely parallel our version of Gelfond's criterion. Philippon's proof is not as easy to sketch as the zero estimate of section 3 because the argument forks at every step in the reduction of dimension. Finally to treat the dimension zero case, an elaboration of Gelfond's proof is developed. We note that a detailed sketch for the case $D_N = S_N = N$ is given in [B6].

C. Two Useful Properties

Nesterenko's basic inequality above for resultants says roughly that if a Cayley-Chow form and an ordinary form are both small at a point of $\mathbb{P}_n$, then so is their resultant. Philippon discovered [P2] that the resultant is also small even if the form is only small compared to the distance to the zeros of a prime ideal underlying the Cayley-Chow form. To state this result more precisely, we define for representatives $\omega = [\omega_0:\dots:\omega_n]$, $\theta =$

$[\theta_0:\ldots:\theta_n]$ of points in $\mathbf{P}_n$ the *projective distance* between them:

$$d(\omega,\theta) = \frac{\max_{i<j} \{|\omega_i\theta_j - \omega_j\theta_i|\}}{(\max |\omega_i|)(\max |\theta_i|)}$$

Proposition (Philippon). *If* F *is the Cayley-Chow form of a homogeneous prime ideal of* $\mathbf{Z}[x_0,\ldots,x_n]$ *intersecting* $\mathbf{Z}$ *only in* 0 *and if for each of its zeros* β,

$$\|F\|_\omega \leq d(\omega,\beta)^\mu,$$

where $0 < \mu \leq 1$, *then*

$$\|R(F,Q)\|_\omega \leq \|F\|^\mu H(F)^{\deg Q} H(Q)^{\delta(F)} \exp(8n(\deg F)(\deg Q)).$$

By continuity, it is clear that a near-by zero forces a Cayley-Chow form to be small. Philippon noticed [P2] that in a certain sense the converse is also true.

Proposition (Philippon). *If* F *is a Cayley-Chow form of a homogeneous prime ideal of* $\mathbf{C}[x_0,\ldots,x_n]$ *of dimension* $d \geq 0$, *then for every* $\omega \in \mathbf{P}_n(\mathbf{C})$, *there is a zero* $\beta \in \mathbf{P}_n(\mathbf{C})$ *of* $\mathfrak{P}$ *such that*

$$d(\omega,\beta)^{\deg F} \leq \|F\|_\omega \exp(3n^2 \deg F).$$

In other words, the Cayley-Chow form of a prime ideal is small only near zeros of the prime ideal. It seems that the exponent $\deg F$ on the left-hand side can be replaced by $\delta(F)$, which we plan to incorporate in a future note.

D. Applications.

1. $N\varepsilon$-independence

In [Ca], J.W.S. Cassels showed a general result which implied in particular that *for any* $\varepsilon > 0$ *and* $n \geq 3$, *there are algebraically independent* $\xi_1,\ldots,\xi_n \in \mathbf{R}$ *such that for each* $N \geq N_0$ *there are* $\nu_1,\ldots,\nu_n \in \mathbf{Z}$ *with* $0 < \max |\nu_i| < N$ *satisfying*

$$\log |\Sigma\nu_i\xi_i| < -N^\varepsilon.$$

Of course this is not typical for n-tuples of real numbers. In fact, to the best of my knowledge, not a single explicit such n-tuple is known. We summarize the fact that $\xi_1,\ldots,\xi_n$ satisfy the inequality of Cassels' result by saying that the numbers $\xi_1,\ldots,\xi_n$ are $N\varepsilon$*-dependent*. This terminology is chosen to call to mind both the exponent N^ε of the strong inequality being satisfied and the notion that such an inequality means that the numbers are quantitatively "Nεarly dependent."

2. Algebraic Independence

As a consequence of his generalization of Gelfond's criterion, Philippon obtained his remarkable results on algebraic independence. The method of proof was essentially an elaboration of the classical Gelfond-Schneider approach involving the Thue-Siegel Lemma to construct an auxiliary function with many zeros and the Schwarz Lemma (Maximum Modulus Principle) to conclude smallness of the values. In order to be able to use his criterion, he had to be able to construct auxiliary functions whose coefficients had no common zeros in small balls about the point whose coordinates were the values whose independence was under investigation. For this he invented the technique of redundant variables and extended Tijdeman's results [Tij] to exponential functions of several variables. Consequently he showed, in our terminology, the following as one of a trio of results.

Theorem (Philippon). *Let* $u_1,\dots,u_m$ *and* $v_1,\dots,v_n$ *be linearly independent sets of complex numbers. If, for every* $\varepsilon > 0$, *these two sets of numbers are not* $N\varepsilon$*-dependent, then*

$$\text{tr. deg. } \mathbb{Q}(e^{u_i v_j}) \geq \frac{mn}{m+n} - 1.$$

This result was subsequently extended to elliptic curves without complex multiplication [W2] and to algebraic groups [W3] by M. Waldschmidt, who also wrote the survey [W1] of the field before Philippon's criterion was available. Since then this result has been improved by G. Diaz [D], who has replaced the right hand side by $\left[\frac{mn}{m+n}\right]$. Philippon has given a general measure of algebraic independence in given dimension once an analogue of the hypotheses i)-iii) of his criterion are satisfied. Nesterenko used essentially the same ideas in his effective measure [N5] of algebraic independence of the numbers appearing in the complex versions of the three theorems covered by Philippon's original results. E.M. Jabbouri has developed measures of algebraic independence in the setting of commutative algebraic group varieties [J].

V. Generalized Liouville Inequalities and Applications

A. Some Examples

Given polynomials $Q_1,\dots,Q_k \in \mathbb{Z}[x_1,\dots,x_n]$, having $\deg Q_i \leq D$, $H(Q_i) \leq H$, and a point $\omega \in \mathbb{C}^n$ (or in $\mathbb{C}_p{}^n$), it is a natural to ask whether one can give a lower bound for $\max |Q_i(\omega)|$. Of course put in this generality, zero is the best lower bound, since ω might well be a common zero of the given polynomials. However useful lower bounds were established classically for the case that ω is not a common zero but all coordinates of ω are algebraic numbers. Such an inequality is called a *Liouville inequality*.

We propose to establish a much more general inequality in Paragraph

V.B. But first we consider in more detail the examples introduced in [B5] to ascertain which ingredients must necessarily appear in a lower bound for $\max |Q_i(\omega)|$. These examples are all variants of an example [B2] devised by Masser and Philippon to give lower bounds for degrees in the Nullstellensatz.

i) The polynomials

$$x_1^D,\ x_1x_n^{D-1}-x_2^D,\dots,x_{n-2}x_n^{D-1}-x_{n-1}^D,\ 1-x_{n-1}x_n^{D-1},$$

with all but first and last terms of the form $Q_i = x_{i-1}x_n^{D-1}-x_i^D$, have

$$\max |Q_i(\tau)| = |Q_1(\tau)| = t^{-D^n+D}$$

at $\tau = (t^{-D^{n-1}+1},\dots,t^{-D+1},t)$ $t > 1$. Note that the Q_i have no common zeros in $\mathbb{C}^n$ to account for the smallness. Since $H = 1$, we see that any lower bound must contain a factor essentially as small as $\|\omega\|^{cD^n}$ with $c \geq 1$, where $\|\omega\| = \max\{1, |\omega_j|\}$. Our example is an improvement of the one devised by Masser and Philippon, which had $\max |Q_i(\tau)| = t^{-D^n+D^{n-1}}$.

ii) An example in which no coordinate tends to infinity is given by

$$x_1^D,\ x_1-x_2^D,\dots,x_{n-1}-x_n^D,\ HX_n-1$$

at $\tau = (H^{-D^{n-1}},\dots,H^{-1})$, $H \geq 1$. Here also the polynomials have no common zeros in $\mathbb{C}^n$. The fact that

$$\max |Q_i(\tau)| = |Q_1(\tau)| = H^{-D^n}$$

shows that any lower bound must contain a factor essentially as small as H^{-cD^n} with $c \geq 1$.

iii) Finally we consider a sequence of examples

$$x_1^D,\ x_1-x_2^D,\dots,x_{n-2}-x_{n-1}^D,\ x_{n-1}-(qx_n-p)^D,$$

where $\{p/q\}$ are a sequence of good rational approximations to the fixed Liouville number ξ with $0 < \xi < 1$, i.e. $\xi \in \mathbb{R}\backslash\mathbb{Q}$, but the ratio $(-\log |q\xi-p|)/\log q$ is unbounded. For our points we take $\tau = (\varepsilon^{D^{n-1}},\dots,\varepsilon^D,\xi)$, where $\varepsilon = q\xi-p$. The polynomials in this example have precisely one (finite) zero in common: $(0,\dots,0,p/q)$. Note that $\varepsilon = H\rho$, where $\rho < 1$ is the distance from τ to the common zero. In our example we have

$$\max |Q_i(\tau)| = |Q_1(\tau)| = (H\rho)^{D^n}.$$

Since ξ is a Liouville number, no power of H to a fixed power of D can produce such smallness, and $\|\tau\|$ tends neither to ∞ nor to 0. Therefore the lower bound sought must contain a factor of the form $(\min\{1,\rho\})^{cD^n}$, where ρ is the distance from ω to the set of common zeros of the Q_i.

Viewed projectively the contributions from i) above are of the same form as in iii), since the polynomials do have common zeros at infinity and the projective distance from ω to the points in $\mathbb{P}_n \setminus \mathbb{C}^n$ is essentially $1/\|\omega\|$.

B. The Local Liouville Inequality

The method which we used in Section III applies just as readily here to establish the following result, which is a strong local version of the inequality which was first established by Masser and G. Wüstholz [MW] as a corollary to their effective version of the Hilbert Nullstellensatz. As in the previous section, for an affine prime ideal $\mathfrak{P}$ of $\mathbb{Q}[x_1,\ldots,x_n]$ we will denote by $\|\mathfrak{P}\|_\omega$ the absolute value $\|F\|_\omega$ of the Cayley-Chow form of the homogenization of $\mathfrak{P}$ and similarly the degree and size of $\mathfrak{P}$ are identified with those of its homogenization.

Theorem. *Let* $\mathfrak{P}$ *be an ideal of* $T = \mathbb{Z}[x_1,\ldots,x_n]$ *of dimension* d, $\mathfrak{P} \cap \mathbb{Z} = \{0\}$. *Let* $\delta(\mathfrak{P}) = D_d$, size $\mathfrak{P} = \sigma_d$. *Let* $Q_1,\ldots,Q_k \in T$ *have* $\deg Q_i \le D$, size $Q_i \le \sigma$. *If* $\omega \in \mathbb{C}^n$ *and* $\mathfrak{P}, Q_1,\ldots,Q_k$ *have no common zero in the ball of radius* $\rho \le 1$ *centered at* ω, *then*

$$\log \max \{\|\mathfrak{P}\|_\omega, |Q_1(\omega)|,\ldots,|Q_k(\omega)|\} \ge$$
$$-cD^{d+1}\sigma_d - cD^d D_d \sigma - cD_d D^d \log \|\omega\| + cD_d D^d \log \rho,$$

for $c = 11(n+1)^5$.

In case there is no ideal $\mathfrak{P}$ given, we can take $\mathfrak{P} = 0$, $D_n = \sigma_n = 1$ to obtain a lower bound

$$\log \max \{|Q_1(\omega)|,\ldots,|Q_k(\omega)|\} \ge -cD^n\sigma - cD^n \log \|\omega\| + cD^d \log \rho,$$

which was previously unknown even for $n = 1$.

In the situation of the classical Liouville inequalities, $d = 0$, $k = 1$, and $\mathfrak{P}$ vanishes at ω. Since $Q_1(\omega) \ne 0$, $Q_1(x)$ and $\mathfrak{P}$ have <u>no</u> common zeros and we can set $\rho = 1$. Of course sharper values of c are known in that case, and in fact the value of c can be decreased somewhat in our general case as well.

The examples given in V.A show that, up to reducing the value of c, all the terms are best possible, with one exception. Hidden within the first two terms is one of the form $-cD^{d+1}D_d$. Although its occurence is natural from the point of view of the proof (Gelfond's Lemma II, p. 135 [G]), no example is known to imply its necessity. The analogous remark applies to Philippon's criterion. Also since we used the sparse sequence of good approximations to a fixed Liouville number to demonstrate the necessity of the appearance of $D_d D^d \log \rho$, it would be interesting to determine the optimality of the term $D \log \rho$ in Philippon's criterion.

C. Application

Just as Philippon's criterion extends Gelfond's method for algebraic independence to higher dimensions, our inequality extends the classical transcendence methods of Gelfond and Schneider to higher dimension. In fact the two approaches yield rather comparable results. As an example, we mention the following proposition:

Theorem. *Let* $u_1, \ldots, u_m$ *and* $v_1, \ldots, v_n$ *be complex numbers and let* $\varepsilon = 1/24(m+n)$. *Then there is an integer* N_0, *depending on the* u_i *and* v_j *such that if the sets are each not* $N\varepsilon$*-dependent for* <u>any single</u> $N \geq N_0$, *then*

$$\operatorname{tr\,deg} \mathbb{Q}(\exp(u_i v_j)) \geq \frac{mn}{m+n} - 1.$$

The main differences between the applications of Philippon's criterion versus the local Liouville inequality lie in

i) the necessity of an infinite, or at least sufficiently long, sequence of ideals to apply the criterion,

ii) the necessity for a zero-free region, rather than isolated zeros, for the local Liouville inequality, and finally

iii) the better dependence on ρ in the criterion.

It is the latter point which thus far prevents one from obtaining Diaz' results [D] via the inequality. Only in dimension 2 does point ii) seem to matter, thanks to the sharp algebraic zero estimates now available [P5].

D. Outline of Proof of Inequality

One constructs recursively a sequence $L_d, \ldots, L_0 \in \Sigma_i \mathbb{Z} Q_i$ and a sequence of Cayley-Chow forms $F_{d+1}, \ldots, F_0$ of descending dimension $d, \ldots, 1, 0, -1$, such that i) F_{d+1} is the Cayley-Chow form of $\mathbb{P}$, ii) for $d \geq i \geq 1$, F_{i-1} is obtained by omitting from $R(F_i, {}^hL_i)$ all factors whose asssociated primes have no zero within ρ of ω, and of course iii) $F_{-1} = R(F_0, L_0) \in \mathbb{N}$. For this purpose of course the magnitude of the coefficients from $\mathbb{Z}$ must be controlled using, e.g., Lemma 1, p.438 of [MW]. Then one verifies from the properties that Nesterenko established for resultants and Philippon's result on the existence of zeros near points where Cayley-Chow forms are small that:

1) $\delta(F_i) \leq D_d D^{d\ i}$, $\log \operatorname{ht} F_i \ll D^{d-i-1}(D\sigma_d + D_d\sigma)$,

2) $\log \|F_i\|_{\omega^*} \leq \log \max \{\|F_d\|_{\omega^*}, |Q_j(\omega)|\} + cD^{d-i-1}(D_d\sigma + D\sigma_d) + cD^{d-i-1} \log \|\omega\|/\rho$,

where $\omega^* = (1, \omega_1, \ldots, \omega_n)$ is the point in projective space corresponding to ω and $\|\omega\| = \max\{1, |\omega_i|\}$. The lower bound follows from the case $i = -1$ and the remark that $F_{-1} \in \mathbb{N}$ and hence $|F_{-1}| \geq 1$.□

E. Nullstellensatz

The method of proof of the local Liouville inequality works just as well over $\mathbb{C}$, although the notion of height of the Cayley-Chow form is not so relevant. One obtains the following result [B2]:

Basic Proposition. *Let* $P_1,\ldots,P_k \in \mathbb{C}[x_1,\ldots,x_n]$ *have no common zeros in* $\mathbb{C}^n$, $\deg P_j \le D$, $D \ge 1$. *Then there is a constant* $C > 0$ *such that*

$$\max \{|P_j(\omega)|/\|\omega\|^{\deg P_j}\} \ \ge \ \|\omega\|^{-(n-1)D^n}.$$

C. Berenstein and A. Yger kindly indicated that an effective version of the Nullstellensatz follows from this type of inequality using some results from the theory of several complex variables. The sharpest bounds so far follow by invoking the following special case of a more general result [S]:

Theorem. (Skoda). *Let* $Q_1,\ldots,Q_m \in \mathbb{C}[\underline{x}] = \mathbb{C}[x_1,\ldots,x_n]$ *and for* $\varepsilon > 0$, *the constant* $K \ge 0$ *be large enough so that*

$$\int_{\mathbb{C}^n} |Q|^{-2(1+\varepsilon)q-2}\|z\|^{-2K} d\lambda \ = \ I \ < \ \infty,$$

where $q = \min\{n,m-1\}$, $|Q|^2 = \Sigma|Q_i|^2$, $\|z\|^2 = \Sigma|z_i|^2$. *Then there exist holomorphic* $A_1,\ldots,A_m$ *such that*

$$1 \ = \ A_1Q_1 + \ldots + A_mQ_m \tag{*}$$

and

$$\int_{\mathbb{C}^n} |A|^2|Q|^{-2(1+\varepsilon)q}\|z\|^{-2K} d\lambda \ \le \ \frac{1+\varepsilon}{\varepsilon}\, I < \infty.$$

On combining these two results one obtains the following proposition [B2]:

Nullstellensatz. *Let* $Q_1,\ldots,Q_m \in \mathbb{C}[\underline{x}]$ *have no common zero in* $\mathbb{C}^n$ *and* $\deg Q_i \le D$, $D \ge 1$. *Then there exist* $A_1,\ldots,A_m \in \mathbb{C}[\underline{x}]$ *satisfying* (*) *such that*

$$\deg A_i \le (n-1)(q+1)D^\mu + qD,$$

for $\mu = \min\{m,n\}$.

It seems likely that the factor involving n can be omitted, and we hope to return to this question in the near future. As pointed out by Masser and Philippon, the first example above shows that the coefficient of D^μ cannot be essentially less than 1. Using a refinement of Rabinowitsch's technique, our result has been extended [B3] to give a sharp effective version of the full Hilbert Nullstellensatz. However many interesting open questions remain [B4]. Perhaps the Cayley-Chow form will be useful in resolving them as well.

Bibliography

[BB] Bertrand, D. and Beukers, F. Equations differentielles et majorations de multiplicités, Ann. Sci. Ecole Norm. Sup. 18 1985, 181-192.

[BM] Brownawell, W.D. and Masser, D.W. Multiplicity estimates for analytic functions II, Duke Math J. 47(1980), 273-295.

[B1] Brownawell, W.D. Sequences of Diophantine approximations, J. Number Th. 6(1974), 11-21.

[B2] ---------------. Bounds on the degree in the Nullstellensatz, Annals of Math., 126(1987), 577-591.

[B3] ---------------. Borne effective pour l'exposant dans le théorème des zéros, C. R. Acad. Sci. Paris, Sér. I., 305(1987), 287-290.

[B4] ---------------. Aspects of the Hilbert Nullstellensatz, in *New Advances in Transcendence*, A. Baker, ed, Cambridge University Press, Cambridge, to appear.

[B5] ---------------. Local Diophantine Nullstellen inequalities, J. Am. Math. Soc., to appear.

[B6] ---------------. Large transcendence degree revisited I. Exponential and CM cases, in *Bonn Workshop on Transcendence*, G. Wüstholz, ed, Springer Lecture Notes, to appear.

[B7] ---------------. Note on a paper of P. Philippon, Mich. Math. J., in press.

[BT] Brownawell, W.D. and Tubbs, R. Large transcendence revisited II. The CM case, in *Bonn Workshop on Transcendence*, G. Wüstholz, ed, Springer Lecture Notes, to appear.

[Ca] Cassels, J.W.S. *An Introduction to Diophantine Approximation*, Cambridge University Press, Cambridge, 1957.

[C] Chudnovsky, G.V. Some analytic methods in the theory of transcendental numbers, Inst. of Math., Ukr. SSR Acad. Sci., Preprint IM 74-8 and IM 74-9, Kiev, 1974 = Chapter 1 in *Contributions to the Theory of Transcendental Numbers*, Am. Math. Soc., Providence, R.I. 1984.

[D] Diaz, G. Grands degrés de transcendance pour des familles d'exponentielles, C. R. Acad. Sci. Paris 305(1987), 159-162.

[G] Gelfond, A.O. Transcendental and Algebraic Numbers, GITTL Moscow 1952 = Dover, New York, 1960.

[J] Jabbouri, E.M. Mesures d'indépendance algébriques sur les groupes algébriques commutatifs, manuscript.

[MW] Masser,D.W. and Wüstholz, G. Fields of large transcendence degree generated by values of elliptic functions, Invent. Math. 72(1983), 407-463.

[N1] Nesterenko, Yu.V. On the algebraic dependence of the components of solutions of a system of linear differential equations, Izv. Akad. Nauk SSR Ser. Mat. 38(1974), 495-512 = Math. USSR Izv. 8(1974), 501-518.

[N2] ----------------. Estimates for the orders of zeros of functions of a certain class and applications in the theory of transcendental numbers, Izv. Akad. Nauk SSSR Ser. Mat. 41 (1977), 253-284 = Math. USSR Izv. 11(1977), 239-270.

[N3] ----------------. Bounds for the characteristic function of a prime ideal, Mat. Sbornik 123, No. 1(1984), 11-34 = Math. USSR Sbornik 51(1985), 9-32.

[N4] ----------------. On algebraic independence of algebraic powers of algebraic numbers, Mat. Sbornik 123, No. 4(1984), 435-459 = Math. USSR Sbornik 51(1985), 429-454, brief version in *Approximations Diophantiennes et Nombres Transcendants*, D. Bertrand and M. Waldschmidt, eds, Birkhäuser Verlag, Verlag, Boston-Basel-Stuttgart, 1983, pp.199-220.

[N5] ----------------. On a measure of the algebraic independence of the values of some functions, Mat. Sbornik 128, No.4(1985), 545-568 = Math USSR Sbornik 56(1986), 545-567.

[N6] ----------------. On bounds of measures of algebraically independent numbers, pp.65-76 in *Diophantine Approximations*, P.L. Ulnova, ed., Moscow Univ. Press, Moscow, 1985 (Russian).

[N7] ----------------. On a measure of algebraic independence of values of the exponential function, Doklady Akad. Nauk SSSR 286, No.4, 1986, 817-821 = Soviet Math. Doklady 33(1986), 20-203.

[N8] ----------------. On a measure of algebraic independence of values of elliptic functions at algebraic points, Uspehi Mat. Nauk 40(1985), 221-222 = Russian Math. Surveys 40, No. 4, 1985, 237-238.

[N9] ----------------. Measures of algebraic independence of numbers and functions, pp 141-149 in *Journées Arithmétiques de Besancon*, Asterisque, Vol. 147-148, Soc. Math. France, Paris, 1987.

[O] Osgood, C.F. Nearly perfect systmes and effective generalizations of Shidlovski's theorem, J. Number Th. 13(1981), 515-540.

[P1] Philippon, P. Indépendance algébrique de valeurs des fonctions exponentielles p-adiques, J. reine angew. Math. 329 (1981), 42-51.

[P2] ------------. Pour une théorie de l'indépendance algébrique, Thèse, Université de Paris XI, 1983.

[P3] ------------. Sur les mesures d'indépendance algébrique, pp.219-233 in *Seminaire de Theorie des Nombres*, Catherine Goldstein, ed, Birkhäuser, Boston-Basel-Stuttgart, 1985.

[P4] ------------. Critères pour l'indépendance algébrique, Inst. Hautes Etudes Sci. Publ. Math. No. 64, 1986, 5-52.

[P5] ------------. Lemmes de zéros dans les groupes algébriques commutatifs, Bull. Soc. Math. France.114(1986), 355=383.

[P6] ------------. Elimination effective, Chap. XXVIII in *Journées Algorithmiques-Arithmétiques*, Univ. St.Etienne, 1983.

[Sh] Shidlovsky, A.B. *Transcendental Numbers*, Nauka, Moscow 1987 (in Russian).

[S] Skoda, H. Applications des techniques L^2 à la théorie des idéaux algèbre de fonctions holomorphes avec poids, Ann. Sci. Ecole Norm. Sup. 5(1972), 545-579.

[Tij] Tijdeman, R. On the number of zeros of general exponential polynomials, Indag. Math. 33(1971),1-7.

[VdW] Van der Waerden, B.L. Zur algebraischen Geometrie 19, Grundpolynom und zugeordnete Form, Math. Ann. 136(1958), 139-155.

[W1] Waldschmidt, M. Algebraic independence of transcendental numbers. Gel'fond's method and its developments, pp.551-571, in *Perspectives in Mathematics, Anniversary of Oberwolfach*, W. Jager, J. Moser, R. Remmert, eds, Birkhäuser Verlag, Boston-Basel-Stuttgart, 1984.

[W2] --------------. Algebraic independence of values of exponential and elliptic functions, J. Indian Math. Soc. 48 (1984), 215-228.

[W3] --------------. Groupes algébriques et grands degrés de transcendance, Acta Math. 156(1986), 253-302.

PARTITIONS INTO PARTS WHICH ARE UNEQUAL AND LARGE.

by P. Erdös, J.L. Nicolas and M. Szalay * .

1. Introduction. Let us denote by $p(n)$ the number of partitions of n, by $q(n)$ the number of partitions of n into unequal parts (or into odd parts), by $r(n, m)$ the number of partitions of n into parts $\geq m$, and by $\rho(n, m)$ the number of partitions of n into unequal parts $\geq m$.

In [Erd] two of us gave the following asymptotic relation

$$(1) \qquad \rho(n, m) = (1 + o(1)) \frac{q(n)}{2^{m-1}}, \qquad m = o(n^{1/5})$$

and in [Dix], a quite different result is given for $r(n, m)$, for $m = O(n^{1/4})$

$$r(n, m) = (m-1)! \left(\frac{\pi}{\sqrt{6n}}\right)^{m-1} p(n) \; (1 + O(m^2/\sqrt{n})) .$$

Using a tauberian theorem, J. Herzog (cf. [Her]) has proved, for $m = O(n^{3/8} (\log n)^{1/4})$:
$\log r(n, m) = \pi\sqrt{2n/3} - (1/2) m \log n + m \log m - m(1 + \log(\sqrt{6}/\pi)) + O(n^{1/4}\sqrt{\log n})$.

The aim of this paper is to prove the following three theorems.

Theorem 1. For all $n \geq 1$, and m, $1 \leq m \leq n$, we have

$$(i) \qquad \frac{1}{2^{m-1}} q(n) \leq \rho(n, m) \leq \frac{1}{2^{m-1}} q\left(n + \frac{m(m-1)}{2}\right)$$

and

$$(ii) \qquad \rho(n, m) \leq \frac{1}{2^{m-2}} q\left(n + \left[\frac{m(m-1)}{4}\right]\right)$$

where $[x]$ is the integral part of x.

Theorem 2. When n tends to infinity, and $m = o\left(\frac{n}{\log n}\right)^{1/3}$, we have

$$\rho(n, m) = (1 + o(1)) \frac{1}{2^{m-1}} q\left(n + \left[\frac{m(m-1)}{4}\right]\right).$$

---

* Research partially supported by Hungarian National Foundation for Scientific Research grant n° 1811, and by Centre National de la Recherche Scientifique, Greco "Calcul Formel" and PRC Math. Info. .

Theorem 3. For fixed ε, with $0 < \varepsilon < 10^{-2}$ and for $m = m(n)$, $1 \le m \le n^{3/8 - \varepsilon}$, and $n \to +\infty$, the relation

$$\rho(n, m) = (1 + o(1)) \frac{q(n)}{\prod_{1 \le j \le m-1} \left(1 + \exp\left(-\frac{\pi j}{2\sqrt{3n}}\right)\right)}$$

holds.

The proof of Theorem 1 is very simple and elementary, and gives immediately (1) when $m = o(n^{1/4})$, with the classical asymptotic estimation of $q(n)$ (cf. (2) and Lemma 3 below).

The proof of Theorem 2 follows the same idea as the proof of theorem 1, but with sharper estimations.

The proof of Theorem 3 is analytic, and uses Cauchy's formula for the generating function of $\rho(n, m)$, which was already used to prove (1). It follows easily from Lemma 3 below that Theorem 3 implies Theorem 2.

At the end of the paper, a table of $\rho(n, m)$ is given. It has been calculated with the recurrence formula $\rho(n, m) = \rho(n, m+1) + \rho(n-m, m+1)$ and $\rho(n, m) = 1$ for $m \ge n/2$.

We shall also need the following asymptotic formula of $q(n)$:

$$(2) \qquad q(n) \sim \frac{1}{4\,(3n^3)^{1/4}} \exp\left(\pi \sqrt{n/3}\right).$$

Actually, it is possible to give a more precise expansion, using the result of Hardy and Ramanujan (cf. [Har] and [Hua]) :

$$q(n) = \frac{1}{\sqrt{2}} \frac{d}{dn} J_0\left(i\pi \sqrt{\frac{1}{3}\left(n + \frac{1}{24}\right)}\right) + O\left(\exp\left(\frac{\pi}{3}\sqrt{n/3}\right)\right).$$

By the classical results $J'_0(z) = -J_1(z)$ and $I_1(z) = -i\,J_1(iz)$ on Bessel's functions, the main term of $q(n)$ is equal to

$$\frac{\pi}{2\sqrt{6}\,\sqrt{n + 1/24}}\, I_1\left(\frac{\pi}{\sqrt{3}} \sqrt{n + 1/24}\right).$$

For $m \ge 1$, let us define

$$a_m = \frac{(-1)^m}{2^{3m}\, m!} \prod_{j=1}^{m} \left(4 - (2j-1)^2\right).$$

We have

$$a_1 = -\frac{3}{8} \quad ; \quad a_2 = -\frac{15}{128} \quad ; \quad a_3 = -\frac{105}{1024}.$$

For a real z tending to $+\infty$, we have the asymptotic expansion (cf. [Wat], p. 203) :

$$I_1(z) = \frac{e^z}{\sqrt{2\pi z}} \left(1 + \sum_{m=1}^{M} a_m z^{-m} + O(z^{-M-1})\right),$$

and if we set $c = \dfrac{\pi}{\sqrt{3}}$, $\lambda = 1/24$ and

$$g_M(\lambda, t) = \left(\exp\left(c\,\frac{(1 + \lambda t^2)^{1/2} - 1}{t}\right)\right) \left((1 + \lambda t^2)^{-\frac{3}{4}} + \sum_{m=1}^{M} \frac{a_m}{c^m}\, t^m\, (1 + \lambda t^2)^{-\frac{m}{2} - \frac{3}{4}}\right)$$

then the function g_M is analytic in t in a neighbourhood of 0, thus it has a Taylor expansion

$$g_M(\lambda, t) = 1 + \sum_{m=1}^{M} b_m t^m + O(t^{M+1}).$$

We conclude that

$$(3) \qquad q(n) = \frac{1}{4\,(3n^3)^{1/4}} \left(\exp\left(\frac{\pi}{\sqrt{3}} \sqrt{n}\right) \right) \left(1 + \sum_{m=1}^{M} b_m n^{-m/2} + O(n^{-\frac{M+1}{2}}) \right).$$

The first coefficients b_m have been calculated by the algebraic computer system MACSYMA :

$$b_1 = \frac{\pi}{48\sqrt{3}} - \frac{3\sqrt{3}}{8\pi} = -0.16896 \qquad b_2 = \frac{\pi^2}{13824} - \frac{5}{128} - \frac{45}{128\,\pi^2} = -0.07397$$

$$b_3 = \frac{\pi^3}{1990656\sqrt{3}} - \frac{35\,\pi}{36864\sqrt{3}} + \frac{35\sqrt{3}}{2048\,\pi} - \frac{315\sqrt{3}}{1024\,\pi^3} = -0.009475$$

$$b_4 = \frac{\pi^4}{1146617856} - \frac{7\,\pi^2}{1769472} + \frac{105}{65536} + \frac{315}{16384\,\pi^2} - \frac{42525}{32768\pi^4} = -0.009812.$$

We are pleased to thank J.P. Massias for calculating both the table of $\rho\,(n, m)$ and the asymptotic expansion of q.

2. Proof of Theorem 1. Setting $q\,(0) = \rho\,(0, m) = 1$, we shall consider generating functions :

$$\sum_{n \geq 0} q\,(n)\, x^n = \prod_{n \geq 1} (1 + x^n)$$

and

$$(4) \qquad \sum_{n \geq 0} \rho\,(n, m)\, x^n = \prod_{n \geq m} (1 + x^n).$$

Let us define

$$(5) \qquad P_{m-1}(x) = \prod_{k=1}^{m-1} (1 + x^k) = \sum_{k=0}^{m(m-1)/2} q\,(k, m-1)\, x^k.$$

We observe that $q(k, m-1) \geq 0$ and that

$$\sum_{k=0}^{m(m-1)/2} q\,(k, m-1) = 2^{m-1}.$$

We now write

$$\sum_{n=0}^{\infty} q(n)\, x^n = \left(\sum_{n=0}^{\infty} \rho\,(n, m)\, x^n \right) \left(\sum_{k=0}^{m(m-1)/2} q\,(k, m-1)\, x^k \right)$$

and

$$(6) \qquad q\,(n) = \sum_{k=0}^{m(m-1)/2} q\,(k, m-1)\; \rho\,(n-k, m)$$

where we set $\rho\,(n, m) = 0$ for $n < m$ and $n \neq 0$. Now, it is easy to see that ρ is non decreasing in n, therefore, $\rho(n-k, m) \leq \rho(n, m)$, and then (6) gives $q(n) \leq 2^{m-1} \rho(n, m)$. In the same way,

$$q\left(n + \frac{m(m-1)}{2}\right) = \sum_{k=0}^{m(m-1)/2} q(k, m-1)\, \rho\left(n - k + \frac{m(m-1)}{2}, m\right)$$
$$\geq 2^{m-1} \rho(n, m)$$

and this achieves the proof of (i). To prove (ii), we set $M = [\,(m(m-1)/4\,]$ and get

$$q(n+M) = \sum_{k=0}^{m(m-1)/2} q(k, m-1)\, \rho(n-k+M, m)$$
$$\geq \rho(n, m) \sum_{k=0}^{M} q(k, m-1) \geq 2^{m-2} \rho(n, m).$$

3. Proof of Theorem 2.

We first need a few lemmas :

Lemma 1. For $0 \leq u \leq 1/2$, we have

(i) $-\log(1-u) \leq u + u^2$.

For $m \geq 3$ and $0 \leq u \leq 1$, we have

(ii) $(1-u)^m \geq 1 - mu + \frac{m(m-1)}{2} u^2 - \frac{m(m-1)(m-2)}{6} u^3.$

Proof : (i) is easy. To prove (ii) use Taylor's formula for the function $u \mapsto (1-u)^m$.

Lemma 2. Let $q(r, m-1)$ be defined by (5). If $m \geq 3$, R is an integer, $0 \leq R \leq \frac{m(m-1)}{4}$ and $t = \frac{m(m-1)}{4} - R$, then we have

$$\sum_{r=0}^{R} q(r, m-1) \leq 2^{m-1} \exp\left(-\frac{3t^2}{m^3}\right).$$

Proof : For $x \in [1/2, 1]$ we set

$$P = P(x, R, m) = x^{-R} \prod_{r=1}^{m-1} (1 + x^r)$$

and $x = 1 - u$. So we have $0 \leq u \leq 1/2$ and

$$\log P = -R \log(1-u) + (m-1)\log 2 + \sum_{r=1}^{m-1} \log\left(1 + \frac{(1-u)^r - 1}{2}\right)$$
$$\leq -R \log(1-u) + (m-1)\log 2 + \sum_{r=1}^{m-1} \frac{(1-u)^r - 1}{2}$$

$$= \begin{cases} -R\log(1-u) + (m-1)\log 2 + \dfrac{(1-u)-(1-u)^m}{2u} - \dfrac{m-1}{2}, & \text{if } u > 0; \\ (m-1)\log 2, \text{ if } u = 0. \end{cases}$$

Using Lemma 1, (i) and (ii), we obtain that

$$\log P \le (m-1)\log 2 + Ru + Ru^2 - \frac{m(m-1)}{4} u + \frac{m(m-1)(m-2)}{12} u^2$$

$$\le (m-1)\log 2 - t u + \frac{m^3}{12} u^2$$

because $R \le \frac{m(m-1)}{4}$. We now choose $u = \frac{6t}{m^3}$. As $0 \le t \le \frac{m^2}{4}$, we have $0 \le u \le \frac{3}{2m} \le \frac{1}{2}$ for $m \ge 3$, and we obtain that $\log P \le (m-1)\log 2 - 3t^2/m^3$.

The lemma follows from this inequality, because

$$\sum_{r=0}^{R} q(r, m-1) \le \sum_{r=0}^{R} q(r, m-1) \frac{x^r}{x^R} \le P.$$

Lemma 3. When $n \to +\infty$ and $h = o(n^{3/4})$, we have

$$q(n+h) \sim q(n) \exp\left(\frac{Ah}{\sqrt{n}}\right),$$

where $A = \pi/(2\sqrt{3}) = 0.9069...$

Proof: From (2) we have

$$q(n+h) \sim \frac{1}{4(3(n+h)^3)^{1/4}} \exp(2A\sqrt{n+h})$$

and

$$\sqrt{n+h} = \sqrt{n} + \frac{h}{2\sqrt{n}} + O\left(\frac{h^2}{n^{3/2}}\right) = \sqrt{n} + \frac{h}{2\sqrt{n}} + o(1).$$

Proof of Theorem 2. We first assume that $m \equiv 0$ or $1 \bmod 4$, in order that $m(m-1)/4$ should be an integer. The case $m \equiv 2$ or $3 \bmod 4$ can be treated similarly. We then choose R and t as in lemma 2, and set

$$R' = \frac{m(m-1)}{2} - R = \frac{m(m-1)}{4} + t.$$

We cut the summation in (6) into three parts:

$$S_1 = \sum_{r=0}^{R-1} \; ; \quad S_2 = \sum_{r=R}^{R'} \; ; \quad S_3 = \sum_{r=R'+1}^{m(m-1)/2},$$

and we shall prove that S_1 and S_3 are $o(q(n))$, if we choose conveniently t.

First of all, it is easy to see that $S_3 \le S_1$. Then we consider

$$S_1 = \sum_{r=0}^{R-1} \rho(n-r, m)\, q(r, m-1).$$

We set $s = \frac{m(m-1)}{4} - r$. Theorem 1, (ii) gives:

$$\rho(n-r, m) \le \frac{1}{2^{m-2}} q(n+s)$$

and by Lemma 3,

$$\rho(n-r, m) \ll \frac{1}{2^{m-1}}\, q(n)\, \exp\left(\frac{A s}{\sqrt{n}}\right).$$

By Lemma 2,

$$q(r, m-1) \le 2^{m-1} \exp\left(-\frac{3 s^2}{m^3}\right),$$

thus

$$S_1 \ll q(n) \sum_{s=t+1}^{m(m-1)/4} \exp\left(\frac{As}{\sqrt{n}} - \frac{3s^2}{m^3}\right) \ll q(n) \sum_{s \ge t+1} \exp\left(-\frac{3}{m^3}\left(s - \frac{m^3 A}{6\sqrt{n}}\right)^2\right)$$

and, if we choose $t > \dfrac{m^3 A}{6\sqrt{n}}$, we shall obtain that

$$S_1 \ll q(n) \int_t^{+\infty} \exp\left(-\frac{3}{m^3}\left(u - \frac{m^3 A}{6\sqrt{n}}\right)^2\right) du$$

$$\ll q(n) \frac{m^3}{6\left(t - \dfrac{m^3 A}{6\sqrt{n}}\right)} \exp\left(-\frac{3}{m^3}\left(t - \frac{m^3 A}{6\sqrt{n}}\right)^2\right).$$

Choosing

(7) $$t = \left[\frac{m^3 A}{6\sqrt{n}} + m^{3/2}\sqrt{\log n}\right]$$

implies $S_1 = o(q(n))$ and $S_2 = (1 + o(1))\, q(n)$.

From the definition of S_2, we see that

$$S_2 \le \rho(n-R, m) \sum_{r=0}^{m(m-1)/2} q(r, m-1) = 2^{m-1} \rho(n-R, m),$$

and

$$S_2 \ge \rho(n-R', m) \sum_{r=R}^{R'} q(r, m-1)$$

$$\ge \rho(n-R', m)\, 2^{m-1}\left(1 - 2\exp\left(-\frac{3(t+1)^2}{m^3}\right)\right)$$

by Lemma 2. This implies that

$$\frac{q(n+R)}{2^{m-1}}(1+o(1)) \le \rho(n, m) \le \frac{q(n+R')}{2^{m-1}}(1+o(1))$$

and Theorem 2 follows from Lemma 3, just observing that the hypothesis and (7) imply that $t = o(\sqrt{n})$.

4. Proof of Theorem 3. Let $k = m-1 \ge 1$ and $q_k(n) = \rho(n, m)$.

Let us observe that the relation

$$1 + \sum_{n=1}^{\infty} q_k(n)\, w^n = \prod_{\nu=k+1}^{\infty} (1 + w^\nu)$$

holds for $|w| < 1$. Cauchy's formula gives the representation

$$q_k(n) = \frac{1}{2\pi i} \int_{|w|=r} w^{-n-1} \prod_{\nu=k+1}^{\infty} (1 + w^{\nu})\, dw$$

for $0 < r < 1$. For $\operatorname{Re} z > 0$, let us define $h_k(z)$ by

$$h_k(z) = \prod_{\nu=k+1}^{\infty} (1 + \exp(-\nu z)) .$$

Then we may write

$$q_k(n) = \frac{1}{2\pi} \int_{-\pi}^{\pi} h_k(x + iy) \exp(nx + iny)\, dy$$

for $x > 0$.

Let C_0 be a sufficiently large constant, further, $1 \le k \le n^{\frac{3}{8} - \varepsilon}$. We choose $x = x_0 = \pi / (2\sqrt{3n})$, $y_1 = n^{-3/4 + \varepsilon/3}$, $y_2 = C_0 x_0$, and it will be convenient to set

$$D = \left\{ \prod_{\nu=1}^{k} (1 + \exp(-\nu x_0)) \right\}^{-1} .$$

Observe that, with our choices of x_0 and k, theorem 3 becomes $\rho(n, m) = (1 + o(1))\, D\, q(n)$.
We investigate $q_k(n)$ as

$$q_k(n) = \frac{1}{2\pi} \int_{-\pi}^{\pi} h_k(x_0 + iy) \exp(nx_0 + iny)\, dy$$

$$= \frac{1}{2\pi} \left\{ \int_{-\pi}^{-y_2} + \int_{-y_2}^{-y_1} + \int_{-y_1}^{y_1} + \int_{y_1}^{y_2} + \int_{y_2}^{\pi} \right\}.$$

For $|y| \le y_2$ (and $n \to +\infty$), we can apply (4.3) - (4.4) of [Erd] and get

$$\prod_{\nu=1}^{\infty} (1 + \exp(-\nu(x_0 + iy))) = \exp\left(\frac{\pi^2}{12(x_0 + iy)} - \frac{1}{2}\log 2 + o(1) \right),$$

further

$$(8) \qquad \prod_{\nu=1}^{k} (1 + \exp(-\nu(x_0 + iy)))^{-1} = D \exp\left(-\sum_{\nu=1}^{k} \log\left(1 - \frac{1 - \exp(-\nu i y)}{1 + \exp(\nu x_0)} \right) \right).$$

For $|y| \le y_1$, we deduce from (8)

$$\prod_{\nu=1}^{k} (1 + \exp(-\nu(x_0 + iy)))^{-1} = D \exp(O(k \cdot k y_1)) = D \exp(o(1)).$$

Therefore (cf. [Erd], pp. 435-437),

$$\frac{1}{2\pi}\int_{-y_1}^{y_1} \qquad = (1+o(1))\ D\ q(n).$$

Next, for $y_1 \le |y| \le y_2$, it follows from (8) that

$$\left|\prod_{\nu=1}^{k}(1+\exp(-\nu(x_0+iy)))^{-1}\right| = D\exp\left(-\sum_{\nu=1}^{k}\log\left(\left|1-\frac{1-\exp(-\nu i y)}{1+\exp(\nu x_0)}\right|\right)\right).$$

Here,

$$\left|1-\frac{1-\exp(-\nu i y)}{1+\exp(\nu x_0)}\right| \ge \left|1-\frac{i\nu y}{1+\exp(\nu x_0)}\right| - \frac{|1-\exp(-\nu i y)-i\nu y|}{1+\exp(\nu x_0)} \ge$$

$$\ge 1-\frac{1}{2}\cdot\frac{\nu^2 y^2}{1+\exp(\nu x_0)} \ge 1-\frac{\nu^2 y^2}{4} = 1-O(k^2 y_2^2) = 1-o(1).$$

If $y_1 \le |y| \le y'_1 := n^{-9/16}$, then

$$\sum_{\nu=1}^{k} -\log\left(\left|1-\frac{1-\exp(-\nu i y)}{1+\exp(\nu x_0)}\right|\right) \le \sum_{\nu=1}^{k} O(\nu^2 y_1'^2) = O(k^3 y_1'^2) = O(n^{-3\varepsilon}) = o(1).$$

Thus (cf. [Erd], p. 438),

$$\frac{1}{2\pi}\left|\int_{y_1\le|y|\le y'_1}\right| = O(1)\ D\exp\left(\frac{\pi}{\sqrt{3}}\sqrt{n}-\frac{2\sqrt{3}}{\pi}n^{2\varepsilon/3}\right) = o(1)\ D\ q(n).$$

If $y'_1 \le |y| \le y_2$ $(= C_0 x_0)$ then

$$\left|\prod_{\nu=1}^{k}(1+\exp(-\nu(x_0+iy)))^{-1}\right| \le D\exp(O(k^3 y_2^2)) = D\exp(O(n^{1/8-3\varepsilon})),$$

consequently (cf. [Erd], p. 438),

$$\frac{1}{2\pi}\left|\int_{y'_1\le|y|\le y_2}\right| \le D\exp\left(\frac{\pi^2 x_0}{12(x_0^2+(y'_1)^2)}+n x_0+O(n^{\frac{1}{8}-3\varepsilon})\right).$$

Here,

$$\frac{\pi^2 x_0}{12(x_0^2+y_1'^2)} = \frac{\pi^2}{12x_0}\cdot\frac{1}{1+\frac{(y'_1)^2}{x_0^2}} = \frac{\pi^2}{12x_0}\left(1-\frac{(y'_1)^2}{x_0^2}+O\left(\frac{(y'_1)^4}{x_0^4}\right)\right) \le$$

$$\le \frac{\pi^2}{12x_0}-\frac{\pi^2}{24}\frac{(y'_1)^2}{x_0^3} = \frac{\pi^2}{12x_0}-c_1 n^{\frac{3}{8}}.$$

Thus,

$$\frac{1}{2\pi}\left|\int_{y'_1 \le |y| \le y_2}\right| \le D\ q(n)\ \exp\left(-c_1 n^{\frac{3}{8}} + O(n^{\frac{1}{8}-3\varepsilon}) + O(\log n)\right) = o(1)\ D\ q(n).$$

Finally, for $C_0 x_0 \le |y| \le \pi$,

$$\frac{1}{2\pi}\left|\int_{C_0 x_0 \le |y| \le \pi}\right| \le q(n)\ \exp(-c_2\sqrt{n})$$

with a suitable positive constant c_2 (cf. [Erd] , pp. 439 - 440).

Since

$$D \le 2^k \le \exp(n^{3/8-\varepsilon}) = o(\exp(c_2\sqrt{n})),$$

Theorem 3 is proved.

Remark. In the same way, one can prove Theorem 3 with the factor

$$1 + O\left(n^{-\frac{1}{4}+\varepsilon}\right) + O\left(m^2\, n^{-\frac{3}{4}+\frac{\varepsilon}{3}}\right)$$

instead of $1 + o(1)$.

Table of ρ(n , m)

n= \ m=	1	2	3	4	5	6	7	8	9	10	11	12
1	1											
2	1	1										
3	2	1	1									
4	2	1	1	1								
5	3	2	1	1	1							
6	4	2	1	1	1	1						
7	5	3	2	1	1	1	1					
8	6	3	2	1	1	1	1	1				
9	8	5	3	2	1	1	1	1	1			
10	10	5	3	2	1	1	1	1	1	1		
11	12	7	4	3	2	1	1	1	1	1	1	
12	15	8	5	3	2	1	1	1	1	1	1	1
13	18	10	6	4	3	2	1	1	1	1	1	1
14	22	12	7	4	3	2	1	1	1	1	1	1
15	27	15	9	6	4	3	2	1	1	1	1	1
16	32	17	10	6	4	3	2	1	1	1	1	1
17	38	21	12	8	5	4	3	2	1	1	1	1
18	46	25	15	9	6	4	3	2	1	1	1	1
19	54	29	17	11	7	5	4	3	2	1	1	1
20	64	35	20	12	8	5	4	3	2	1	1	1
21	76	41	24	15	10	7	5	4	3	2	1	1
22	89	48	28	17	11	7	5	4	3	2	1	1
23	104	56	32	20	13	9	6	5	4	3	2	1
24	122	66	38	23	15	10	7	5	4	3	2	1
25	142	76	44	27	17	12	8	6	5	4	3	2
26	165	89	51	31	20	13	9	6	5	4	3	2
27	192	103	59	36	23	16	11	8	6	5	4	3
28	222	119	68	41	26	17	12	8	6	5	4	3
29	256	137	78	47	30	20	14	10	7	6	5	4
30	296	159	91	55	35	23	16	11	8	6	5	4
31	340	181	103	62	39	26	18	13	9	7	6	5
32	390	209	118	71	45	29	20	14	10	7	6	5
33	448	239	136	81	51	34	23	17	12	9	7	6
34	512	273	155	93	58	38	26	18	13	9	7	6
35	585	312	176	105	66	43	29	21	15	11	8	7
36	668	356	201	120	75	49	33	23	17	12	9	7
37	760	404	228	135	84	55	37	26	19	14	10	8
38	864	460	259	154	96	62	42	29	21	15	11	8
39	982	522	294	174	108	70	47	33	24	18	13	10
40	1113	591	332	197	122	79	53	36	26	19	14	10
41	1260	669	375	221	137	88	59	41	29	22	16	12
42	1426	757	425	251	155	100	67	46	33	24	18	13
43	1610	853	478	281	173	111	74	51	36	27	20	15
44	1816	963	538	317	195	125	83	57	40	29	22	16
45	2048	1085	607	356	219	140	93	64	45	33	25	19
46	2304	1219	681	400	245	157	104	71	50	36	27	20
47	2590	1371	764	447	274	174	115	79	55	40	30	23
48	2910	1539	858	502	307	196	129	88	62	44	33	25
49	3264	1725	961	561	342	217	143	97	68	49	36	28
50	3658	1933	1075	628	383	243	160	109	76	54	40	30

n= \ m=	1	2	3	4	5	6	7	8	9	10	11	12
51	4097	2164	1203	701	427	270	177	120	84	60	44	34
52	4582	2418	1343	782	475	301	197	133	93	66	48	36
53	5120	2702	1499	871	529	333	218	147	102	73	53	40
54	5718	3016	1673	972	589	372	243	164	114	81	59	44
55	6378	3362	1863	1081	654	411	268	180	125	89	64	48
56	7108	3746	2073	1202	727	457	297	200	138	98	71	52
57	7917	4171	2308	1336	807	506	329	220	152	108	78	58
58	8808	4637	2564	1483	894	561	364	244	168	119	86	63
59	9792	5155	2847	1645	991	619	401	268	184	130	94	69
60	10880	5725	3161	1825	1098	687	444	297	204	144	104	76
61	12076	6351	3504	2021	1214	757	489	325	223	157	113	83
62	13394	7043	3882	2237	1343	837	540	360	246	173	125	91
63	14848	7805	4301	2476	1485	924	595	395	270	189	136	100
64	16444	8639	4757	2736	1638	1019	655	435	297	208	149	109
65	18200	9561	5260	3023	1809	1122	721	477	325	227	163	119
66	20132	10571	5814	3338	1995	1238	794	526	358	250	179	131
67	22250	11679	6419	3683	2198	1361	872	575	391	272	194	142
68	24576	12897	7083	4060	2422	1498	958	633	429	299	213	155
69	27130	14233	7814	4476	2667	1648	1053	693	470	326	232	169
70	29927	15694	8611	4928	2933	1811	1156	761	515	358	254	185
71	32992	17298	9484	5424	3226	1988	1267	832	562	389	276	200
72	36352	19054	10443	5967	3545	2184	1390	913	616	427	302	219
73	40026	20972	11488	6560	3893	2395	1523	997	672	464	328	237
74	44046	23074	12631	7207	4274	2626	1668	1093	735	508	359	259
75	48446	25372	13884	7917	4691	2880	1827	1194	803	553	390	281
76	53250	27878	15247	8687	5142	3154	1998	1305	876	604	425	306
77	58499	30621	16737	9530	5637	3453	2186	1425	955	656	462	331
78	64234	33613	18366	10449	6175	3780	2390	1558	1043	717	504	362
79	70488	36875	20138	11451	6760	4134	2611	1698	1136	778	546	391
80	77312	40437	22071	12541	7399	4519	2851	1854	1238	849	595	426
81	84756	44319	24181	13732	8095	4941	3114	2021	1349	922	646	461
82	92864	48545	26474	15023	8848	5395	3397	2203	1468	1004	702	502
83	101698	53153	28972	16431	9671	5891	3705	2400	1597	1089	761	542
84	111322	58169	31695	17963	10564	6430	4040	2615	1739	1186	827	590
85	121792	63623	34651	19628	11533	7014	4403	2845	1890	1286	896	637
86	133184	69561	37866	21435	12587	7646	4795	3097	2054	1398	973	692
87	145578	76017	41366	23403	13732	8337	5223	3369	2233	1516	1054	748
88	159046	83029	45163	25535	14971	9080	5683	3662	2424	1646	1142	811
89	173682	90653	49287	27852	16319	9889	6184	3981	2632	1783	1237	875
90	189586	98933	53770	30367	17780	10766	6726	4326	2858	1936	1341	950
91	206848	107915	58628	33093	19361	11715	7312	4697	3100	2096	1450	1024
92	225585	117670	63900	36048	21077	12740	7945	5100	3361	2272	1570	1109
93	245920	128250	69622	39255	22936	13856	8633	5536	3646	2460	1699	1197
94	267968	139718	75818	42725	24945	15056	9373	6004	3950	2664	1837	1295
95	291874	152156	82534	46486	27125	16359	10175	6513	4280	2882	1986	1396
96	317788	165632	89814	50559	29482	17767	11041	7060	4636	3120	2147	1510
97	345856	180224	97690	54965	32029	19289	11977	7651	5019	3373	2319	1627
98	376256	196032	106218	59732	34787	20931	12986	8289	5431	3648	2506	1758
99	409174	213142	115452	64893	37768	22712	14079	8979	5879	3943	2706	1895
100	444793	231651	125433	70468	40986	24627	15254	9718	6357	4261	2920	2045

References

[Dix] J. DIXMIER et J.L. NICOLAS, Partitions sans petits sommants, preprint of I.H.E.S. may 1987, to be published in the proceedings of the colloquium in number theory, Budapest, July 1987.

[Erd] P. ERDÖS and M. SZALAY, On the statistical theory of partitions ; in :Coll. Math. Soc. Jànos Bolyai 34. Topics in Classical Number Theory (Budapest, 1981), pp. 397-450, North - Holland / Elsevier.

[Har] G.H. HARDY and S. RAMANUJAN, Asymptotic formulae in combinatory analysis, Proc. London Math. Soc. (2) 17 (1918), pp. 75-115. (Also in Collected Papers of S. Ramanujan, pp. 276-309. Cambridge Univ. Press., Cambridge, 1927 ; reprinted by Chelsea, New York, 1962).

[Her] J. HERZOG, Gleichmässige asymptotische Formeln für parameterabhängige Partitionenfunktionen, Thesis, University J. W. Goethe, Frankfurt am Main, 1987.

[Hua] L. K. HUA, On the number of partitions of a number into unequal parts, Trans. Amer. Math. Soc. 51 (1942), pp. 194-201.

[Wat] G.N. WATSON, A treatise on the theory of Bessel functions, Cambridge, at the University Press, 1962.

P. ERDÖS
Mathematical Institute of the
Hungarian Academy of Sciences
H - 1053 BUDAPEST, Realtanoda u.13-15
HUNGARY

J.L. NICOLAS
Département de Mathématiques
Université de Limoges
123 av. A. Thomas
F-87060 LIMOGES cédex
FRANCE

M. SZALAY
Department of Algebra and Number Theory
Eötvös Lorànd University
H-1088 BUDAPEST, Muzeum Körut 6-8
HUNGARY

LINKS BETWEEN SOLUTIONS OF A-B = C AND ELLIPTIC CURVES

Gerhard Frey

Fachbereich 9 Mathematik
Universität des Saarlandes
D-6600 Saarbrücken

In the following paper we want to relate conjectures about solutions of the equation A-B = C in global fields with conjectures about elliptic curves. To do this we use the simple idea to interprete A and B as X-coordinates of points of order 2 of elliptic curves $E_{(A,B)}$, and then by using the discriminant of $E_{(A,B)}$ we transfer arithmetical properties of A,B and A-B to properties of $E_{(A,B)}$. In this way the (A-B-C)-conjecture of Masser-Oesterlé (cf. section 1) motivates the height conjecture (H) about elliptic curves (cf. section 2) which implies (A-B-C) (and is true over function fields). An interesting application of (A-B-C) (and so of (H)) is the Asymptotic Fermat Conjecture (cf. 1). In the case K = $\mathbb{Q}$ which is considered in section 4 to 7 the concept of modular elliptic curves is very useful. It turns out that the height of such curves is closely related to the degree of the modular parametrization. (H) is true if this degree is polynomially bounded by the conductor of the elliptic curve (cf. section 4). It becomes obvious that congruence primes play an important role if one tries to prove the height conjecture for modular elliptic curves. They are very important for modular representations of $G(\overline{\mathbb{Q}}/\mathbb{Q})$ too: Serre has formulated very far reaching conjectures about the minimal level of such representations (cf. [21]) and Ribet succeeded to prove an important special case of these conjectures in [19] which is stated in theorem 5.3. This theorem has the remarkable consequence that Taniyama's conjecture for elliptic curves implies Fermat's last theorem.

An overview over various conjectures and implications discussed in this paper can be found at the end of section 7, it should show how ideas of many mathematicians come together to find relations which could give a new approach towards Fermat's conjecture; especially it should emphasize the extremely valuable contributions of B. Mazur, J.F. Mestre, J. Oesterlé, K. Ribet, J.P. Serre and L. Szpiro.

1. The A-B-C-conjecture

We begin by fixing some notation:
K is a global field, i.e. K is a finite extension of $\mathbb{Q}$ or K is a finite extension of $K_o(X)$ with K_o algebraically closed in K and X transcendental over K_o.

We are mainly interested in the number field case, and hence it may be justified to simplify the situation in the function field case by assuming that $\mathrm{char}(K_o) = 0$.

Denote by n_K the degree of K over $\mathbb{Q}$ resp. $K_o(X)$. Σ_K are the non archimedean places of K, ∞ denotes the set of archimedean places of K (which is empty of course if $K \supset K_o(X)$). In each $\mathfrak{p} \in \Sigma_K \cup \infty$ we choose two distinguished valuations:

First case: $\mathfrak{p} \in \Sigma_K$. Then $v_\mathfrak{p} \in \mathfrak{p}$ is determined by $v_\mathfrak{p}(K^\times) = \mathbb{Z}$ (the normed valuation belonging to $\mathfrak{p}$) and $w_\mathfrak{p}$ is defined by

$$w_\mathfrak{p} := N_\mathfrak{p} \cdot v_\mathfrak{p} \quad \text{with}$$

$$N_\mathfrak{p} := \begin{cases} \log \# k_\mathfrak{p} & \text{if } K \text{ is a number field} \\ [k_\mathfrak{p} : K_o] & \text{if } K \text{ is a function field over } K_o \end{cases}$$

with $k_\mathfrak{p}$ the residue field of $\mathfrak{p}$.

Second case: $\mathfrak{p} \in \infty$. Corresponding to $\mathfrak{p}$ there is an imbedding

$$\iota_\mathfrak{p} : K \longrightarrow \mathbb{C} .$$

Define

$$w_\mathfrak{p}(x) := \begin{cases} -\log|\iota_\mathfrak{p}(x)| & \text{if } \iota_\mathfrak{p}(K) \subset \mathbb{R} \\ -2\log|\iota_\mathfrak{p}(x)| & \text{if } \iota_\mathfrak{p}(K) \not\subset \mathbb{R} \end{cases} .$$

It is well known that for $x \in K^\times$ one has the sum formula

$$\sum_{\mathfrak{p} \in \Sigma_K \cup \infty} w_\mathfrak{p}(x) = 0 .$$

To measure the "size" of $x \in K^\times$ one introduces the height: For $x \in K^\times$ define

$$h(x) := \frac{1}{n_K} \sum_{\mathfrak{p} \in \Sigma_K \cup \infty} \mathrm{Max}\{0, w_\mathfrak{p}(x)\} = \frac{1}{n_K} \sum_{\mathfrak{p} \in \Sigma_K \cup \infty} \mathrm{Max}\{0, -w_\mathfrak{p}(x)\} .$$

A divisor D of K is a product

$$D = \prod_{\mathfrak{p}\in\Sigma_K} \mathfrak{p}^{z_\mathfrak{p}} \quad \text{with } z_\mathfrak{p} \in \mathbb{Z} \text{ and } z_\mathfrak{p} = 0 \text{ for almost all } \mathfrak{p} .$$

The degree of D is defined by

$$\deg(D) := \sum_{\mathfrak{p}\in\Sigma_K} z_\mathfrak{p} N_\mathfrak{p} = \sum_{\mathfrak{p}\in\Sigma_K} w_\mathfrak{p}(D) \quad \text{where } w_\mathfrak{p}(D) := z_\mathfrak{p} N_\mathfrak{p} .$$

The support of D is

$$\operatorname{supp}(D) := \prod_{w_\mathfrak{p}(D)\neq 0} \mathfrak{p} ,$$

and for $x \in K^\times$ the principal divisor associated to x is

$$(x) := \prod_{\mathfrak{p}\in\Sigma_K} \mathfrak{p}^{v_\mathfrak{p}(x)} .$$

Now we are ready to state a conjecture which should be contributed to Masser and Oesterle:

<u>CONJECTURE (A-B-C). There is a constant</u> $d = d(K)$ <u>and a constant</u> $c(K,d)$ <u>such that for all</u> $x \in K^\times\setminus\{1\}$ <u>one has</u>

$$(1.1) \qquad h(x) \leq c(K,d) + \frac{d}{n_K} (\deg(\operatorname{supp}(x(x-1)))) .$$

<u>Remarks.</u> 1. Of course one will ask for minimal values of $d(K)$. We will see that $d(K) = 1$ is allowed in the function field case.
For $K = \mathbb{Q}$ $d(\mathbb{Q})$ has to be larger than 1 as the following example shows. A very optimistic guess would be: $d = 1+\epsilon$ is allowed for all $\epsilon > 0$.

<u>Example:</u> For $m \in \mathbb{N}$ take $x = 4^{3^m} = 3^m \cdot x_0 + 1$ with $x_0 \in \mathbb{N}$. Hence $h(x) = 3^m \log 4$, $\operatorname{supp}(x(x-1)) = \operatorname{supp}(3^m \cdot x_0 \cdot 4^{3^m})$ divides $2\cdot 3 \operatorname{supp}(x_0)$, and hence

$\deg(\operatorname{supp}(x(x-1))) \leq \log 2 + \log 3 + \log(4^{3^m} - 1) - m \log 3$.

So $h(x) - \deg(\operatorname{supp}(x(x-1))) \geq c + m \log 3$ and hence (1.1) cannot hold for $d = 1$ and for sufficiently large m.

2. The formulation (1.1) doesn't motivate the name of the conjecture. But: Take $A,B,C \in K^\times$ with $A-B = C$. Take $x = \frac{A}{C}$. Then $h(x) = h(\frac{A}{C})$ and $\operatorname{supp}(x(x-1)) = \operatorname{supp}(ABC^{-2})$ and hence (1.1) is equivalent with

$$\sum_{\substack{\mathfrak{p}\in\Sigma_K\cup\infty \\ w_\mathfrak{p}(A)>w_\mathfrak{p}(C)}} (w_\mathfrak{p}(A)-w_\mathfrak{p}(C)) \leq$$

(1.2)

$$n_K c(K,d) + d\Big(\sum_{\substack{\mathfrak{p}\in\Sigma_K \\ \mathrm{Min}\{v_\mathfrak{p}(A),v_\mathfrak{p}(B),v_\mathfrak{p}(C)\}\neq \\ \mathrm{Max}\{v_\mathfrak{p}(A),v_\mathfrak{p}(B),v_\mathfrak{p}(C)\}}} N_\mathfrak{p}\Big)$$

3. One can go one step further: Let $S_o \subset \Sigma_K$ be such that $O_{S_o} = \{x \in K;$ $v_\mathfrak{p}(x) \geq 0$ for $\mathfrak{p} \notin S_o\}$ is a principal domain. Given $x \in K^\times$ we can write $x = \frac{A}{C}$ with $A,C \in O_{S_o}$ and A,C relatively prime in O_{S_o}. Then $x-1 = \frac{A}{C} - 1 = \frac{A-C}{C} =: \frac{B}{C}$, and A,B,C are elements in O_{S_o} which are relatively prime. We use this to get the following formulation for (<u>A-B-C</u>): For all finite $S_o \subset \Sigma_K$ such that O_{S_o} is a principal domain there is a $d \in \mathbb{R}_{>0}$ and a constant $c(K,d,S_o)$ such that for relatively prime elements $A,B \in O_{S_o}$ one has

$$\sum_{\substack{\mathfrak{p}\in S_o\cup\infty \\ w_\mathfrak{p}(A)>w_\mathfrak{p}(B)}} (w_\mathfrak{p}(A)-w_\mathfrak{p}(B)) + \sum_{\mathfrak{p}\in\Sigma_K\setminus S_o} w_\mathfrak{p}(A) \leq$$

(1.3)

$$n_K c(K,d,S_o) + d\sum_{\substack{\mathfrak{p}\in\Sigma_K\setminus S_o \\ w_\mathfrak{p}(AB(A-B))>0}} N_\mathfrak{p}$$

4. Special case: $K = \mathbb{Q}$. In this case $S_o = \emptyset$ is possible and hence (1.3) yields: For $A,B \in \mathbb{Z}$ relatively prime with $|A| \geq |B|$ one has

(1.4) $$|A| \leq c(d)\Big(\prod_{p\,|\,AB(A-B)} p\Big)^d .$$

(<u>A-B-C</u>) has rather remarkable consequences for ternary diophantine equations. We give one example: Fix $a_1, a_2 \in K^\times$ and define

$$F(a_1,a_2) := \{(z_1,z_2) \in K^2;\ \exists\, n \geq 4 \text{ such that } a_1z_1^n - a_2z_2^n = 1\}.$$

ASYMPTOTIC FERMAT CONJECTURE. *$F(a_1,a_2)$ is finite in the number field case and $F(a_1,a_2)\backslash K_o^2$ is finite in the function field case.*

PROPOSITION 1.1. *(A-B-C) implies the Asymptotic Fermat conjecture.*

Proof. Since for $n \geq 4$ the genus of the curve C_n: $a_1 Z_1^n - a_2 Z_2^n = 1$ is larger than 1, we can use Falting's theorem to conclude that the proposition is true if there is a $n_o \in \mathbb{N}$ such that $\bigcup_{n \geq n_o} C_n(K)$ (resp. $(\bigcup C_n(K)\backslash K_o^2)$ in the function field case) is finite.

Now take (z_1, z_2) with $a_1 z_1^n - a_2 z_2^n = 1$. (A-B-C) implies that for $i = 1,2$ we have $h(a_i z_i^n) \leq c(K,d) + \frac{d}{n_K} \deg \operatorname{supp}(a_1 z_1 \cdot a_2 z_2)$. Hence $n \cdot h(z_i) \leq \tilde{c} + \frac{d}{n_K} \deg \operatorname{supp}(z_1 z_2)$ for some constant $\tilde{c}$ depending on d, K, a_1, a_2.

Now $2n_K h(z_i) \geq \deg \operatorname{supp}(z_i) + c'(K)$ for $i = 1,2$, and so $n(h(z_1)+h(z_2)) \leq c''(K,d,a_1,a_2) + 4d(h(z_1)+h(z_2))$ and so for all $\epsilon \in \mathbb{R}_{>0}$ there is a $n_o \in \mathbb{N}$ such that for all $n \geq n_o$ and $(z_1, z_2) \in C_n(K)$ one has $h(z_i) \leq \epsilon$.

But for ϵ small enough this implies that z_i is a root of unity in the number field case and a constant in the function field case.

2. The height conjecture for elliptic curves

Elliptic curves E/K can be given by Weierstraß equations

$$(2.1) \qquad Y^2 + a_1 XY + a_3 Y = X^3 + a_2 X^2 + a_4 X + a_6 \quad \text{with } a_i \in K .$$

The behaviour of E at $\mathfrak{p} \in \Sigma_K$ is determined by arithmetical properties of some functions in $a_1, \ldots, a_6$ (cf. [26]): Define

$$c_4 := (a_1^2 + 4a_2)^2 - 24(a_1 a_3 + 2a_4) ,$$

$$c_6 := -(a_1^2 + 4a_2)^3 + 36(a_1^2 + 4a_2)(a_1 a_3 + 2a_4) - 216(a_3^2 + 4a_6) .$$

Then $\Delta_E := 12^{-3}(c_4^3 - c_6^2)$ is the discriminant of E (depending on the choice of the equation (2.1) in an obvious way), and $j_E := c_4^3/\Delta_E$ (the j-invariant of E) resp. $\delta_E :\equiv -c_4 \cdot c_6 \bmod K^{\times 2}$ for $c_4 \cdot c_6 \neq 0$ (the Hasse invariant of E) are independent of this choice, (j_E, δ_E) determines E up to K-isomorphy.

<u>DEFINITION.</u> (2.1) <u>is $\mathfrak{p}$-minimal at</u> $\mathfrak{p} \in \Sigma_K$ <u>if</u>

i) $v_{\mathfrak{p}}(a_i) \geq 0$ <u>and</u>

ii) $v_{\mathfrak{p}}(\Delta_E)$ <u>is minimal</u>.

Let $n_{\mathfrak{p}}$ be the $v_{\mathfrak{p}}$-value of the discriminant of a $\mathfrak{p}$-minimal equation for E. Then

$$\mathfrak{D}_E := \prod_{\mathfrak{p} \in \Sigma_K} \mathfrak{p}^{n_{\mathfrak{p}}}$$

is the <u>discriminant divisor</u> of E/K.

Let $\mathcal{E}$/K be the Néron model of E/K (i.e. for all $\mathfrak{p} \in \Sigma_K$ with ring of $\mathfrak{p}$-adic integers $O_{\mathfrak{p}}$ $\mathcal{E} \otimes O_{\mathfrak{p}}$ is the Neron model of $E \otimes O_{\mathfrak{p}}$). Then one has an exact sequence

$$0 \longrightarrow (\mathcal{E} \otimes k_{\mathfrak{p}})^{o} \longrightarrow \mathcal{E} \otimes k_{\mathfrak{p}} \longrightarrow C_{\mathfrak{p}} \longrightarrow 0$$

where $(\mathcal{E} \otimes k_{\mathfrak{p}})^{o}$ is the connected component of the unity of $\mathcal{E} \otimes k_{\mathfrak{p}}$.

<u>DEFINITION.</u>

i) E <u>has good reduction at</u> $\mathfrak{p}$ <u>if</u> $\mathcal{E} \otimes k_{\mathfrak{p}}$ <u>is an elliptic curve</u>.

ii) E <u>has multiplicative reduction at</u> $\mathfrak{p}$ <u>if</u> $(\mathcal{E} \otimes k_{\mathfrak{p}})^{o}$ <u>is a torus</u>.

iii) E <u>has semistable reduction at</u> $\mathfrak{p}$ <u>if</u> E <u>has either good or multiplicative reduction at</u> $\mathfrak{p}$.

We have the following well-known criterion for semistable reduction: E has semistable reduction at $\mathfrak{p}$ if either $v_{\mathfrak{p}}(j_E) \geq 0$ and $v_{\mathfrak{p}}(\mathfrak{D}_E) = 0$ or $v_{\mathfrak{p}}(j_E) < 0$ and $K(\delta_E^{1/2})/K$ is unramified at $\mathfrak{p}$.

Hence there is a finite extension field L of K such that $E \otimes L$ has semistable reduction at all $\mathfrak{p} \in \Sigma_L$.

A measure for the type of reduction is the <u>conductor</u> N_E of E:

$$N_E = \prod_{\mathfrak{p} \in \Sigma_K} \mathfrak{p}^{n_{\mathfrak{p}}}$$

with $n_{\mathfrak{p}} = 0$ if E has good reduction at $\mathfrak{p}$, $n_{\mathfrak{p}} = 1$ if E has multiplicative reduction at $\mathfrak{p}$, and $n_{\mathfrak{p}} \geq 2$ else (see [26] for the exact definition in this case).

The geometric conductor $N_{E,geom}$ of E is defined by

$$N_{E,geom} := \prod_{v_{\mathfrak{p}}(j_E)<0} \mathfrak{p} .$$

One sees that N_E divides $\mathfrak{D}_E$. It was remarked by Parshin and Szpiro that in the function field case $\mathfrak{D}_E$ is not "too far away" from N_E, and Szpiro conjectured that this should be so in the general case too.

We'll state a conjecture about heights of elliptic curves which is closely related to (but stronger than) Szpiro's conjecture and which has the advantage that it has an obvious generalization to abelian varieties.

Firstly we recall the definition of heights of abelian varieties given by Faltings (for details cf. [4]):
Let A be an abelian variety of dimension d defined over K, let $\mathcal{G}$ be its Neron model with respect to all places of K, and $\omega(A) := \overset{d}{\wedge} (\mathrm{Lie}(\mathcal{G})^{\vee})$. Then $\omega(A)$ is a projective module of rank 1 over the ring of integers of K (in the number field case) resp. over the curve defined by K in the function field case, i.e. $\omega(A)$ defines a divisor of K. This module gets a hermitian structure at all archimedean places of K in the following way:
For $\iota: K \longrightarrow \mathbb{C}$ and for $\alpha \in \overset{d}{\wedge} (\mathrm{Lie}(\iota A)^{\vee})$ define

$$\|\alpha\|_{\iota}^2 = \langle \alpha, \alpha \rangle := \frac{1}{(2\pi)^d} \int_{\iota A} |\alpha \wedge \bar{\alpha}| .$$

The degree of $\omega(A)$ is defined in the following way: For all $\mathfrak{p} \in \Sigma_K$ choose an isomorphism $\varphi_{\mathfrak{p}}: \omega(A) \otimes O_{\mathfrak{p}} \longrightarrow O_{\mathfrak{p}}$. Take $\alpha \in \omega(A)\setminus\{0\}$. Then

$$\deg(\omega(A)) = \sum_{\iota: K \to \mathbb{C}} -\log\|\alpha\|_{\iota} + \sum_{\mathfrak{p}\in\Sigma_K} w_{\mathfrak{p}}(\varphi_{\mathfrak{p}}(\alpha)) .$$

<u>DEFINITION.</u> $h(A) := \frac{1}{n_K} \deg \omega(A)$.

<u>Example.</u> Take $K = \mathbb{Q}$. Then $\overset{d}{\wedge} (\mathrm{Lie}\ \mathcal{G})^{\vee}$ is a $\mathbb{Z}$-module of rank 1. Let α be a generator (α is the "Neron d-form" of A). Then

$$h(A) = -\frac{1}{2} \log \frac{1}{(2\pi)^d} \int_{A\otimes\mathbb{C}} |\alpha\wedge\bar{\alpha}| .$$

As in the theory of elliptic curves one has the notion of semi-stability of A at $\mathfrak{p} \in \Sigma_K$: The connected component of the unity of $\mathcal{G} \otimes k_{\mathfrak{p}}$ is a finite extension of an abelian variety by a torus. There is a finite extension L of K such that $A \otimes L$ is semistable at all $\mathfrak{p} \in \Sigma_L$.

DEFINITION. $h_{geom}(A) := h(A \otimes L)$.

One has: $h_{geom}(A) \leq h(A)$, and equality holds if and only if A is semi-stable over K.

In the case of elliptic curves everything can be computed explicitely (cf. [4]):
For $\iota: K \longrightarrow \mathbb{C}$ let τ_ι be an element with $\mathrm{Im}(\tau_\iota) \geq \frac{\sqrt{3}}{2}$ such that $\iota E \cong \mathbb{C}/\mathbb{Z}+\mathbb{Z}\tau_\iota$. With $q_\iota = e^{2\pi i \tau_\iota}$ one has

$$\iota j_E = \frac{1}{q_\iota} + 744 + \ldots$$

(the usual q-expansion of the j-function).
Then

$$h_{geom}(E) = \frac{1}{12 n_K} \{-\log \Big| \prod_{\iota: K \to \mathbb{C}} ((4\pi \,\mathrm{Im}\, \tau_\iota)^6 \cdot q_\iota \prod_{n \in \mathbb{N}} (1-q_\iota^n)) \Big| + \sum_{\mathfrak{p} \in \Sigma_K} \mathrm{Max}(0, -w_{\mathfrak{p}}(j_E))\} .$$

Since

$$-\log|q_\iota| = 2\pi \,\mathrm{Im}(\tau_\iota) \geq 2\pi \frac{\sqrt{3}}{2} = \pi\sqrt{3}$$

and

$$\frac{2\pi \,\mathrm{Im}(\tau_\iota)}{\log|\iota j_E|} \sim 1 \quad \text{for } |\iota j_E| \longrightarrow \infty$$

one has

$$h_{geom}(E) \sim \frac{1}{12} h(j_E) .$$

By using the definition of the height of E by differentials it is easy to see how to compare $h(E)$ with $h_{geom}(E)$:

$$h(E) = h_{geom}(E) + \frac{1}{12 n_K} \Big(\sum_{v_{\mathfrak{p}}(j_E) \geq 0} w_{\mathfrak{p}}(\mathcal{D}_E) + 6 \sum_{\substack{v_{\mathfrak{p}}(j_E)<0 \\ v_{\mathfrak{p}}(\delta_E) \equiv 1 \bmod 2}} N_{\mathfrak{p}} \Big) .$$

Hence

$$h(E) \leq h_{geom}(E) + \frac{1}{2 n_K} \deg\Big(\frac{N_E}{N_{E,geom}}\Big) .$$

We are now ready to state height conjectures for elliptic curves:

(H_{geom}). There is a number d *and for all finite* $S \subset \Sigma_K$ *there is a constant* $c(K,d,S)$ *such that for all elliptic curves* E/K *which are semi-stable outside* S *one has*

$$(2.2) \qquad h_{geom}(E) \leq c(K,d,S)+ \frac{d}{2n_K} \deg(N_{E,geom}) \ .$$

Remark. As stated above this is equivalent with

$$h(j_E) \leq c'(K,d,S,\varepsilon)+ \frac{6(d+\varepsilon)}{n_K} \ (\sum_{w_{\mathfrak{p}}(j_E)<0} N_{\mathfrak{p}})$$

for all $\varepsilon > 0$, and $\varepsilon = 0$ in the function field case.

One could hope that $c(K,d,S)$ depends linearly on $\sum_{\mathfrak{p}\in S} N_{\mathfrak{p}}$, and this would motivate the height conjecture

(H). There is a $d \in \mathbb{R}$ *and a constant* $c(K,d)$ *such that for all* E/K *one has*

$$h(E) \leq c(K,d)+ \frac{d}{2n_K} \deg(N_E) \ .$$

It is obvious how to generalize (*H*) to abelian varieties:

(H). For all* A/K *one has*

$$h(A) \leq c(K,d,\dim A)+ \frac{d(\dim A,K)}{n_K} \deg(N_A)$$

where N_A *is the conductor of* A.

Let us give a first application of (H_{geom}):

PROPOSITION 2.1. *For given* $c,d \in \mathbb{R}$ *let* $\mathcal{E}_{c,d}$ *be the set of all elliptic curves* E/K *with*

$$h_{geom}(E) \leq c+ \frac{d}{2n_K} \deg(N_{E,geom})$$

with j*-invariant not in* K_o *in the function field case. Then*

$$\# E(K)_{tor} \leq M(c,d,K) \qquad \text{for all } E \in \mathcal{E}_{c,d}$$

where $M(c,d,K)$ *can be effectively computed.*

Proof. Assume that E(K) contains a point P of order m. For m large enough (depending on n_K only) it follows that E and $E' := E/\langle P\rangle$ are semistable at all $\mathfrak{p} \in \Sigma_K$ and that for all places $\mathfrak{p}$ with $v_{\mathfrak{p}}(2) > 0$ one has $v_{\mathfrak{p}}(j_E) < 0$, hence $N_E = N_{E,geom} = N_{E'}$ is non trivial in the number field case, and since $j_E \notin K_o$ by assumption this follows in the function field case too.
Now take $\mathfrak{p} \in \Sigma_K$ dividing N_E. Let $m_o \in \mathbb{N}$ be minimal such that m_oP lies in the connected component of $\mathcal{E}^{(\mathfrak{p})}$ (the fiber of the Neron model of E at $\mathfrak{p}$). The theory of Tate curves (cf. [20]) implies that $m_o | v_{\mathfrak{p}}(j_E)$ and $v_{\mathfrak{p}}(j_{E'}) = \frac{m}{m_o^2} v_{\mathfrak{p}}(j_E)$, hence

$$\mathrm{Max}\{-v_{\mathfrak{p}}(j_E), -v_{\mathfrak{p}}(j_{E'})\} \geq m^{1/2} ,$$

and so

$$\frac{1}{12n_K} m^{1/2} \deg N_E \leq h_{geom}(E)+h_{geom}(E') \leq 2c+ \frac{d}{n_K} \deg N_E + \log m$$

since $|h_{geom}(E)-h_{geom}(E')| \leq \log m$ (cf. [4]), and hence the proposition follows.

Especially one gets

COROLLARY 2.2. *If* (H_{geom}) *holds in* K *then the torsion of all elliptic curves over* K *is uniformly bounded by a number depending on* n_K, d(K) *and* c(K,d).[1]

A good test for conjectures for global fields is the function field case. We have

PROPOSITION 2.3. *Let* K *be a function field of genus* g. *Let* E/K *be an elliptic curve which is semistable outside* $S \subset \Sigma_K$. *Then*

$$h_{geom}(E) \leq \frac{1}{n_K} [((g-1)+ \frac{1}{2} \sum_{\mathfrak{p}\in S} N_{\mathfrak{p}})+ \frac{1}{2} \deg N_{E,geom}] .$$

Hence (*H*) *holds in* K *with* d = 1 *and* $c(K,1) = \frac{1}{n_K}(g-1)$.

Proof. Without loss of generality we can assume that K_o is algebraically closed and $j_E \notin K_o$.

---

1) as conjectured in general and proved for function fields and K = $\mathbb{Q}$.

$n_K \cdot h(j_E) = 12 n_K\, h_{geom}(E) = [K:K_o(j_E)] =: d$ is equal to the degree of the zero divisor of j_E resp. $(j_E - 12^3)$. For all places $\mathfrak{p} \in \Sigma_K$ one has

$$v_\mathfrak{p}(\mathfrak{D}_E) \equiv v_\mathfrak{p}(j_E^2(j_E-12^3)^3) \bmod 6,$$

and hence for $\mathfrak{p} \notin S \cup \operatorname{supp} N_{E,geom}$ one has

$$0 \equiv v_\mathfrak{p}(j_E^2(j_E-12^3)^3) \bmod 6 .$$

Hence all $\mathfrak{p} \notin S$ with $v_\mathfrak{p}(j_E) > 0$ (resp. $v_\mathfrak{p}(j_E-12^3) > 0$) are ramified in $K/K_o(j_E)$ of order $e_\mathfrak{p} \geq 3$ (resp. $e_\mathfrak{p} \geq 2$) and

$$\#\{\mathfrak{p} \in \Sigma_K,\ v_\mathfrak{p}(j_E(j_E-12^3)) > 0\} \leq \frac{d}{3} + \frac{d}{2} + \#S = \frac{5}{6}d + \#S .$$

We apply the Hurwitz genus formula and get

$$2g-2 \geq -2d + \sum_{v_\mathfrak{p}(j_E)<0} (-v_\mathfrak{p}(j_E)-1) + \sum_{v_\mathfrak{p}(j_E(j_E-12^3))>0} (e_\mathfrak{p}-1)$$

$$\geq d - \deg N_{E,geom} - \frac{5}{6}d - \#S$$

or:

$$\frac{n_K}{6}\, h(j_E) \leq 2g - 2 + \#S + \deg N_{E,geom}$$

which proves the proposition.

We end this section by discussing a "relative" situation. Let $\pi: C_1 \longrightarrow C_2$ be a K-morphism of curves. If $g(C_2) \geq 2$ then $\deg \pi$ is bounded by the genus of C_1 due to the Hurwitz genus formula. If $g(C_2) = 1$ the situation is not so easy to describe, and surely $\deg \pi$ cannot be bounded if one does not impose a maximality condition for C_2. So we define for elliptic curves E/K and curves C/K with Jacobian J(C) the number

$$e_K(E,C) := \begin{cases} 0 \text{ if } E \text{ is not K-isogenous to a factor of } J(C) \\ \operatorname{Min}\{\deg \pi;\ \pi: J(C) \xrightarrow[K]{} E^*\} \\ E^* \text{ K-isogenous to } E \\ E^* \text{ a factor of } J(C) \end{cases}$$

Using this definition we formulate

<u>QUESTION (D).</u> Are there $d(K) \in \mathbb{R}_{>0}$ and for all finite sets $S \subset \Sigma_K$ numbers $c(K,S)$ with the following property:

For all curves C/K such that the Jacobian J(C) of C is semistable outside of S one has

$$\log(e_K(E,C)) \leq c(K,d,S)+d(K)\ \deg(\prod_{\substack{\mathfrak{p}\in\Sigma_K \\ J(C) \text{ has bad reduction mod } \mathfrak{p}}} \mathfrak{p})\ \log(g(C)) \quad .$$

We'll see that coverings of elliptic curves arise in a rather natural way if $K = \mathbb{Q}$, and so a special case of (D) plays an important role if one wants to prove (H) over $\mathbb{Q}$.

3. The curves $E_{(A,B)}$

The purpose of this section is to relate the height conjectures about elliptic curves with the A-B-C-conjecture. The tool we use are the elliptic curves $E_{(A,B)}$ discussed in various papers by Hellegouarch ([10]) and the author ([7], [8]).

We take $S_o \subset \Sigma_K$ such that S_o contains all divisors of 2 and O_{S_o} is a principal domain. For $x \in K^\times\setminus\{1\}$ we choose relatively prime $A, C \in O_{S_o}$ with $x = \frac{A}{C}$. Then $x-1 = \frac{A-C}{C} =: \frac{B}{C}$ with $B \in O_{S_o}$ and B relatively prime to A too.

The elliptic curve

$$E_{(A,B)}: Y^2 = X(X-A)(X-B)$$

has j-invariant

$$j_{(A,B)} = 2^8 \frac{(A^2+B^2-AB)^3}{A^2B^2(A-B)^2} = 2^8 \frac{(x^2-x+1)^3}{x^2(x-1)^2} ,$$

the Hasse invariant of $E_{(A,B)}$ is

$$\delta_{(A,B)} \equiv \frac{1}{2}(A^2+B^2-AB)(A+B)(2A^2+2B^2-5AB) \bmod K^{\times 2} .$$

So $E_{(A,B)}$ is semistable outside S_o and its conductor is

$$N_{(A,B)} = N'_{S_o} \prod_{\substack{v_\mathfrak{p}(AB(A-B))>0 \\ \mathfrak{p}\notin S_o}} \mathfrak{p} \quad \Big|\ N'_{S_o}\ \operatorname{supp}(x(x-1))$$

with $\deg N'_{S_o}$ bounded by

$$c + 2 \sum_{\mathfrak{p}\in S_o} N_\mathfrak{p}$$

where $c = c(K)$ is a constant reflecting the behaviour of 2 in K. Hence conjecture ($\underline{H}_{geom}$) implies: There is a $\tilde{d} \in \mathbb{R}_{>0}$ and a constant $c(K,\tilde{d})$ such that for all $x \in K^{\times}\setminus\{1\}$ one has

$$(3.1) \qquad h(j_{(A,B)}) = h\left(2^8 \frac{(x^2-x+1)^3}{x^2(x-1)^2}\right) \le c(K,\tilde{d}) + \frac{6\tilde{d}}{n_K} \deg(\mathrm{supp}(x(x-1))) .$$

Since $h\left(2^8 \frac{(x^2-x+1)^3}{x^2(x-1)^2}\right) = c' + 6h(x)$ we have:

$$h(x) \le \tilde{c}(K,\tilde{d}) + \frac{\tilde{d}}{n_K} \deg(\mathrm{supp}(x(x-1))) .$$

<u>Result.</u> The (A-B-C)-conjecture is true if ($\underline{H}_{geom}$) holds for $E_{(A,B)}$ with $A,B \in O_{S_0}$ relatively prime. (Note that the number $\tilde{d}$ in (3.1) can be taken equal to $d+\varepsilon$ ($\varepsilon > 0$) if d occurs in formula (2.1).)

<u>Remark.</u> For $K = \mathbb{Q}$ one sees conversely that the validity of (1.1) for all $d = 1+\varepsilon$ implies (2.1) with $d = 1+\varepsilon$ for all $\varepsilon > 0$.

The (A-B-C)-conjecture predicts that the "average exponent" of prime divisors of A,B and A-B is equal to d, and we have seen above how this conjecture can be translated into the language of elliptic curves using $E_{(A,B)}$. But $E_{(A,B)}$ gives information about the exponent of a fixed $\mathfrak{q} \in \Sigma_K$ in AB(A-B) too:

For a prime p let $\mathcal{E}_p$ be the group scheme which is equal to the kernel of the multiplication by p of the Neron model of E/K, $\mathcal{E}$, and let $\rho_{E,p}$ be the representation of $G(\overline{K}/K)$ induced by the action of $G(\overline{K}/K)$ on $\mathcal{E}_p \otimes \overline{K} = E(\overline{K})_p$, hence

$$\rho_{E,p} : G(\overline{K}/K) \longrightarrow Gl(2,\mathbb{Z}/p) .$$

Define

$$R_{\rho_{E,p}} := \prod_{\substack{\mathfrak{q}\in\Sigma_K \\ \mathcal{E}_p\otimes O_{\mathfrak{q}} \text{ is \underline{not} finite over } O_{\mathfrak{q}}}} \mathfrak{q} .$$

If $\rho_{E,p}$ is finite at p then $R_{\rho_{E,p}}$ is the support of $N_{\rho_{E,p}}$, the prime-to-p part of the conductor of $\rho_{E,p}$. One has:

$$R_{\rho_{E,p}} \mid N_E .$$

But $R_{\rho_{E,p}}$ is smaller than N_E in many cases: If $\mathfrak{q} \in \Sigma_K$ such that E is semistable at $\mathfrak{q}$ (and p big enough if $v_{\mathfrak{q}}(p) > 0$) then

$$\mathfrak{q} \mid R_{\rho_{E,p}} \text{ if and only if } \mathrm{Min}\{0, v_{\mathfrak{q}}(j_E)\} \not\equiv 0 \bmod p .$$

Example. Take $A, B \in O_{S_o}$ relatively prime and $E_{(A,B)}$ as above. Then

$$R_{\rho_{E_{(A,B)},p}} = R'_{S_o} \cdot \prod_{\substack{v_{\mathfrak{q}}(AB(A-B)) \not\equiv 0 \bmod p \\ \mathfrak{q} \notin S_o}} \mathfrak{q}$$

with $R'_{S_o} \mid \prod_{\mathfrak{q} \in S_o} \mathfrak{q}$.

Especially: Take (z_1, z_2) as solution of

$$a_1 Z_1^p - a_2 Z_2^p = 1 .$$

Let $A, C \in O_{S_o}$ be relatively prime with $\frac{A}{C} = z_1$, $B = z_2^p C^p$. Then

$$R_{\rho_{E_{(a_1 A^p, a_2 B^p)},p}} \mid R_o(a_1, a_2, K)$$

where $R_o(a_1, a_2, K)$ is independent of p.
Hence to prove that $a_1 Z_1^p - a_2 Z_2^p = 1$ has no non trivial solution in K one can try to prove that no representation of $G(\overline{K}/K)$ on $\mathbb{Z}/p \times \mathbb{Z}/p$ of the type described above (and hence with rather small conductor) exists.

For general number fields K this seems to be far out of range today, but for $K = \mathbb{Q}$ we shall see that this point of view has very interesting consequences.

4. The height conjecture for modular elliptic curves

For $N \in \mathbb{N}$ let $X_o(N)/\mathbb{Q}$ be a canonical model over $\mathbb{Q}$ of the Riemann surface $\mathbb{H}^*/\Gamma_o(N)$ with

$$\mathbb{H}^* = \{z \in \mathbb{C};\ \mathrm{Im}(z) > 0\} \cup \mathbb{Q} \cup \{i\infty\} \quad \text{and}$$

$$\Gamma_o(N) = \{\begin{pmatrix} a & b \\ c & d \end{pmatrix} \in \mathrm{Sl}(2,\mathbb{Z});\ N \mid c\} .$$

Hence the points of $X_o(N)$ parametrize elliptic curves with cyclic isogeny of degree N.

$J_o(N)/\mathbb{Q}$ is the Jacobian of $X_o(N)/\mathbb{Q}$, its Néron model over $\mathbb{Q}$ is denoted by $\mathcal{J}_o(N)$.

Let l be a prime number dividing N. Then one has a natural map

$$\varphi_l : X_o(N) \longrightarrow X_o(\tfrac{N}{l})$$

(use the modular interpretation of $X_o(N)$) and an involution ω_l operating on $X_o(N)$ which is the identity on $X_o(\frac{N}{l})$. Hence we can apply φ_l^* resp. $(\varphi_l \circ \omega_l)^*$ to various geometric objects like differentials on $X_o(\frac{N}{l})$ or the Jacobian of $X_o(\frac{N}{l})$ to lift these objects to corresponding objects of $X_o(N)$. These lifted objects are called "old" (or "coming from a lower level").

An important fact we have to recall is that one knows a commutative subring of the endomorphism of $J_o(N)$, namely for primes there is a Hecke operator T_l (for definition cf. [22]), the algebra generated by $\{T_l;\ l \in \mathbb{P}\}$ and Fricke involutions $\omega_l(l|N)$ is called the Hecke algebra $\mathbb{T}_N$. $\mathbb{T}_N$ operates on various objects, for instance it operates on the space of holomorphic differentials of $X_o(N)$. Let ω be such a differential. Then $\omega = f \cdot \frac{dq}{q}$ with $q = e^{2\pi iz}$ and f has a Fourier expansion at $i\infty$:

$$f(z) = \sum_{i=1}^{\infty} a_i q^i \quad \text{with } a_i \in \mathbb{C}\ .$$

<u>DEFINITION.</u> 1. <u>f is a cusp form of weight 2 and level N. For all rings $R \subset \mathbb{C}$ let $S_2(N)(R)$ be the R-module of cusp forms of weight 2 and level N for which $a_i \in R$ for all i.</u>
2. <u>f is a new form of level N if f is an eigenfunction under $\mathbb{T}_N$, and if f is orthogonal under the Petersson scalar product to the space generated by all old forms.</u>

One knows that $J_o(N)$ has good reduction outside of divisors of N and semistable reduction for all primes l dividing the square-free part N_{sf} of N.

<u>DEFINITION. An elliptic curve E/$\mathbb{Q}$ is a modular curve if there is a non trivial $\mathbb{Q}$-morphism $\varphi: X_o(N) \longrightarrow E$.</u>

If we choose N minimal then $N = N_E$ ([2]). Hence E is modular if and only if E is a $\mathbb{Q}$-factor of $J_o(N_E)$.

We want to compute the height of E. Let ω be a Néron differential of E (i.e. $\langle\omega\rangle = \Omega^1(\mathcal{E})/\mathbb{Z}$). Then by definition we have

$$h(E) = -\frac{1}{2}\log\frac{1}{2\pi}\int_{E\otimes\mathbb{C}} |\omega\wedge\bar{\omega}| .$$

Define $\omega^* := \varphi^*(\omega) = \tilde{f}_E \cdot \frac{dq}{q}$ with $\tilde{f}_E \in S_2(N_E)(\mathbb{Z})$.[1)]

So

$$\omega^* = c_\varphi \cdot (q + \sum_{i=2}^{\infty} a_i q^i) \frac{dq}{q} \quad \text{with } c_\varphi, a_i \in \mathbb{Z} .$$

Define $f_E := \tilde{f}_E \cdot c_\varphi^{-1}$.

We'll have to use the following facts:

1. For $l \in \mathbb{P}$ let $\sigma_l \in G(\bar{\mathbb{Q}}/\mathbb{Q})$ be a Frobenius element for l. Then for $p \in \mathbb{P}$ one has:

$$\mathrm{Tr}_{\rho_{E,p}}(l) := \mathrm{Tr}(\rho_{E,p}(\sigma_l)) \equiv a_l \bmod p \quad \text{for all } l \nmid p\cdot N_E .$$

2. For $l \nmid N_E$ we have:

$$|a_l = (l+1) - \#(\mathcal{E} \otimes \mathbb{Z}/l)(\mathbb{Z}/l)| \leq 2\sqrt{l} .$$

3. f_E is a new form, and so $|a_n| \leq c\cdot n$ for some constant c independent of N_E.

We use 3. to estimate h(E):

$$h(E) = -\frac{1}{2}\log\frac{1}{2\pi}\int_{E\otimes\mathbb{C}} |\omega\wedge\bar{\omega}|$$

$$= -\frac{1}{2}\log\frac{1}{2\pi \deg\varphi}\int_{X_o(N_E)\otimes\mathbb{C}} |\tilde{f}_E^2|\,dv$$

with $dv = 4\pi dxdy$. So

$$h(E) = \frac{1}{2}\log(\deg\varphi) - \log|c_\varphi| - \frac{1}{2}\log\frac{1}{2\pi}\int_{X_o(N_E)\otimes\mathbb{C}} |f_E|^2\,dv .$$

We choose b large enough (independent of N_E) such that

$$|f_E(z)| \geq \left||q| - \sum_{i=2}^{\infty}|a_i q^i|\right| \geq \frac{1}{2}|q| \quad \text{for } \mathrm{Im}(z) \geq b ,$$

and a fundamental domain in $\mathbb{H}$ for $\Gamma_o(N_E)$ containing

$$U := \{z \in \mathbb{C};\ \mathrm{Im}(z) \geq b,\ |\mathrm{Re}(z)| < \tfrac{1}{2}\}$$

---

1) It is clear that $\tilde{f}_E \in S_2(N_E)(\mathbb{Q})$, and using results of [12] one can show that the Fourier expansion of $\tilde{f}_E$ at $i\infty$ has integral coefficients.

to get:

$$\frac{1}{2\pi}\int_{X_o(N_E)\otimes\mathbb{C}} |f_E|^2\, dv \geq \frac{1}{8\pi}\int_U |q|^2\, dv =: k > 0$$

with k independent of N_E.
Hence

$$h(E) \leq \frac{1}{2}\log(\deg\varphi) - \log|c_\varphi| - \frac{1}{2}\log k.$$

Now it becomes obvious how the question (D) stated in 2. is related to the height conjecture:
Specialize (D) to modular parametrizations of elliptic curves, use Mazur's result that the degree of cyclic isogenies over $\mathbb{Q}$ is bounded by 163 and conclude: There is a $d \in \mathbb{R}_{>0}$ and for all $N' \in \mathbb{N}$ a constant $c(d,N')$ such that for all elliptic curves with conductor N_E with $\frac{N_E}{N_{E,sf}} | N'$ and for all $\varphi: X_o(N_E) \longrightarrow E$ one has

(4.1) $$\frac{\deg\varphi}{c_\varphi^2} \leq c(d,N')\, g(X_o(N_E))^d .$$

<u>CONJECTURE</u> (D_{mod}). <u>For all modular parametrizations</u> $\varphi: X_o(N_E) \longrightarrow E/\mathbb{Q}$ (4.1) <u>holds</u>.

Since $g(X_o(N_E))$ can be bounded by a linear function in N_E we get:

<u>PROPOSITION 4.1.</u> (D_{mod}) <u>implies the geometric height conjecture for modular elliptic curves</u>.

We have seen in section 2 how to compute deg φ inside of $J_o(N_E)$. Assume that $\varphi: X_o(N_E) \longrightarrow E$ has minimal degree and that $p \in \mathbb{P}$ such that E has no $\mathbb{Q}$-rational isogeny of degree p. Let A_E be the connected component of the unity of ker φ_* where $\varphi_*: J_o(N_E) \longrightarrow E$ is induced by φ. Then $p | \deg\varphi$ implies that $A_E \cap \varphi^*E \supset \varphi^*E_p$, so E_p is isomorphic to a subscheme of a simple abelian subvariety A_1 of $J_o(N)$ which is not isogenous to E. Following Shimura we associate an eigenform $f_{A_1} \in S_2(N_E)(\mathcal{O})$ to A_1 where $\mathcal{O}$ is the ring of integers of some number field:

$$f_{A_1}(z) = q + \sum_{i=2}^{\infty} b_i q^i , \quad b_i \in \mathcal{O} .$$

Now we use the relation between $\rho_{E,p}$ (E_p regarded as subscheme of A_1) and the coefficients of f_{A_1}: There is a prime $\mathfrak{p} | p$ in $\mathcal{O}$ such that

$$\mathrm{Tr}\,\rho_{E,p} \equiv b_l \bmod \mathfrak{p}$$

for all l prime to $p\cdot N_E$.
Hence $b_l \equiv a_l \bmod \mathfrak{p}$ for all $l \nmid pN_E$, and this means that <u>p is a congruence prime</u> for f_E (cf. [8]).

It follows that the discussion of the height conjecture for modular elliptic curves is closely related to the discussion of congruence primes.
There are two possible cases: If p is a congruence prime for f_E and "$f \equiv g$" then g can be an l-old form or g has to be a new form. A necessary criterion for the first case is that $\mathcal{E}_p \otimes \mathbb{Z}_l$ is finite over $\mathbb{Z}_l$ if $l^2 \nmid N_E$. We'll see in the next section that under reasonable assumptions this condition is sufficient too!

But unfortunately most of all congruence primes seem to be of the second kind, and this can be the reason for the difficulties of the height conjecture even for modular elliptic curves.

We end this section by presenting an easy method to construct congruence primes.

PROPOSITION 4.2. <u>Assume that E is modular with even conductor. Let p, l be odd primes with $p \geq 5$ such that E has no Q-isogeny of degree p, $l \| N_E$ and $v_l(\mathcal{D}_E) \equiv 0 \bmod p$. Then p is a congruence prime for f_E.</u>

<u>Proof.</u> We use the description of $\mathcal{J}_o(N_E)$ at l which is due to Deligne-Rapoport and Mazur (cf. [13]): Up to elements of 2- or 3-power order the group of connected components of $\mathcal{J}_o(N_E) \otimes (\mathbb{Z}/l) =: \mathcal{J}^{(l)}$ is generated by the image of the Q-rational cusp divisor $c = (0)-(\infty)$.
It follows that all components $C^{(i)}$ of $(\varphi^*\mathcal{E}) \otimes (\overline{\mathbb{Z}/l})$ with $p\cdot C^{(i)}$ equal to the connected component of the unity $C^{(o)}$ are in $\mathcal{J}^{(l)o}$, the connected component of the unity of $\mathcal{J}^{(l)}$, since the norm map from $X_o(N_E)$ to $X_o(\frac{N_E}{2})$ maps φ^*E to 0 but c to $\tilde{c}$ with order $(\tilde{c}) = \dfrac{\text{order } c}{3}$.
Hence $\varphi_*(C^{(i)}) \in (\mathcal{E} \otimes (\overline{\mathbb{Z}/l}))^o$ if $pC^{(i)} = C^{(o)}$.
Now let P be a point of order p in $\varphi^*E(\mathbb{Q})$. Let $\mathcal{E}'$ be the Néron model of E over $\mathbb{Q}(P)$. Semi-stability of $\varphi^*(E)$ at l implies: For all prime divisors $\mathfrak{l} | l$ of $\mathbb{Q}(P)$ we have a natural mapping of the connected components of $(\varphi^*\mathcal{E}) \otimes (\overline{\mathbb{Z}/l})$ to components of $\varphi^*\mathcal{E}' \otimes (\overline{\mathbb{Z}/l})$, and since $v_l(\mathcal{D}_E) \equiv 0$ mod p it follows that components C' of $\varphi^*\mathcal{E}'$ mod l with $pC' = C'^{(o)}$ are images of components $C^{(i)}$ of $\varphi^*\mathcal{E}$ with $pC^{(i)} = C^{(o)}$, hence $\varphi_*(C'^{(i)}) \subset$

$(\mathcal{E}')^o$ mod l.
Since $pP = 0$ it is clear that P is mapped modulo l to a component $C'^{(i)}$ with $pC'^{(i)} = 0$. So $\varphi_*(P)$ is in the connected component of the unity of $\mathcal{E}'$ modulo l and hence $\varphi_*(P)$ is contained in a cyclic group of order p which is independent of the special choice of P. Hence $p \mid \#(\mathrm{Ker}\ \varphi_* \cap \varphi^*E)$ and since φ^*E has no $\mathbb{Q}$-rational isogeny of degree p the proposition follows.

Example. Take $A,B \in \mathbb{Z}$ relatively prime and assume that $E_{(A,B)}$ is modular. (For example take $A = 2^8$, $B = 3^5$. Then $E_{(2^6,3^5)}$ is the curve (78B) in [16].)
Assume that $v_l(AB(A-B)) \equiv 0 \bmod p$ for $p \geq 5$ and $l \geq 3$. Then p is a congruence prime for f_E. (In our example 5 is a congruence prime, and f_E is congruent mod 5 to the cusp form of level 26 corresponding to the curve (26B).)

5. Modular representations of $G(\overline{\mathbb{Q}}/\mathbb{Q})$: Ribet's theorem

Let p be a prime ≥ 5 and $E/\mathbb{Q}$ an elliptic curve over $\mathbb{Q}$ without $\mathbb{Q}$-rational isogeny of degree p. As usual $\mathcal{E}_p/\mathbb{Z}$ is the scheme of points of order p of the Néron model $\mathcal{E}$ of E, and $\rho_{E,p}$ the representation induced by the action of $G(\overline{\mathbb{Q}}/\mathbb{Q})$ on $E_p := \mathcal{E} \otimes \mathbb{Q}$. $\rho_{E,p}$ is irreducible and $\det(\rho_{E,p}) = \chi_p$, the cyclotomic character mod p.

$$R_{\rho_{E,p}} := \prod_{\mathcal{E}_p \otimes \mathbb{Z}_l \text{ is not finite}} l$$

can easily be described.
As an example we take $E = E_{(A,B)}$ with $A,B \in \mathbb{Z}$ relatively prime. $E_{(A,B)}$ is semistable at all primes $l \neq 2$, and hence

$$R_{\rho_{E_{(A,B)},p}} = 2^{\delta} \prod_{\substack{l \neq 2 \\ v_l(AB(A-B)) \not\equiv 0 \bmod p}} l \quad .$$

Moreover $\delta = 0$ if $v_2(AB(A-B)) \equiv 4 \bmod p$, and $\delta = 1$ else.

We use this to characterize solutions of Fermat-type equations:

PROPOSITION 5.1. Let S be a finite set of primes not containing 2 and p where p is a prime ≥ 5. Then the following assertions are equivalent:

i) There are relatively prime integers a_1, a_2, a_3 with $\mathrm{supp}(a_1a_2a_3) \mid \prod_{l \in S} l$ such that $a_1 Z_1^p - a_2 Z_2^p = a_3 Z_3^p$ has a solution in $(\mathbb{Z}\setminus\{0\})^3$.

ii) There is an elliptic curve $E/\mathbb{Q}$ with $E(\mathbb{Q})_2 = \mathbb{Z}/2 \times \mathbb{Z}/2$ which is semistable at all primes and has discriminant $\Delta_E = 2^{-8} d_1 \cdot d_2^p$ with $\mathrm{supp}(d_1) \mid \prod_{l \in S} l$.

iii) There is an elliptic curve $E/\mathbb{Q}$ with $E(\mathbb{Q})_2 = \mathbb{Z}/2 \times \mathbb{Z}/2$, semistable at all primes such that $R_{\rho_{E,p}} \mid 2 \prod_{l \in S} l$ and $v_2(j_E) \equiv 8 \bmod p$.

The proof of this proposition is a straight-forward verification (cf. [9]).

We note the

COROLLARY. $Z_1^p - Z_2^p = Z_3^p$ has a non trivial solution for $p \geq 5$ if and only if there is a semistable elliptic curve $E/\mathbb{Q}$ with $E(\mathbb{Q})_2 = \mathbb{Z}/2 \times \mathbb{Z}/2$, $v_2(j_E) \equiv 8 \bmod p$ and $R_{\rho_{E,p}} = 2$.

A strategy to prove Fermat's conjecture could be to prove that a representation $\rho_{E,p}$ of $G(\overline{\mathbb{Q}}/\mathbb{Q})$ with properties as in the corollary cannot exist.

To proceed in this direction we need some definitions.

DEFINITION. Let F be a finite field of characteristic p, $\rho: G(\overline{\mathbb{Q}}/\mathbb{Q}) \longrightarrow Gl(2,F)$ an irreducible continuous representation with $\det(\rho) = \chi_p$. ρ is modular of weight 2 and level N [1] if there is a number field K with ring of integers $\mathcal{O}$, a prime $\mathfrak{p} \mid p$ and a new form $f = q + \sum_{i=2}^{\infty} a_i q^i \in S_2(N)(\mathcal{O})$ such that for $l \nmid p \cdot N$:

$$\mathrm{Tr}_\rho(l) \equiv a_l \bmod \mathfrak{p} .$$

In other words: There is a maximal ideal $\mathfrak{m} \subset \mathbb{T}_N$ and an embedding $\omega: \mathbb{T}/\mathfrak{m} \longrightarrow \overline{F}$ such that for almost all primes l one has

$$\mathrm{Tr}_\rho(l) = \omega(T_l \bmod \mathfrak{m})$$

(where T_l is the l-th Hecke operator).

---

[1] In the following "modular" always means modular with respect to a fixed prime p.

Example. If E is modular without Q-isogeny of degree p then $\rho_{E,p}$ is modular of weight 2 and level N_E.

Problem. For given modular representation ρ find a minimal level!

We want to give a rough sketch of ideas how to attack this problem.

One considers the group scheme $V := \mathcal{J}_o(N)[\mathfrak{m}]$ (the intersection of kernels of elements in $\mathfrak{m}$.

Let l be an odd prime with $l \| N$ such that $V \otimes \mathbb{Z}_l$ is finite over $\mathbb{Z}_l$: $l \nmid R_\rho$. Let A_l be the part of $J_o(N)$ one gets by "lifting" $J_o(\frac{N}{l})$ and assume that $V \cap A_l = \{0\}$. By arguments similar to those in the proof of proposition 4.2 one concludes that modulo l V lies in the connected component of the unity of $\mathcal{J}_o(N) \otimes (\overline{\mathbb{Z}/l})$ and hence it is mapped injectively to the connected component of the unity of $\mathcal{J}_o(N)/A_l$ modulo l which is a torus T split over an extension k of $\mathbb{Z}/l$ of degree ≤ 2. It follows that there is an extension K of degree ≤ 2 of $\mathbb{Q}_l$ such that

$$V \otimes K \subset (\mu_p)^d \otimes K$$

for some d. So

$$\det(\rho) \otimes \mathbb{Q}_l = \chi(K/\mathbb{Q}_l)^2 \cdot (\chi_p^2 \otimes \mathbb{Q}_l)$$

where $\chi(K/\mathbb{Q}_l)$ is the character belonging to $K/\mathbb{Q}_l$. Hence

$$\chi_p \otimes \mathbb{Q}_l = \det(\rho) \otimes \mathbb{Q}_l = \chi_p^2 \otimes \mathbb{Q}_l$$

and so $l \equiv 1 \mod p$.

So we found a result proven by Mazur (cf. [15]) and in a special case in [9]:

PROPOSITION 5.2. *If ρ is modular of level N and $l \| N$ but $l \nmid R_\rho$ and $l \not\equiv 1 \mod p$ then ρ is modular of level dividing $\frac{N}{l}$.*

This is the "easy case" of lowering the level of ρ. A conjecture of Serre (cf. [21]) predicts that all $l \nmid R_\rho$ can be cancelled in the level, hence the condition $l \not\equiv 1 \mod p$ should not be necessary, and this is essentially the beautiful result of K. Ribet ([19]):

THEOREM 5.3 (Ribet). *Let p be a prime ≥ 3, F a finite field of characteristic p and $\rho: G(\overline{\mathbb{Q}}/\mathbb{Q}) \longrightarrow Gl(2,F)$ an irreducible modular representation of weight 2 and level N with $p \nmid N$. Let l be a prime with $l \| N$, $l \equiv 1 \mod p$ and $l \nmid R_\rho$.*

<u>Then ρ is modular of weight</u> 2 <u>and level dividing</u> $\frac{N}{l}$.

Combining proposition 5.2 with Ribet's theorem one gets the

<u>COROLLARY.</u> If ρ is modular of weight 2 and level N, $p \nmid R_\rho$ and $p^2 \nmid N$ then ρ is modular of weight 2 and level dividing $\prod_{\substack{l \mid R_\rho \\ \text{or } l^2 \mid N}} l$.

In this article it is impossible to sketch Ribet's proof, the only thing we can do is to mention some of the tools he uses.

The basic idea of Ribet is to replace the difficult prime $l \equiv 1 \bmod p$ by an easy prime $q \equiv -1 \bmod p$ with the additional property that $Tr(\rho(\sigma_q)) = 0$ and $\det(\rho(\sigma_q)) = -1$ for a Frobenius element $\sigma_q \in G(\overline{\mathbb{Q}}/\mathbb{Q})$ for q. Čebotarev's density theorem guarantees that there are infinitely many such primes. The first step in Ribet's proof is <u>to raise</u> the level from N to q·N in a non trivial way, i.e. to associate to ρ a modular form of level N·q which is "q-new", i.e. it is not a lifted modular form of level N. The second step is to lower the level and to associate a form of level $\frac{N}{l}\cdot q$ to ρ, and then proposition 5.2 can be applied.

The main tool in Ribet's proof is the Shimura curve of level $l\cdot q\cdot(\frac{N}{l}) = l\cdot q\cdot N'$:
Take $\mathcal{O}$ as Eichler order of level N' in the indefinite Quaternion algebra B over $\mathbb{Q}$. Let Γ_∞ be the elements of reduced norm 1 in B. An embedding $B \longrightarrow M(2,\mathbb{R})$ induces an embedding $\Gamma_\infty \longrightarrow Sl(2,\mathbb{R})$. Now take $C_{p,q}$ as canonical model of $\mathbb{H}^*/\Gamma_\infty$. In an earlier paper Ribet showed that the "bad fibers" of C at the primes l and q describe the "primes of fusion" of cusp forms of weight $l\cdot q\cdot N'$, i.e. congruence primes which relate modular forms of level $l\cdot N'$ with forms of level $q\cdot N'$. Ribet uses the geometric description of these bad fibers given by Jordan-Livné (cf. [11]), the l- (resp. q-) adic description of C_{pq} due to Čerednic-Drinfeld [3] and a very careful study of the operation of Hecke operators on C_{pq} and their relation with Hecke operators on $J_o(l\cdot q\cdot N')$ to prove his theorem.

6. Consequences of Ribet's theorem

<u>PROPOSITION 6.1. Assume that</u> $z_1^p - z_2^p = z_3^p$ <u>with</u> $z_i \in \mathbb{Z}\setminus\{0\}$, $\gcd(z_1,z_2,z_3) = 1$ <u>and</u> $p \geq 3$. <u>Then</u>

$$E_{(z_1^p, z_2^p)} : Y^2 = X(X-z_1^p)(X-z_2^p)$$

is not a modular elliptic curve.

Proof. Assume without loss of generality that $p > 163$ and that $2|z_1$ and $z_2 \equiv 1 \bmod 4$. Then $E_{(z_1^p, z_2^p)}$ is semistable at all primes and $\rho = \rho_{E_{(z_1^p, z_2^p)}, p}$ has a conductor $R_\rho = 2$. If ρ would be modular then ρ would be modular of level 2 by Ribet's theorem. But the genus of $X_o(2)$ is 0, and hence we get a contradiction.

PROPOSITION 6.2. Let $\{p_i\}$ be an infinite set of primes such that for each i there exists a modular elliptic curve E_i with square free conductor N_i such that $R_{\rho_{E_i, p_i}}$ has only divisors in a finite set S of primes. Then there exists a infinite subset $\{p_{ij}\}$ and a modular elliptic curve E with

i) $N_E | N_{E_{ij}}$

ii) $l | N_E \Rightarrow l \in S$

iii) $\rho_{E, p_{ij}} = \rho_{E_{ij}, p_{ij}}$

iv) If $E_{ij}(\mathbb{Q})_2 = \mathbb{Z}/2 \times \mathbb{Z}/2$ then $E = E_{(A,B)}$ with $\mathrm{supp}(A \cdot B(A-B)) \mid \prod_{l \in S} l$.

Proof. We use a trick of Mazur. Firstly we can assume that $p_i \notin S$ for all i. By Ribet's theorem we find for each i a new form $f_i \in S_2(\tilde{N}_i, \mathfrak{O}_i)$ associated with $\rho_i := \rho_{E_i, p_i}$ such that $\tilde{N}_i$ has only prime divisors in S. Since there are only finitely many possibilities for f_i we can and shall assume that there is a ring $\mathfrak{O}$ of integers in a number field, a new form $f \in S_2(\tilde{N}, \mathfrak{O})$ and for each i a prime $\mathfrak{p}_i | p_i$ of $\mathfrak{O}$ such that $f = \sum_{j=1}^{\infty} a_j a^j$ and for fixed i one has for primes l prime to $p_i\tilde{N}$:

$$a_l \equiv \mathrm{Tr}_{\rho_i}(l) \bmod \mathfrak{p}_i .$$

Now $\mathrm{Tr}_{\rho_i}(l)$ is congruent modulo p_i to an integer with absolute value bounded by $l+1$, namely the trace of the Frobenius operation on the Tate module of p-power order of E_i: For $l \nmid N_{E_i}$ one has

$$\mathrm{Tr}_{\rho_i}(l) \equiv 1+l-\#(\mathcal{E}_i \otimes \mathbb{Z}/l)(\mathbb{Z}/l) \bmod p_i$$

and for $l \mid N_i$:

$$\mathrm{Tr}_{\rho_i}(l) \equiv \pm(l+1) \bmod p_i \ .$$

So for infinitely many primes p_i the element a_l is congruent to a fixed rational integer and hence $a_l \in \mathbb{Z}$ for almost all (and hence for all) l. Since $f \in S_2(\tilde{N},\mathbb{Z})$ there is a modular elliptic curve E_f associated with f and this proves i)-iii). If E_{ij} has all points of order 2 rational over $\mathbb{Q}$ then $\#(E \otimes \mathbb{Z}/l)(\mathbb{Z}/l) \equiv 0 \bmod 4$ for almost all l and hence $4 \mid \# E(\mathbb{Q})_{tor}$. It follows that E is isogenous to a curve E' with $E'(\mathbb{Q})_2 = \mathbb{Z}/2 \times \mathbb{Z}/2$ and this proves iv).

<u>Application.</u> Fix relatively prime integers a_1, a_2, a_3 and assume that for infinitely many primes p_i the equation

$$a_1 Z_1^{p_i} - a_2 Z_2^{p_i} = a_3 Z_3^{p_i}$$

has (relatively prime) solutions $(z_{1,i}, z_{2,i}, z_{3,i})$. Assume moreover that the curves

$$E_i := E_{(a_1 z_{1,i}^{p_i},\, a_2 z_{2,i}^{p_i})}$$

are modular. Then there are integers a_1', a_2', a_3' with $a_1' - a_2' = a_3'$ and $\mathrm{supp}(a_1' \cdot a_2' \cdot a_3') \mid 2a_1 a_2 a_3$.
(Clearly this result aims into the direction of the Asymptotic Fermat Conjecture.)

<u>Special case (Mazur).</u> For $a_1 = 2^4$, $a_2 = 1$ and $a_3 = q \in \mathbb{P}$ we get: If the equations

$$2^4 Z_1^{p_i} - Z_2^{p_i} = q Z_3^{p_i}$$

have solutions for infinitely many primes such that $E_{(2^4 z_{1,i}^{p_i},\, z_{2,i}^{p_i})}$ are modular then $q = 17$. For: $E_{(2^4 z_{1,i}^{p_i},\, z_{2,i}^{p_i})}$ has good reduction in 2, and hence we must have $2^4 \pm 1 = q$. More general take $a_1 = 2^m$, $a_2 = \pm 1$ and

$a_3 = q \in \mathbb{P}$. Then under the assumptions made above we get $2^k \pm 1 = q$, and hence q is a Mersenne or a Fermat prime.

PROPOSITION 6.3. *Let E/Q be a modular elliptic curve with prime conductor q. Then* $\mathfrak{D}_E | q^6$.

Sketch of the proof. Assume that E has a Q-rational isogeny of degree l and that $v_q(j_E) \equiv 0 \bmod l$. Then E has a Q-rational point of order l. Now assume that

$$\mathfrak{D}_E = p^{2^{\alpha_2} \cdot 3^{\alpha_3}} \prod_{l \nmid 6} l^{\alpha_l} .$$

If $\alpha_2 > 0$ then E has a point of order 2 over Q. Exclude $q = 17$ (for which 6.3 is true). Then $E(\mathbb{Q})_2 \neq \mathbb{Z}/2 \times \mathbb{Z}/2$. Hence if $\alpha_2 \geq 2$ then E has a point of order 4 over Q, and we would find a curve E' isogenous to E with $E'(\mathbb{Q})_2 = \mathbb{Z}/2 \times \mathbb{Z}/2$ which is impossible. Hence $0 \leq \alpha_2 \leq 1$. In the same way one sees that $0 \leq \alpha_3 \leq 1$.
Now assume that $l \geq 5$ and $\alpha_l > 0$. Ribet's result implies that E has Q-rational isogeny of degree l, and hence a Q-rational point of order l. Now we use Riemann's hypothesis over $\mathbb{Z}/2$ and get: $l \leq 5$ and $\alpha_5 = 0$ if $\alpha_2 > 0$ or $\alpha_3 > 0$.
The last step is to show that $\alpha_5 \leq 1$, and then 6.3 is proven.

Proposition 6.3 states that the following conjecture due to Szpiro is true for modular elliptic curves with prime conductor.

(SZ). There are constants d *and* c = c(d) *such that for all elliptic curves over* Q *one has*

$$\deg(\mathfrak{D}_E) \leq c + 6d \deg(N_E) .$$

(It is clear how (SZ) has to be formulated over arbitrary global fields.)

Obviously (SZ) is weaker than the height conjecture but it is sufficiently strong to give results about torsion points and the Asymptotic Fermat Conjecture we got by applying (H). It could be regarded as "finite part" of (H).

So the question arises: Can we conclude that ($\underline{H}$) holds for modular elliptic curves with prime conductor.

It turns out that we need an additional information about the absolute value of j: Look at the discriminant equation: $4g_2^3-27g_3^2 = \Delta_E$ (which we assume to be minimal with respect to short Weierstraß equations for E for which $g_2, g_3 \in \mathbb{Z}$). Since

$$j_E = 12^3 \cdot 4 \frac{g_2^3}{\Delta_E}$$

we get:

$$h(j_E) = \log|\Delta_E| + \operatorname{Max}\{0, 3\log|g_2| - \log|\Delta_E|\} + \log(12^3 \cdot 4) .$$

Now there is a conjecture about the size of integral points on elliptic curves which we apply to the discriminant equation to get

HALL'S CONJECTURE. $\log|g_2| \leq c'(\varepsilon)+(2+\varepsilon)\log|\Delta_E|$ for $\varepsilon \in \mathbb{R}_{>0}$ (cf. [23]).

Hence assuming that Hall's conjecture is true we get

$$h(j_E) \leq c'+\log|\Delta_E|+(5+\varepsilon)\log|\Delta_E| = c'+(6+\varepsilon)\log|\Delta_E| .$$

This would give ($\underline{H}$) for E as in 6.3 with $d = 6+\varepsilon$. But we can do better: By looking at the proof of 6.3 we can conclude that there is an elliptic curve E' isogenous to E with $\Delta_{E'} = q$. Since $h(E) \leq h(E')+\log 6$ we get indeed:

$$h(j_E) \leq c(\varepsilon)+6(1+\varepsilon)\log|q| ,$$

the height conjecture for E with $d = 1+\varepsilon$.

7. Taniyama's conjecture

In the last three sections it became obvious that modular elliptic curves over $\mathbb{Q}$ are good objects to test various conjectures about elliptic curves. One could have the impression that modular elliptic curves are rather exotic. But there is a famous conjecture saying that the contrary is true:

CONJECTURE (T). Every elliptic curve over $\mathbb{Q}$ is modular.

Essentially this conjecture was stated by Taniyama 1955 in [25]. Its precise formulation is strongly influenced by results of Shimura who proved that elliptic curves over $\mathbb{Q}$ with complex multiplication are modular and who systematically related eigenfunctions in $S_2(N)(\mathbb{C})$ to quotients of $J_o(N)$ (cf. [22], 7.5). A very important criterion for elliptic curves to be modular is due to Weil ([28]).

Before giving two motivations for (T) we state the results about elliptic curves we got under the assumption that (T) is true:

<u>THEOREM 7.1. Assume that (T) is true. Then it follows that</u>

1. <u>Fermat's conjecture is true, and</u>
2. <u>(D_{mod}) (the degree conjecture for modular parametrizations) implies the geometric height conjecture for elliptic curves, and so the A-B-C-conjecture and the Asymptotic Fermat conjecture too.</u>

Why should one believe that (T) is reasonable?
First of all define the L-series of elliptic curves over $\mathbb{Q}$ by the following Dirichlet series:

$$L_E(s) := \prod_{l\in\mathbb{P}} \frac{1}{L_E^{(l)}(l^{-s})} =: \sum_{n\geq 1} a_n n^{-s} \quad \text{with}$$

$$L_E^{(l)}(T) := \begin{cases} 1-(l+1-\#(\mathcal{E}^{(l)}(\mathbb{Z}/l)))T+lT^2 & \text{if } l\nmid N_E \\ 1+T & \text{if } \mathcal{E}^{(l)o} \cong G_m \\ 1-T & \text{if } \mathcal{E}^{(l)o} \text{ is a non split torus} \\ 1 & \text{else} \end{cases}$$

where $\mathcal{E}^{(l)o}$ is the connected component of the Néron model of E modulo l.

For a Dirichlet character χ of $\mathbb{Z}$ define

$$L_{E\otimes\chi}(s) := \sum_{n=1}^{\infty} \chi(n) a_n n^{-s} .$$

Then $L_{E\otimes\chi}$ is a Dirichlet series which converges for s with Re(s) sufficiently large.

<u>(GENERALIZED) CONJECTURE OF HASSE-WEIL. For all</u> m <u>prime to</u> N_E <u>and all primitive Dirichlet characters</u> χ <u>with conductor</u> m <u>the series</u>

$$L_{E\otimes\chi}(s)\cdot N_E^{s/2}\cdot(\frac{m}{2\pi})^s\ \Gamma(s) =: \Lambda_{E,\chi}(s)$$

has an analytic extension to $\mathbb{C}$ bounded in vertical stripes satsifying the functional equation

$$\Lambda_{E,\chi}(s) = w_E\ \frac{g(\chi)}{g(\overline{\chi})}\ \chi(-N_E)\ \Lambda_{E,\overline{\chi}}(2-s)$$

with $w_E = \pm 1$ and $g(\chi)$ the Gauss sum to χ.

A beautiful theorem of Weil [28] states that then $L_E(s)$ is the Mellin transform of an eigenform $f \in S_2(N)(\mathbb{Z})$. So there is a modular curve E_f (Shimura) with conductor N_E (Carayol) with $L_{E_f}(s) = L_E(s)$ (Eichler-Shimura), hence E is isogenous to E_f (Faltings).

So we get

PROPOSITION 7.2. (T) is true if and only if the conjecture of Hasse-Weil is true.

Let us give a second motivation for (T) coming from representation theory of $G(\overline{\mathbb{Q}}/\mathbb{Q})$.

Serre formulated a very far reaching conjecture which could be called a "mod p-Langlands-conjecture" (cf. [21]); a very special case of this conjecture is

(S). Let $E/\mathbb{Q}$ be an elliptic curve, p a prime ≥ 3 such that E has no $\mathbb{Q}$-rational isogeny of degree p, $\mathcal{E}_p$ the kernel of the multiplication by p of the Néron model of $E/\mathbb{Q}$. Assume that $\mathcal{E}_p \otimes \mathbb{Z}_p$ is finite over $\mathbb{Z}_p$. Then $\rho_{E,p}$ is modular of weight 2 and level $N_{\rho_{E,p}}$.

One sees that Ribet's theorem proves an essential part of this conjecture in the case that E is modular. But assuming (S) one gets

PROPOSITION 7.3. If for given E (S) is true for infinitely many primes p_i then E is modular, hence if (S) is true for all $E/\mathbb{Q}$ and all p then (T) is true.

The proof of 7.3 is essentially a repetition of arguments given in the proof of proposition 6.2: One uses the congruences between the coefficients of cusp forms associated to ρ_{E,p_i} and the Dirichlet series $L_E(s)$ to conclude that $L_E(s)$ is the Mellin transform of a cusp form, and then as above Faltings' result implies that E is modular.

We end by mentioning that there are very interesting approaches toward (A-B-C) by Vojta (cf. [27]) and by transcendental methods (cf. [24]).

The following diagram gives an overview of conjectures and implications we discussed more or less extensively; it should be noticed that the degree conjecture for modular curves is the link between representation theory of $G_{\mathbb{Q}}$ (Serre's conjecture) and height conjectures.

POSTSCRIPT (April 1988)

A.N. Parshin started in [31] a very promising new line to attack conjecture (H) and hence (A-B-C): For algebraic surfaces V which are not ruled surfaces with second Chern class $c_2(V) < 0$ there is the Bogomolev-Miyaoka-Yau inequality

$$c_1^2(V) \leq 3c_2(V) \qquad \text{(cf. [30])} .$$

Parshin formulates an analogy of this inequality for arithmetical surfaces V/B whose general fiber has genus $g \geq 1$, where B is the ring of integers in a number field K: Let $\omega_{V/B}$ be the relative dualizing sheaf of V/B as metrized vector bundle in the sense of Arakelov-Faltings. Then

$$\omega_{V/B}^2 \leq 3 \sum_{v \in \Sigma_K \cup \infty} \delta_v' N_v + (2g-2) \log|D_{K/\mathbb{Q}}| + c(K)$$

where δ_v' is the number of singular points in the fiber of a stable model of V over v if $v \in \Sigma_K$, and δ_v is introduced by Faltings in [29] for $v \in \infty$. Then Parshin shows that this inequality implies (H).

Recently (March 1988) Miyaoka announced results which are valuable steps towards the proof of the inequality but in the moment it seems that they are not strong enough to prove the inequality.

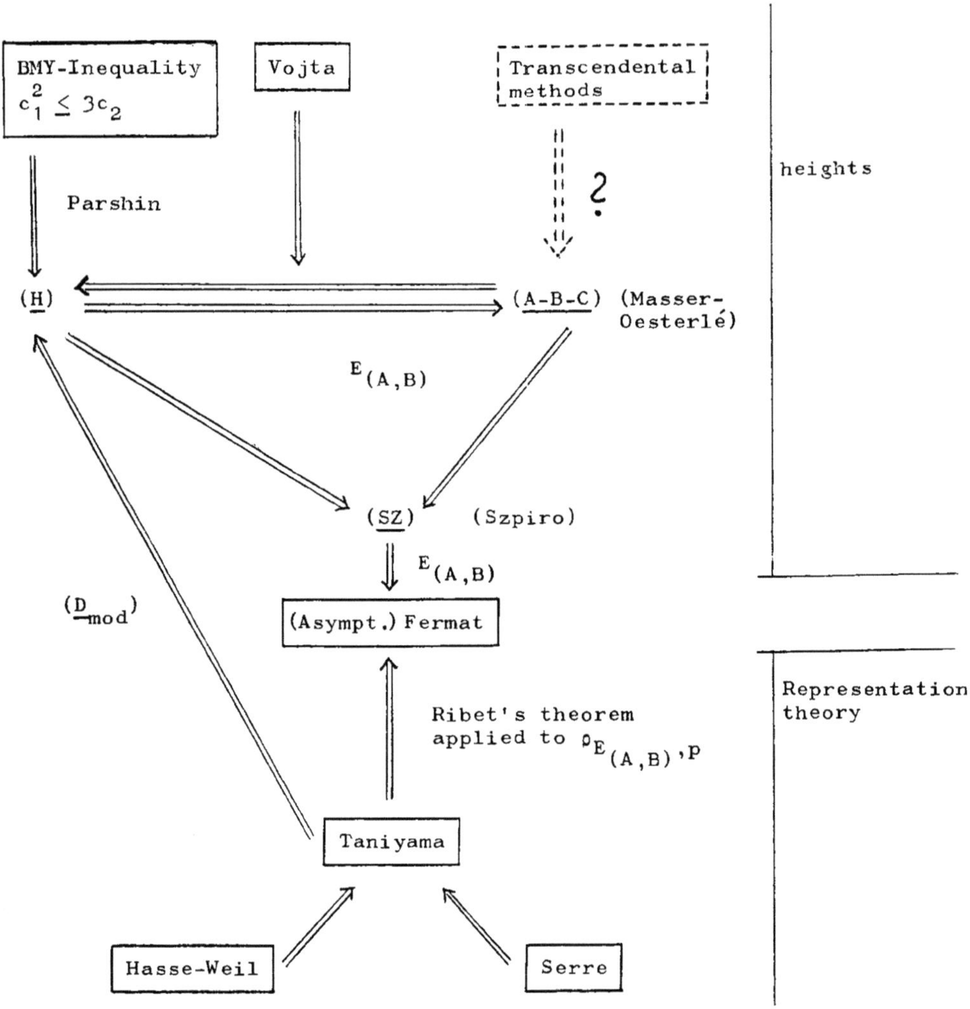
BMY-Inequality
$c_1^2 \leq 3c_2$
Vojta
Transcendental methods
?
heights
Parshin
(H)
(A-B-C) (Masser-Oesterlé)
$E_{(A,B)}$
(SZ) (Szpiro)
$E_{(A,B)}$
(D_{mod})
(Asympt.) Fermat
Ribet's theorem applied to $\rho_{E_{(A,B)},p}$
Representation theory
Taniyama
Hasse-Weil
Serre

[1] BROWNAWELL, W.D., MASSER, D.W.: Vanishing sums in function fields. Math. Proc. Cambr. 1986.

[2] CARAYOL, H.: Sur les représentation l-adiques associées aux formes modulaires de Hilbert. Ann. Sci. ENS 19, 409-468 (1986).

[3] CEREDNIK, I.V.: Uniformization of algebraic curves by discrete arithmetic subgroups of $PGl_2(k_w)$ with compact quotients. Transl. in Math. USSR Sb. 29, 55-78 (1976).

[4] DELIGNE, P.: Preuve des conjectures de Tate et de Shafarevitch (d'après G. Faltings). Sém. Bourbaki 616 (1983).

[5] DELIGNE, P., RAPOPORT, M.: Les schémas de modules de courbes elliptiques. in Modular Functions of One Variable II, Springer Lecture Notes in Math. 349, 143-316 (1972).

[6] FALTINGS, G.: Endlichkeitssätze für abelsche Varietäten über Zahlkörpern. Invent. Math. 73, 349-366 (1983).

[7] FREY, G.: Some remarks concerning points of finite order on elliptic curves over global fields. Ark. f. Mat. 15, 1-19 (1977).

[8] FREY, G.: Rationale Punkte auf Fermatkurven und getwisteten Modulkurven. J. Reine u. Angew. Math. 33, 185-191 (1982).

[9] FREY, G.: Links between stable elliptic curves and certain diophantine equations. Ann. Univ. Sarav. Math. Ser. Vol. 1, 1-40 (1986).

[10] HELLEGOUARCH, Y.: Points d'ordre $2p^h$ sur les courbes elliptiques. Acta Arith. 26, 253-263 (1975).

[11] JORDAN, B., LIVNÉ, R.: Local diophantine properties of Shimura curves. Math. Ann. 270, 235-248 (1985).

[12] KATZ, N., MAZUR, B.: Arithmetic moduli of elliptic curves. Princeton Univ. Press (1985).

[13] MAZUR, B.: Modular curves and the Eisenstein ideal. Publ. Math. IHES 47, 33-186 (1977).

[14] MAZUR, B.: Rational isogenies of prime degree. Invent. Math. 44, 129-162 (1978).

[15] MAZUR, B.: Letter to J.F. Mestre (16. August 1985).

[16] Modular Functions of One Variable IV. Springer Lecture Notes in Math. 476 (1975).

[17] RIBET, K.: Mod p Hecke operators and congruences between modular forms. Invent. Math. 71, 193-205 (1983).

[18] RIBET, K.: Congruence relations between modular forms. Proc. Int. Congr. Math., 503-514 (1983).

[19] RIBET, K.: On modular representations of Gal($\overline{\mathbb{Q}}/\mathbb{Q}$) arising from modular forms. Math. Sc. Research Institute Berkeley, CA, Preprint # 06420-87 (1987).

[20] ROQUETTE, P.: Analytic theory of elliptic functions over local fields. Hamb. Math. Einzelschriften, N.F. Heft 1 (1969).

[21] SERRE, J.P.: Sur les représentations modulaires de degré 2 de Gal($\overline{\mathbb{Q}}/\mathbb{Q}$). Duke Math. J. 54, 179-230 (1987).

[22] SHIMURA, G.: Introduction to the arithmetic of automorphic functions. Princeton Univ. Press (1971).

[23] SILVERMAN, J.H.: The arithmetic of elliptic curves. New York-Berlin-Heidelberg-Tokyo (1986).

[24] STEWART, C.L., TIJDEMAN, R.: On the Oesterlé-Masser conjecture. Monatsh. f. Math. 102, 251-257 (1986).

[25] TANIYAMA, Y.: in: Problem session of the Tokyo-Nikko conference on number theory; problem 12, 1955.

[26] TATE, J.: Algorithm for finding the type of a singular fiber in an elliptic pencil. in [16], 33-52.

[27] VOJTA, P.: Diophantine approximation and value distribution theory. Springer Lecture Notes in Math. 1239 (1987).

[28] WEIL, A.: Über die Bestimmung Dirichletscher Reihen durch Funktionalgleichungen. Math. Ann. 168, 149-156 (1967).

[29] FALTINGS, G.: Calculus on arithmetical surfaces. Ann. Math. 119, 387-424, 1984.

[30] MIYAOKA, Y.: On the Chern numbers of surfaces of general type. Inv. Math. 42, 225-237 (1977).

[31] PARSHIN, A.N.: The Bogomolov-Miyaoka-Yau inequality for the arithmetical surfaces and its applications. Preprint 1988.

FACTORIZATIONS OF ALGEBRAIC INTEGERS

Alfred Geroldinger

1. Introduction

Let R be the ring of integers of an algebraic number field K with ideal class group $C\ell(K)$. Every element $a \in R^{\#} = R \setminus (R^{\times} \cup \{0\})$ has a (not necessarily unique) factorization $a = u_1 \dots u_k$ into irreducible elements $u_1, \dots, u_k \in R^{\#}$; k is called the length of the factorization and $L(a) = \{k \,/\, a$ has a factorization of length $k\}$ is called the set of lengths of a. It is well known, that R is factorial if and only if $\# C\ell(K) = 1$. One possible measure for the non-uniqueness of factorization is the structure of the sets $L(a)$ for various $a \in R^{\#}$. The main results concerning sets of lengths are the following:

(i) $\# L(a) = 1$ for every $a \in R^{\#}$ if and only if $\# C\ell(K) \le 2$ (Carlitz [1]).

(ii) If $\# C\ell(K) \ge 3$, then for every $m \in \mathbb{N}$ there is an element $a \in R^{\#}$ with $\# L(a) = m$ (Śliwa [4]).

(iii) There are constants $M(C\ell(K))$ and $D(C\ell(K))$ such that every set of lengths $L(a)$ has the following form:

$$\begin{aligned} L(a) = \{x_1, \dots, x_\alpha, y_o, y_1, \dots\dots\dots\dots y_\mu, y_o + d,\\ y_1 + d, \dots\dots\dots\dots, y_o + 2d,\\ \dots\dots\dots\dots\dots\\ y_1 + (k-1)d, \dots\dots, y_o + kd, z_1, \dots, z_\beta\} \end{aligned}$$

with $x_1 < \dots < x_\alpha < y_o < \dots < y_\mu < y_o + d \le y_o + kd < z_1 < \dots < z_\beta$, $\alpha \le M(C\ell(K))$, $\beta \le M(C\ell(K))$ and $d \le D(C\ell(K)) - 2$ (Geroldinger [3]).

In this paper I establish a property concerning elements $a \in R^{\#}$ which makes the sets $L(a)$ as simple as possible, i. e. of the form $L(a) = \{y, y+1, \dots, y+k\}$ with $y, k \in \mathbb{N}$ (Theorem 1'). Furthermore I obtain that there is a rational integer λ such that $L(\lambda n)$ has this form for every $n > 1$ in $\mathbb{N}$. In certain Galois number fields $\lambda = 1$.

2. Notations

As in [3] I translate the arithmetical problem of sets of lengths into a combinatorial problem on abelian groups.

Let G be an additively written abelian group with zero element 0.

Two sequences $b = (g_1,\dots,g_m)$ and $b' = (g'_1,\dots,g'_m)$ of elements of G are called equivalent, if for some permutation π $g_i = g'_{\pi(i)}$ for every $i \in \{1,\dots,m\}$. An equivalence class of sequences $B = \langle g_1,\dots,g_m\rangle$ is called a block, if $g_1+\dots+g_m = 0$. The equivalence class consisting of the empty sequence is called empty block. Let $v_g(B)$ denote the multiplicity of g in B and let $\ell(B) = \sum_{g\in G} v_g(B) = m$ denote the length of B. The set $\mathcal{B}(G)$ of all blocks has a natural semigroup structure, defined as follows:

$$\langle a_1,\dots,a_m\rangle * \langle b_1,\dots,b_n\rangle = \langle a_1,\dots,a_m,b_1,\dots,b_n\rangle$$

(The empty block is the unit element). A block $B \in \mathcal{B}(G)$ is called irreducible, if $B = B_1 * B_2$ with $B_1,B_2 \in \mathcal{B}(G)$ implies that B_1 or B_2 is the empty block. For every block $B \in \mathcal{B}(G)$ its set of lengths is defined as $L(B) = \{k \,/\, B$ has a factorization into k irreducible blocks$\}$.

The relationship between factorizations of blocks $B \in \mathcal{B}(C\ell(K))$ and factorizations of elements $a \in R^{\#}$ is as follows: let $(a) = \mathfrak{p}_1\cdot\ldots\cdot\mathfrak{p}_r$ be the prime ideal decomposition of (a) and let $[\mathfrak{p}_i]$ denote the ideal class of $\mathfrak{p}_i$, then $B(a) = \langle[\mathfrak{p}_1],\dots,[\mathfrak{p}_r]\rangle$ is a block in $\mathcal{B}(C\ell(K))$ and $L(a) = L(B(a))$ (see Proposition 1 in [3]).

3. A combinatorial result

Let G be an abelian group, $B \in \mathcal{B}(G)$ a non-empty block and $B = \mathop{*}_{k=1}^{r} B_k$ a factorization of B into r irreducible blocks. For $1 \le \ell < m \le r$ let $B_\ell = \langle a_1,\dots,a_i,a_{i+1},\dots,a_u\rangle$, $B_m = \langle c_1,\dots,c_j,c_{j+1},\dots,c_v\rangle$ with $u \ge i+1 \ge 2$, $v \ge j+1 \ge 2$ and $a_1+\dots+a_i = c_1+\dots+c_j$. I transform B_ℓ, B_m into $B'_\ell = \langle c_1,\dots,c_j,a_{i+1},\dots,a_u\rangle$ and $B'_m = \langle a_1,\dots,a_i,c_{j+1},\dots,c_v\rangle$; B'_ℓ and B'_m are still blocks but not necessarily irreducible. I call this process an (exchange) transformation of the form $((a_1,\dots,a_i) \longleftrightarrow (c_1,\dots,c_j))$ [or briefly $(i \longleftrightarrow j)$] between B_ℓ and B_m. Since $B_\ell * B_m = B'_\ell * B'_m$ I get $B = \mathop{*}_{k\neq\ell,m} B_k * B'_\ell * B'_m$. If B'_ℓ and B'_m are irreducible, I will say, the exchange transformation between B_ℓ and B_m leaves the length r of the factorization of B invariant. If $B'_\ell = \mathop{*}_{k=1}^{s} B_{\ell,k}$ and $B'_m = \mathop{*}_{k=1}^{t} B_{m,k}$ are factorizations of B'_ℓ and B'_m into irreducible blocks $B_{\ell,k}$ and $B_{m,k}$ with $s,t \ge 1$, $s+t > 2$, then I will say, the exchange transformation between B_ℓ and B_m gives rise to a factorization of B of length $r-2+s+t$.

PROPOSITION 1. *Let* $B = \overset{r}{\underset{k=1}{*}} B_k$ *be a factorization of* $B \in \mathcal{B}(G)$ *into* r *irreducible blocks. Then every exchange transformation of the form* $(2 \longleftrightarrow 1)$ *between two of the blocks* $B_1, \ldots, B_r$ *either leaves the length* r *of the factorization of* B *invariant or gives rise to a factorization of* B *of length* $r+1$.

PROOF. Let $1 \leq \ell < m \leq r$, $B_\ell = \langle a_1, a_2, \ldots, a_u \rangle$, $B_m = \langle c_1, \ldots, c_v \rangle$ with $u \geq 3$, $v \geq 2$ and $a_1 + a_2 = c_1$. Then $B'_\ell = \langle c_1, a_3, \ldots, a_u \rangle$ is irreducible and $B'_m = \langle a_1, a_2, c_2, \ldots, c_v \rangle$ can be represented as a product of at most two irreducible blocks. ◇

LEMMA 1. *Let* $A = \langle a_1, \ldots, a_i, a_{i+1}, \ldots, a_u \rangle$ *be an irreducible block with* $u \geq i+1$ *and* $i \geq 3$. *Then for some* $\ell \in \{2, \ldots, i-1\}$ *there are indices* $1 \leq k_1 < \ldots < k_\ell \leq i$ *such that* $a_{k_1} + \ldots + a_{k_\ell} \notin \{a_1, \ldots, a_i\}$.

PROOF. Suppose that for every $k \in \{1, \ldots, i\}$ $a_1 + \ldots + a_{k-1} + a_{k+1} + \ldots + a_i \in \{a_1, \ldots, a_i\}$; so $a_1 + \ldots + a_{k-1} + a_{k+1} + \ldots + a_i = a_k$, as A is irreducible. If $i \geq 4$ this implies $a_1 + \ldots + a_{i-2} \notin \{a_1, \ldots, a_i\}$, so the assertion holds for $\ell = i-2$. If $i = 3$, it follows that $a_1 + a_2 + a_3 = 2(a_1 + a_2 + a_3) = 0$, a contradiction to the irreducibility of A . ◇

PROPOSITION 2. *Let* $B \in \mathcal{B}(G)$ *be a block such that* $\{g \in G \,/\, v_g(B) > 0\} \cup \{0\}$ *is a subgroup of* G *and let* $r \in L(B)$. *For every factorization of* B *into* r *irreducible blocks let the length* r *be invariant under all exchange transformations of the form* $(2 \longleftrightarrow 1)$. *Then for every factorization of* B *into* r *irreducible blocks the length* r *is invariant under all exchange transformations.*

PROOF. I have to consider exchange transformations of the form $(i \longleftrightarrow j)$ and I assume without restriction $i \geq j \geq 1$. I prove the assertion by induction on $i+j$. By assumption it is correct for $i+j = 3$.

To do the induction step, let $B = \overset{r}{\underset{k=1}{*}} B_k$ be a factorization of B into r irreducible blocks. I consider a transformation of the form $(i \longleftrightarrow j)$ with $i+j \geq 4$ between two of the blocks $B_1, \ldots, B_r$. Without restriction let B_1, B_2 be these two blocks; let $B_1 = \langle a_1, \ldots, a_i, a_{i+1}, \ldots, a_u \rangle$, $B_2 = \langle c_1, \ldots, c_j, c_{j+1}, \ldots, c_v \rangle$ with $u \geq i+1$, $v \geq j+1$ and $a_1 + \ldots + a_i = c_1 + \ldots + c_j$. Then I must show that the blocks $B''_1 = \langle c_1, \ldots, c_j, a_{i+1}, \ldots, a_u \rangle$ and $B''_2 = \langle a_1, \ldots, a_i, c_{j+1}, \ldots, c_v \rangle$ are still irreducible, or equivalently, that $B = B''_1 * B''_2 * B_3 * \ldots * B_r$ is a factorization of B into r irreducible blocks.

<u>Case 1.</u> $j = 1$. This implies $i \geq 3$; according to Lemma 1 I assume without restriction that $e_1 = a_1+\dots+a_k \notin \{a_1,\dots,a_i\}$ for some $k \in \{2,\dots,i-1\}$. Since $\{g \in G / v_g(B) > 0\} \cup \{0\}$ is a subgroup of G there is an index $n \in \{1,\dots,r\}$ with $v_{e_1}(B_n) > 0$.

<u>Case 1.1.</u> $n = 1$. Since $e_1 \notin \{a_1,\dots,a_i\}$ I assume without restriction that $e_1 = a_{i+1}$. I transform B_1, B_2 into $B_1' = \langle a_1,\dots,a_k,c_1, a_{i+2},\dots,a_u\rangle$ and $B_2' = \langle a_{k+1},\dots,a_i,a_{i+1},c_2,\dots,c_v\rangle$. By induction hypothesis B_1' and B_2' are irreducible and $B = B_1' * B_2' * B_3 * \dots * B_r$ is a factorization of B into r irreducible blocks. Next I transform B_1', B_2' into B_1'' and B_2'' . By induction hypothesis again B_1'' and B_2'' are irreducible and $B = B_1'' * B_2'' * B_3 * \dots * B_r$ is a factorization of B into r irreducible blocks. Thus the length r of the factorization $B = \mathop{*}\limits_{k=1}^{r} B_k$ is invariant under the transformation $((a_1,\dots,a_i) \longleftrightarrow (c_1,\dots,c_j))$ between B_1 and B_2 .

<u>Case 1.2.</u> $n = 2$. Since $e_1 = a_1+\dots+a_k \neq a_1+\dots+a_i = c_1$ I assume without restriction that $e_1 = c_2$. First I transform B_1, B_2 into $B_1' = \langle c_2,a_{k+1},\dots,a_i,\dots,a_u\rangle$, $B_2' = \langle c_1,a_1,\dots,a_k,c_3,\dots,c_v\rangle$ and by induction hypothesis $B = B_1' * B_2' * B_3 * \dots * B_r$ is a factorization into r irreducible blocks. Then I transform B_1', B_2' into B_1'', B_2'' and by induction hypothesis again $B = B_1'' * B_2'' * B_3 * \dots * B_r$ is a factorization into r irreducible blocks.

<u>Case 1.3.</u> $n \geq 3$. Without restriction I assume $n = 3$. Let $B_3 = \langle e_1,\dots,e_w\rangle$ with $w \geq 2$. First I transform B_1, B_3 into $B_1' = \langle e_1,a_{k+1},\dots,a_i,a_{i+1},\dots,a_u\rangle$, $B_3' = \langle a_1,\dots,a_k,e_2,\dots,e_w\rangle$ and by induction hypothesis $B = B_1' * B_2 * B_3' * B_4 * \dots * B_r$ is a factorization into r irreducible blocks. Then I transform B_1', B_2 into B_1'', $B_2' = \langle e_1,a_{k+1},\dots,a_i,c_2,\dots,c_v\rangle$ and by induction hypothesis $B = B_1'' * B_2' * B_3' * B_4 * \dots * B_r$ is a factorization into r irreducible blocks. Finally I transform B_2', B_3' into B_2'', $B_3'' = B_3$ and by induction hypothesis again $B = B_1'' * B_2'' * B_3 * \dots * B_r$ is a factorization into r irreducible blocks.

<u>Case 2.</u> $j \geq 2$. Let $e_1 = a_1+\dots+a_i$; since $\{g \in G / v_g(B) > 0\} \cup \{0\}$ is a subgroup of G there is an index $n \in \{1,\dots,r\}$ with $v_{e_1}(B_n) > 0$.

<u>Case 2.1.</u> $n = 1$. Since $e_1 \notin \{a_1,\dots,a_i\}$ I assume without restriction that $e_1 = a_{i+1}$. First I transform B_1, B_2 into $B_1' = \langle a_1,\dots,a_i, c_1,\dots,c_j,a_{i+2},\dots,a_u\rangle$, $B_2' = \langle a_{i+1},c_{j+1},\dots,c_v\rangle$ and by induction hypothesis $B = B_1' * B_2' * B_3 * \dots * B_r$ is a factorization into r irreducible blocks. Then I transform B_1', B_2' into B_1'', B_2'' and by induction hypothesis again $B = B_1'' * B_2'' * B_3 * \dots * B_r$ is a factorization into r irreducible blocks.

<u>Case 2.2.</u> $n = 2$. Analogous to Case 2.1.

Case 2.3. $n \geq 3$. Without restriction I assume $n = 3$. Let $B_3 = \langle e_1, \ldots, e_w \rangle$ with $w \geq 2$. First I transform B_1, B_3 into $B_1' = \langle e_1, a_{i+1}, \ldots, a_u \rangle$, $B_3' = \langle a_1, \ldots, a_i, e_2, \ldots, e_w \rangle$ and by induction hypothesis $B = B_1' * B_2 * B_3' * B_4 * \ldots * B_r$ is a factorization into r irreducible blocks. Then I transform B_1', B_2 into B_1'', $B_2' = \langle e_1, c_{j+1}, \ldots, c_v \rangle$ and by induction hypothesis $B = B_1'' * B_2' * B_3' * B_4 * \ldots * B_r$ is a factorization into r irreducible blocks. Finally I transform B_2', B_3' into B_2'', $B_3'' = B_3$ and by induction hypothesis again $B = B_1'' * B_2'' * B_3 * \ldots * B_r$ is a factorization into r irreducible blocks. ◇

PROPOSITION 3. *Let* $B = \mathop{*}\limits_{k=1}^{r} B_k$ *be a factorization of* $B \in \mathbf{B}(G)$ *into* r *irreducible blocks. If the length* r *is invariant under all exchange transformations, then, for every* $1 \leq \ell < m \leq r$, $B_\ell * B_m$ *cannot be represented as a product of more than two irreducible blocks.*

PROOF. Let $B_\ell = \langle a_1, \ldots, a_u \rangle$, $B_m = \langle c_1, \ldots, c_v \rangle$ and let $B_\ell * B_m = \mathop{*}\limits_{k=1}^{s} C_k$ be a factorization of $B_\ell * B_m$ into s irreducible blocks with $s \geq 2$. If one of the C_k equals B_ℓ or B_m, then $s = 2$. Otherwise I assume without restriction that $C_1 = \langle c_1, \ldots, c_j, a_{i+1}, \ldots, a_u \rangle$; thus $c_1 + \ldots + c_j = -(a_{i+1} + \ldots + a_u) = a_1 + \ldots + a_i$ and the exchange transformation $((a_1, \ldots, a_i) \longleftrightarrow (c_1, \ldots, c_j))$ transforms B_ℓ, B_m into $B_\ell' = C_1$ and $B_m' = \langle a_1, \ldots, a_i, c_{j+1}, \ldots, c_v \rangle = C_2 * \ldots * C_s$. Since B_m' is irreducible $s = 2$ follows. ◇

PROPOSITION 4. *Let* $B \in \mathbf{B}(G)$ *be a block such that* $\{g \in G / v_g(B) > 0\} \cup \{0\}$ *is a subgroup of* G *and let* $r \in L(B)$. *Let* $i \in \{2, \ldots, r\}$ *be maximal in respect to the following condition:*

for every factorization $B = \mathop{*}\limits_{k=1}^{r} B_k$ *into* r *irreducible blocks and for any* j *indices* $1 \leq k_1 < \ldots < k_j \leq r$ *with* $j \leq i$ $B_{k_1} * \ldots * B_{k_j}$ *cannot be represented as a product of more than* j *irreducible blocks.*

Then either $(i = r)$ *and thus* $r = \max L(B)$ *or there is a factorization* $B = \mathop{*}\limits_{k=1}^{r} B_k$ *into* r *irreducible blocks such that* $B_1 * \ldots * B_{i+1}$ *can be represented as a product of* $i + 2$ *irreducible blocks.*

PROOF. Assume $i < r$; let μ be minimal in respect to the following condition: there is a factorization $B = \mathop{*}\limits_{k=1}^{r} B_k$ into r irreducible blocks such that $B_1 * \ldots * B_{i+1}$ is a product of $s \geq i + 2$ irreducible blocks and $\mu = \ell(B_{i+1})$. Now let $B = \mathop{*}\limits_{k=1}^{r} B_k$ be such a factorization, $\mathop{*}\limits_{k=1}^{i+1} B_k = C_1 * \ldots * C_s$ with irreducible C_k , $s \geq i+2$ and $\mu = \ell(B_{i+1})$. It suffices to prove $s \leq i+2$.

First I show $B_{i+1} \neq \langle 0 \rangle$: assume $B_{i+1} = \langle 0 \rangle$; this implies that there is an index $j \in \{1,\ldots,s\}$ with $C_j = \langle 0 \rangle$; therefore I get $\mathop{*}\limits_{k=1}^{i} B_k = \mathop{*}\limits_{k \neq j} C_k$, a contradiction to the definition of i .

Since B_{i+1} is irreducible and $B_{i+1} \neq \langle 0 \rangle$, it follows that $v_0(B_{i+1}) = 0$; therefore $\{g \in G / v_g(B_{i+1}) > 0\}$ is not a subgroup, and because being finite it cannot be closed under addition. Thus there are elements a_1, a_2 with $v_{a_1}(B_{i+1}) > 0$, $v_{a_2}(B_{i+1}) > 0$ and $v_{a_1+a_2}(B_{i+1}) = 0$. Because $v_{a_1+a_2}(B) > 0$ there is an index $j \in \{1,\ldots r\}$ with $v_{a_1+a_2}(B_j) > 0$. I transform B_j, B_{i+1} into $B_j' = \langle a_1, a_2, \ldots \rangle$ and $B_{i+1}' = \langle a_1+a_2, \ldots \rangle$. Since $B_j * B_{i+1} = B_j' * B_{i+1}'$ cannot be represented as a product of more than two irreducible blocks, it follows that B_j' is irreducible. I put $B_k' = B_k$ for $k \in \{1,\ldots,r\} \setminus \{j, i+1\}$.

Next I show that $j \notin \{1,\ldots,i\}$: assuming to the contrary $j \in \{1,\ldots,i\}$ I obtain $B = \mathop{*}\limits_{k=1}^{r} B_k = \mathop{*}\limits_{k=1}^{r} B_k'$, $\mathop{*}\limits_{k=1}^{i+1} B_k = \mathop{*}\limits_{k=1}^{i+1} B_k' = \mathop{*}\limits_{k=1}^{s} C_k$ and $\ell(B_{i+1}') < \ell(B_{i+1}) = \mu$, a contradiction to the minimality of μ .

Without restriction I assume $v_{a_1}(C_{s-1} * C_s) > 0$ and $v_{a_2}(C_{s-1} * C_s) > 0$; let $C_{s-1} * C_s = \langle a_1, a_2, \ldots, a_u \rangle$. I define $C_{s-1}' = \langle a_1+a_2, a_3, \ldots, a_u \rangle$ and I put $C_k' = C_k$ for $k \in \{1,\ldots,s-2\}$. Summing up I obtain $B = \mathop{*}\limits_{k=1}^{r} B_k = \mathop{*}\limits_{k=1}^{r} B_k'$, $\mathop{*}\limits_{k=1}^{i+1} B_k' = \mathop{*}\limits_{k=1}^{s-1} C_k'$ (since $j \in \{i+2,\ldots,r\}$) and $\ell(B_{i+1}') < \ell(B_{i+1}) = \mu$. Thus the minimality of μ implies that C_{s-1}' is irreducible and that $s-1 \leq i+1$. ◇

<u>THEOREM 1.</u> *Let* G *be an abelian group and let* $B \in \mathcal{B}(G)$ *be a non-empty block. If* $\{g \in G / v_g(B) > 0\} \cup \{0\}$ *is a subgroup of* G *, then* $L(B) = \{y, y+1, \ldots, y+k\}$ *with* $y, k \in \mathbb{N}$.

<u>PROOF.</u> Let $r \in L(B)$ be given; by Proposition 1 there are two possibilities:

<u>Case 1.</u> there is a factorization $B = \mathop{*}\limits_{k=1}^{r} B_k$ into r irreducible blocks and there is an exchange transformation of the form $(2 \longleftrightarrow 1)$ which gives rise to a factorization of length $r+1$.

<u>Case 2.</u> for every factorization of B into r irreducible blocks the length r remains invariant under all exchange transformations of the form $(2 \longleftrightarrow 1)$. Then by Proposition 2 it remains invariant under all

exchange transformations. Thus by Proposition 3 there is an $i \in \{2,\dots,r\}$ which satisfies the condition in Proposition 4, which finally states that either $r = \max L(B)$ or $r+1 \in L(B)$. ◇

4. Algebraic integers having simple sets of lengths

The relationship between factorizations of blocks and factorizations of algebraic integers immediately gives rise to the following arithmetic version of Theorem 1.

THEOREM 1'. *Let* R *be the ring of integers of an algebraic number field* K *with ideal class group* $\mathcal{C}\ell(K)$ *and let* C^o *denote the class of principal ideals. If for some element* $a \in R^{\#} = R\setminus(R^{\times} \cup \{0\})$ $\{[\mathfrak{p}] \,/\, \mathfrak{p}$ *is a prime ideal dividing* $(a)\} \cup \{C^o\}$ *is a subgroup of* $\mathcal{C}\ell(K)$, *then* $L(a) = \{y,y+1,\dots,y+k\}$ *with* $y,k \in \mathbb{N}$.

REMARK. According to Proposition 1 in [3] there is an arithmetic analogue to Theorem 1 not only for rings of integers but also for semigroups H having a divisor theory $\partial: H \to D$.

COROLLARY 1. *Let* R *be the ring of integers of an algebraic number field* K . *Then there is a rational integer* λ *such that for every* $a \in R^{\#}$ $L(\lambda a) = \{y,y+1,\dots,y+k\}$ *with* $y,k \in \mathbb{N}$; *in particular,* $L(\lambda n)$ *has this form for every* $n \in \mathbb{Z}^{\#} = \mathbb{Z}\setminus\{-1,0,1\}$.

PROOF. Because there are prime ideals in each ideal class $C \in \mathcal{C}\ell(K)$, there exist primes $p_1,\dots,p_k \in \mathbb{N}$ with $p_iR = \mathfrak{p}_{i,1}\cdots\mathfrak{p}_{i,\ell_i}$, such that $\mathcal{C}\ell(K) = \bigcup_{i=1}^{k} \{[\mathfrak{p}_{i,j}] \,/\, 1 \le j \le \ell_i\}$. Then $\lambda = p_1\cdots p_k$ has the property desired. ◇

COROLLARY 2. *Let* K *be a Galois number field with Galoisgroup* $\mathrm{Gal}(K/Q)$ *and ideal class group* $\mathcal{C}\ell(K)$, *let* C^o *denote the class of principal ideals.*

1. *If* $\mathrm{Gal}(K/Q)$ *operates transitively on* $\mathcal{C}\ell(K)\setminus\{C^o\}$, *then for every* $n \in \mathbb{Z}^{\#}$ $L(n) = \{y,y+1,\dots,y+k\}$ *with* $y,k \in \mathbb{N}$.
2. *If for every ideal class* $C \in \mathcal{C}\ell(K)$ $\{C^{\sigma} \,/\, \sigma \in \mathrm{Gal}(K/Q)\} \cup \{C^o\}$ *is a subgroup of* $\mathcal{C}\ell(K)$, *then for every prime power* $p^n \in \mathbb{Z}$ $L(p^n) = \{y,y+1,\dots,y+k\}$ *with* $y,k \in \mathbb{N}$.

PROOF. For every prime $p \in \mathbb{Z}$ $Gal(K/Q)$ operates transitively on the set of prime ideals dividing pR and therefore both assertions follow from Theorem 1'. ◇

REMARK. If $Gal(K/Q)$ operates transitively on $C\ell(K)\setminus\{C^o\}$, then $C\ell(K)$ is of the type $(p,\ldots,p)$, $p \in \mathbb{Z}$ prime.

I give some examples of Galois algebraic number fields K whose Galoisgroup operates transitively on $C\ell(K)\setminus\{C^o\}$. Let $h(K)$ denote the class number of K and let $Q(\zeta_m)$ denote the mth cyclotomic field. The class numbers of certain cyclotomic fields are from [5], page 353.

To do the first example, I need a lemma.

LEMMA 2. *Let* K/K_o *be a cyclic extension of algebraic number fields with prime degree* $[K:K_o] = \ell$, $\gcd(h(K),h(K_o)) = 1$ *and* $h(K) > 1$. *Then* $Gal(K/K_o)$ *operates transitively on* $C\ell(K)\setminus\{C^o\}$ *if and only if* $h(K) = \ell+1$.

PROOF. Let $Gal(K/K_o) = \langle\sigma\rangle$ and R the ring of integers of K.

(i) For $C \in C\ell(K)$ it is obviously that $\#\{C^{\sigma^k}/1 \leq k \leq \ell\} \in \{1,\ell\}$.

(ii) If $Gal(K/K_o)$ operates transitively on $C\ell(K)\setminus\{C^o\}$, then $h(K) \leq \ell+1$. From (i) $h(K) = \ell+1$ follows.

(iii) Let $h(K) = \ell+1$. Let $\mathfrak{P}_1 \triangleleft R$ be a prime ideal, which is not principal, such that $\mathfrak{P}_1 \cap K_o = \mathfrak{p}$ is unramified. Then $\mathfrak{p}R = \mathfrak{P}_1\ldots\mathfrak{P}_\ell$ and $Gal(K/K_o)$ operates transitively on $G_o = \{[\mathfrak{P}_i] / 1 \leq i \leq \ell\}$. Since $\#G_o > 1$ (assume to the contrary $\#G_o = 1$: let $[\mathfrak{P}_i] = C \neq C^o$ for $1 \leq i \leq \ell$. Then $C^o = [\mathfrak{p}R]^{h(K_o)} = C^{\ell h(K_o)}$, a contradiction), it follows by (i) that $\#G_o = \ell$ and thus $Gal(K/K_o)$ operates transitively on $C\ell(K)\setminus\{C^o\}$. ◇

EXAMPLE 1. Let $K = Q(\zeta_{29})$, $Gal(K/Q) \cong (\mathbb{Z}/29\mathbb{Z})^\times \cong \mathbb{Z}/4\mathbb{Z} \times \mathbb{Z}/7\mathbb{Z}$ and $C\ell(K) \cong \mathbb{Z}/2\mathbb{Z} \times \mathbb{Z}/2\mathbb{Z} \times \mathbb{Z}/2\mathbb{Z}$ (see [5], page 187).

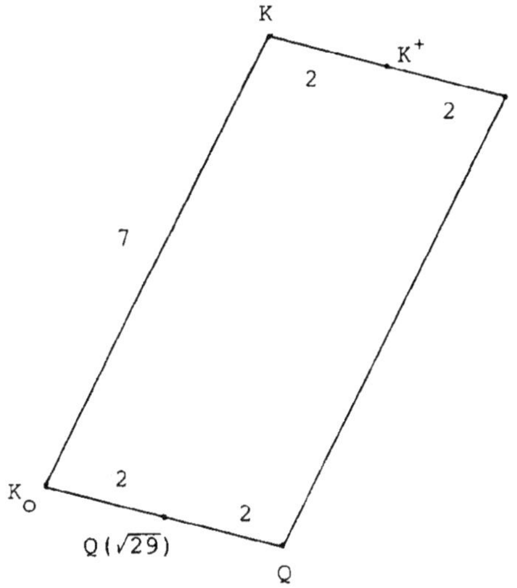

(K^+ denotes the maximal real subfield of K)

K_o is cyclic quartic and $h(K_o) = 1$ (see for example [6]). Thus Lemma 2 implies that $Gal(K/K_o) < Gal(K/Q)$ operates transitively on $C\ell(K)\setminus\{C^o\}$.

To do the next examples, I need a further lemma.

LEMMA 3. *Let* K/K_o *be a cyclic extension of algebraic number fields with* $Gal(K/K_o) = \{1,\sigma,\dots,\sigma^{n-1}\}$. *Let* $C\ell(K) = \{1,C,\dots,C^{h(K)-1}\}$, $h(K) > 1$, $\gcd(n,h(K)) = 1$ *and let* $C^\sigma = C^r$. *If for a field* K_1 *with* $K_o \subset K_1 \subset K$, $[K_1:K_o] = d$ *and* $\gcd(h(K_1),h(K)) = 1$, *it follows* $r^d \not\equiv 1 \bmod h(K)$.

PROOF. I consider the norm map

$$N_{K/K_1}: \begin{cases} C\ell(K) \to C\ell(K_1) \\ C \to C^{1+r^d+\dots+r^{(\frac{n}{d}-1)d}} = C^v \end{cases}$$

Thus $N_{K/K_1}(C^{h(K_1)}) = C^{vh(K_1)}$, which implies $vh(K_1) \equiv 0 \bmod h(K)$ and therefore

$$v = 1+r^d+\dots+(r^d)^{\frac{n}{d}-1} \equiv 0 \bmod h(K).$$

Since $r^d \equiv 1 \bmod h(K)$ implies $\frac{n}{d} \equiv 0 \bmod h(K)$, a contradiction to $\gcd(n,h(K)) = 1$, it follows $r^d \not\equiv 1 \bmod h(K)$. ◇

EXAMPLE 2. Let $K = Q(\zeta_{64})$, $Gal(K/Q) \cong (\mathbb{Z}/64\mathbb{Z})^\times \cong \mathbb{Z}/2\mathbb{Z} \times \mathbb{Z}/16\mathbb{Z}$ and $C\ell(K) = \{C^o,C,C^2,\dots,C^{16}\}$.

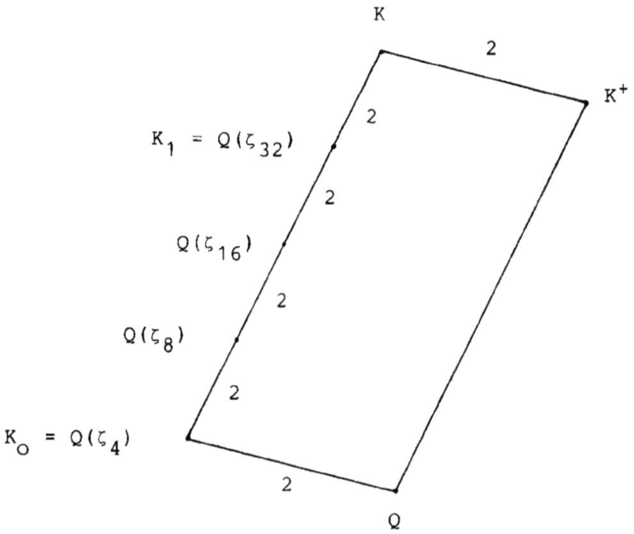

Let $Gal(K/K_o) = \{1,\sigma,\ldots,\sigma^{15}\}$ and $C^\sigma = C^r$. Since $h(K_1) = 1$, Lemma 3 implies $r^8 \not\equiv 1 \bmod 17$. Thus $ord_{17}(r) = 16$ and $Gal(K/K_o) <$ $< Gal(K/Q)$ operates transitively on $C\ell(K)\setminus\{C^o\}$.

<u>EXAMPLE 3.</u> Let $K = Q(\zeta_{51})$, $Gal(K/Q) \cong (\mathbb{Z}/51\mathbb{Z})^x \cong \mathbb{Z}/2\mathbb{Z} \times \mathbb{Z}/16\mathbb{Z}$ and $C\ell(K) = \{C^o, C, \ldots, C^4\}$.

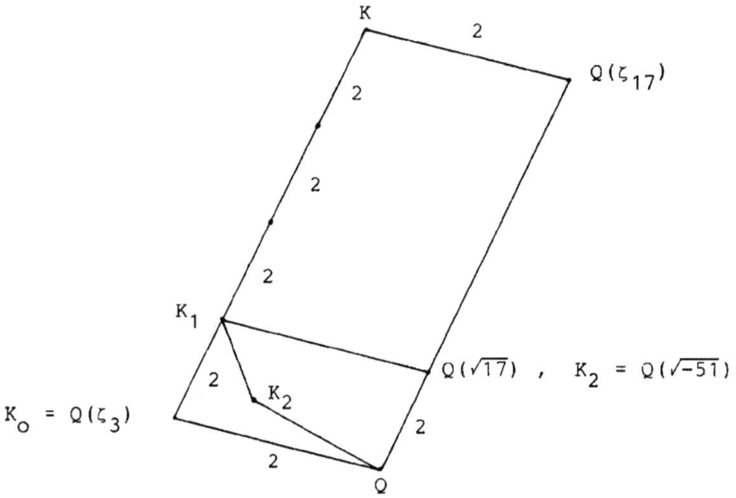

Let $Gal(K/K_o) = \{1,\sigma,\ldots,\sigma^{15}\}$ and $C^\sigma = C^r$. K_1 is bicyclic biquadratic and $h(K_1) = 1$ (see [2]). Lemma 3 implies $r^2 \not\equiv 1 \bmod 5$ and thus $ord_5(r) = 4$. Therefore $Gal(K/K_o) < Gal(K/Q)$ operates transitively on $\mathcal{C}\ell(K)\setminus\{C^o\}$.

Corollary 2 shows that the operation of the Galois group Gal(K/Q) on the ideal class group $\mathcal{C}\ell(K)$ can greatly influence the structure of sets of lengths of rational integers. This may give rise to the hope, that sets of lengths of rational integers have certain nice properties even without assuming that Gal(K/Q) operates transitively on $\mathcal{C}\ell(K)\setminus\{C^o\}$. So if for example Gal(K/Q) operates trivially on $\mathcal{C}\ell(K)$, then $\#L(p^n) = 1$ for every prime power $p^n \in \mathbb{Z}$. In general, however, not every rational integer has a set of lengths of the form $\{y,y+1,\dots,y+k\}$ with $y,k \in \mathbb{N}$; this means that the rational integer λ arising in Corollary 1 need not be equal to 1 (even if K/Q is Galois). I give an example of such a situation.

EXAMPLE. Let K be a quadratic number field whose ideal class group contains a subgroup isomorphic to $\mathbb{Z}/n\mathbb{Z} = \{\overline{0},\overline{1},\dots,\overline{n-1}\}$ with $n \geq 6$. Let $B_1 = \langle\overline{1},\dots,\overline{1}\rangle$, $B_2 = \langle\overline{n-1},\dots,\overline{n-1}\rangle$ and $B_3 = \langle\overline{2},\overline{n-2}\rangle$ be irreducible blocks in $\mathcal{B}(\mathcal{C}\ell(K))$ and let $B_k = B_1^k * B_2^k * B_3$ with $k \in \mathbb{N}_+$. Since the non-trivial automorphism $\sigma \in \mathrm{Gal}(K/Q)$ maps each ideal class onto its inverse, there are (infinitely many) rational integers n_k with $B(n_k) = B_k$. I determine $L(B_k)$:

1. $L(B_1^k * B_2^k) = \{2k, 2k+(n-2),\dots,2k+k(n-2)\}$,
 $L(B_1^{k-1} * B_2^{k-1}) = \{2k-2, 2k-2+(n-2),\dots,2k-2+(k-1)(n-2)\}$.
2. Let $B_4 = \langle\overline{2},\overline{1},\dots,\overline{1}\rangle$, $B_5 = \langle\overline{2},\overline{n-1},\overline{n-1}\rangle$, $B_6 = \langle\overline{n-2},\overline{n-1},\dots,\overline{n-1}\rangle$, $B_7 = \langle\overline{n-2},\overline{1},\overline{1}\rangle$ and $B_8 = \langle\overline{1},\overline{n-1}\rangle$ be irreducible blocks. Then $B_k = B_1^k * B_2^k * B_3 = \mathop{*}\limits_{i=1}^{8} B_i^{r_i}$ with $r_3+r_4+r_5 = 1$ and $r_3+r_6+r_7 = 1$. I discuss the cases $r_3 = 1$, $(r_4 = 1,\ r_6 = 1)$, $(r_4 = 1,\ r_7 = 1)$ and $(r_5 = 1,\ r_7 = 1)$ (The case $(r_5 = 1,\ r_6 = 1)$ is symmetric to $(r_4 = 1,\ r_7 = 1)$).

Case 1. $r_3 = 1$. $B_1^k*B_2^k*B_3 = B_1^{r_1} * B_2^{r_2} * B_3*B_8^{r_8}$; thus $r_1+r_2+1+r_8 \in L(B_k)$ if and only if $r_1+r_2+r_8 \in L(B_1^k*B_2^k)$.

Case 2. $r_4 = 1$ and $r_7 = 1$. $B_k = B_1^{k-1}*B_2^{k-1}*B_2*B_4*B_7 = B_1^{r_1} * B_2^{r_2} * B_8^{r_8} *B_4*B_7$; $nr_1+r_8+n = v_{\overline{1}}(B_k) = v_{\overline{n-1}}(B_k) = nr_2+r_8$ implies $r_2 \geq 1$; thus $r_1+(r_2-1)+r_8+3 \in L(B_k)$ if and only if $r_1+(r_2-1)+r_8 \in L(B_1^{k-1}*B_2^{k-1})$.

Case 3. $r_4 = 1$ and $r_6 = 1$. $B_k = B_1^{k-1}*B_2^{k-1}*B_4*B_6*B_8^2 = B_1^{r_1} * B_2^{r_2} *$

$* B_8^{r_8} * B_4 * B_6$; $v_{\overline{1}}(B_k) = nr_1 + r_8 + (n-2) \equiv 0 \bmod n$ implies $r_8 \geq 2$; thus $r_1 + r_2 + (r_8 - 2) + 4 \in L(B_k)$ if and only if $r_1 + r_2 + r_8 - 2 \in L(B_1^{k-1} * B_2^{k-1})$.

Case 4. $r_5 = 1$ and $r_7 = 1$. $B_k = B_1^{k-1} * B_2^{k-1} * B_5 * B_7 * B_8^{n-2} = B_1^{r_1} * B_2^{r_2} * {}$ $* B_8^{r_8} * B_5 * B_7$; $v_{\overline{1}}(B_k) = nr_1 + r_8 + 2 \equiv 0 \bmod n$ implies $r_8 \geq n-2$; thus $r_1 + r_2 + (r_8 - (n-2)) + n \in L(B_k)$ if and only if $r_1 + r_2 + (r_8 - (n-2)) \in$ $\in L(B_1^{k-1} * B_2^{k-1})$.

3. $L(B_k) = \{$

$$
\begin{array}{lll}
 & 2k+1, & 2k+2, \\
2k+(n-2), & 2k+1+(n-2), & 2k+2+(n-2), \\
 & \ldots\ldots\ldots\ldots\ldots\ldots & \\
2k+(k-1)(n-2), & 2k+1+(k-1)(n-2), & 2k+2+(k-1)(n-2), \\
2k+k(n-2), & 2k+1+k(n-2)\} &
\end{array}
$$

Acknowledgement. I would like to thank Prof. F. Halter-Koch for reading the paper and giving valuable comments.

References

[1] L. Carlitz, A characterization of algebraic number fields with class number two, Proc. Amer. Math. Soc. 11 (1960), 391 - 392.

[2] E. Brown, C. J. Parry, The imaginary bicyclic biquadratic fields with class number 1, J. Reine Angew. Math. 266 (1974), 118 - 120.

[3] A. Geroldinger, Über nicht-eindeutige Zerlegungen in irreduzible Elemente, Math. Z. 197 (1988), 505 - 529.

[4] J. Śliwa, Remarks on factorizations in algebraic number fields, Coll. Math. 46.1 (1982), 123 - 130.

[5] L. C. Washington, Introduction to cyclotomic fields, GTM 83, Springer 1982.

[6] K. S. Williams, Calculation of the classnumber of certain quartic fields, Proceedings of the international conference on classnumbers and fundamental units, June 24 - 28, 1986, Katata, Japan.

Alfred Geroldinger
Institut für Mathematik
Karl-Franzens-Universität Graz
Halbärthgasse 1/1
A-8010 Graz

Etude d'une famille modulaire de variétés abéliennes.

Roland Gillard

Université de Grenoble I, Laboratoire de Mathématiques associé au C.N.R.S.
Institut Fourier, BP 74, F-38402 Saint Martin d'Hères

1. Introduction.

L'exposé oral a résumé [Gi]. Il s'agissait de démontrer, pour un nombre premier p, l'analogue p-adique des relations monomiales de Shimura [Sh] par une méthode trouvant sa source dans [Gr]. Pour cela, on utilise une variété de Shimura Sh (la même que dans [Sh]) mais sur une base entière en p. On a alors besoin de la connexité de la fibre spéciale. Dans [Gi], on s'en tire en observant que la variété Sh est compacte sous certaines conditions. Il m'a semblé plus intéressant de rappeler ici la construction d'un modèle entier en p de Sh (sans l'hypothèse ordinaire) *cf.* §2 et d'indiquer * comment les idées de Faltings, [F], devraient permettre de construire une compactification toroïdale pour Sh dans le cas non compact, *cf.* §3 et 4. On espère que les constructions ci-dessous seront utiles à la généralisation de [Gi] au cas non ordinaire. Notons enfin que le cas des variétés de [D1] 4.9 peut sans doute être traité de la même façon.

Désignons par L un corps CM, ρ la conjugaison complexe et par $\mathcal{O}(L)$ l'anneau des entiers de L. Soit F le sous corps réel maximum de L. Les variétés abéliennes que nous considèrerons seront munies d'une action de $\mathcal{O}(L)$ ainsi que d'une polarisation compatible.

2. Modèle entier en p.

Avec L comme dans le §1, on suppose que $L = F(\xi)$ avec $\xi^\rho = -\xi$ (ρ désignant la conjugaison complexe). Pour chaque plongement μ de F dans $\mathbb{R}$, on choisit un plongement τ de L dans $\mathbb{C}$ le prolongeant et tel que la partie imaginaire de $\tau(\xi)$ soit positive; on désigne par $\overline{\tau}$ le plongement conjugué.

On fixe un nombre premier p non ramifié dans $L/\mathbb{Q}$. On choisit une clôture algébrique $\overline{\mathbb{Q}}$ de $\mathbb{Q}$ et des plongements $\overline{\mathbb{Q}} \hookrightarrow \mathbb{C}$ et $\overline{\mathbb{Q}} \hookrightarrow \overline{\mathbb{Q}}_p$. Ceci définit une place v de $\overline{\mathbb{Q}}$. On note K une sous-extension de $\overline{\mathbb{Q}}$ non ramifiée en p et assez grande: en particulier K contient tous les plongements $\tau(L)$ de L dans $\overline{\mathbb{Q}}$. Désignons par R l'anneau de valuation de v dans K et R_v son complété.

---

* en espérant l'indulgence du lecteur pour les nombreuses lacunes de mon texte!

2.1. Une famille analytique de variétés abéliennes. — On considère une somme formelle Φ de plongements de L dans $\mathbb{C}$; Φ s'écrit donc

2.1.1 $$\Phi = \sum (r_\mu \tau + s_\mu \bar{\tau}) ,$$

et on suppose qu'il existe m tel que

2.1.2 $$\text{pour tout } \mu , r_\mu + s_\mu = m .$$

On considère un réseau $\mathcal{L}$ dans le L-espace vectoriel $V := L^m$. On munit V d'une forme ρ-hermitienne H (i.e. $H(y,x) = H(x,y)^\rho$) telle que pour chaque μ la forme sur $\mathbb{C}^m$ qu'on déduit de H par τ est de signature (r_μ, s_μ). La forme ξH est anti-hermitienne et joue le rôle de la matrice T de [Sh] §4. Elle permet d'obtenir un accouplement sur $\mathcal{L}$ dont l'ensemble des valeurs est exactement $\mathbb{Z}$ si H est bien choisie. Cet accouplement est défini par une forme ρ-alternée, i.e. $E(y,x) = -E(x,y)$ et $E(a\,x, y) = E(x, a^\rho\, y)$ pour tout a dans $\mathcal{O}(L)$:

2.1.3 $$E(x,y) := Tr_{L/\mathbb{Q}}(\xi\, H(x,y)) .$$

Les données de H et E sont équivalentes. On suppose que l'indice de l'image de $\mathcal{L}$ dans son dual $\mathcal{L}^*$ (c'est l'ensemble des y dans L^m tels que pour tout x dans $\mathcal{L}$, $E(x,y)$ soit dans $\mathbb{Z}$) soit premier à p.

Posons $V_\mathbb{R} := V \otimes \mathbb{R}$; à partir de $V_\mathbb{R}$, pour obtenir une variété abélienne telle que E soit une forme de Riemann, on doit munir $V_\mathbb{R}$ d'une structure complexe J telle que la forme

2.1.4 $$(x,y) \to E(x, Jy) \text{ soit définie positive.}$$

On sait qu'il existe une bijection entre les structures complexes vérifiant 2.1.4 et les points $x := (x_\mu)$ dans le produit $\mathbf{H} := \prod \mathbf{H}_\mu$ où $\mathbf{H}_\mu$ est le domaine symétrique des matrices complexes d'ordre $r_\mu \times s_\mu$ telles que $I_{r_\mu} - x_\mu \bar{x}_\mu^t$ soit hermitienne définie positive ($\mathbf{H}_\mu$ est réduit à un point si r_μ ou s_μ est nul). Enfin en munissant $V_\mathbb{R}/\mathcal{L}$ de la structure correspondant au point x de $\mathbf{H}$, on obtient une famille analytique de variétés abéliennes

2.1.5 $$\mathbf{H} \times V_\mathbb{R}/\mathcal{L} \to \mathbf{H} .$$

Si $A = V_\mathbb{R}/\mathcal{L}$ est muni d'un J comme plus haut, on a un isomorphisme canonique

2.1.6 $$\alpha : \mathcal{L} \xrightarrow{\sim} \mathrm{H}_1(A, \mathbb{Z}) ;$$

de plus, A est munie naturellement de la polarisation

2.1.7 $$\lambda : A \to A'$$

déduite de E, A' désignant la variété abélienne duale de A; une telle polarisation munit $\mathrm{H}_1(A,\mathbb{Z})$ d'une forme ρ-alternée. On a de plus un plongement θ de $\mathcal{O}(L)$ dans $\mathrm{End}(A)$. Ainsi θ et λ sont compatibles: l'involution de Rosati définie par λ vérifie: $\theta(a) \to \theta(a^\rho)$.

2.2. La variété de Shimura. — Pour chaque point x de $\mathbf{H}$, la variété abélienne qui lui correspond dans 2.1.5 fait partie d'un quadruplet $(A,\theta,\lambda,\alpha)$, avec la condition que l'action de $\mathcal{O}(L)$ sur $Lie(A)$ vérifie:

2.2.1 $$Tr(a, Lie(A)) = Tr_\Phi\, a\ .$$

Soit $\mathbf{G}$ le groupe algébrique des $\mathcal{O}(L)$-similitudes de la forme H à multiplicateur rationnel défini par ses points à valeurs dans une $\mathbb{Q}$ algèbre variable Λ:

2.2.2
$$\mathbf{G}(\Lambda) = \{g \in Gl(V \otimes \Lambda) | \exists m(g) \in \Lambda^* \ , \forall x,y \in V \otimes \Lambda\ , H(g\,x, g\,y) = m(g)\, H(x,y)\}\ .$$

C'est le groupe algébrique associé au domaine $\mathbf{H}$; $\mathbf{G}(\mathbb{R})$ contient un sous-groupe égal au produit $SU := \prod SU_\mu$ où SU_μ désigne le groupe spécial unitaire de la forme H_μ déduite de H par l'extension des scalaires $\tau : L \to \mathbb{C}$. En raisonnant composante par composante, on voit que SU opère transitivement sur H.

On peut alors regarder le sous-groupe arithmétique $\Gamma \subset SU \subset \mathbf{G}(\mathbb{R})$ des g dans la composante neutre de $\mathbf{G}(\mathbb{R})$ fixant $\mathcal{L} \subset V_\mathbb{R}$: deux points x et x' sont dans la même orbite sous Γ si les quadruplets associés donnent des triplets (A,θ,λ) isomorphes (oubli de α). En général, les triplets (A,θ,λ) ont beaucoup d'isomorphismes. On rigidifie donc la situation en imposant une structure de niveau M, *i.e.* un $\mathcal{O}(L)$-isomorphisme symplectique

2.2.3 $$k : \mathcal{L}/M\,\mathcal{L} \xrightarrow{\sim} \mathrm{H}_1(A,\mathbb{Z}) \otimes \mathbb{Z}/M\,\mathbb{Z}\ ,$$

pour M dans $\mathbb{N}$. Pour éviter tout problème, en 2.3, on suppose que M est premier à p. Dès que $\mathrm{H}_1(A,\mathbb{Z})$ est isomorphe à $\mathcal{L}$, un tel isomorphisme se remonte en un isomorphisme α comme plus haut. Si Γ_M est le sous-groupe de Γ fixant $\mathcal{L}/M\,\mathcal{L}$, deux points x et x' de $\mathbf{H}$ donnent le même quadruplet (A,θ,λ,k) si et seulement s'ils sont dans la même orbite selon Γ_M: si bien que $S_\mathbb{C} := \Gamma_M \setminus \mathbf{H}$ classifie sur $\mathbb{C}$ les quadruplets à isomorphisme près et porte un quadruplet universel $(A^u,\ldots)$ déduit de 2.1.5 par passage au quotient pour l'action de Γ_M. On sait ([D1]) que le morphisme de variétés analytiques $A^u \to S_\mathbb{C}$ provient en fait d'un morphisme de variétés algébriques (noté de la même façon) défini sur un corps de nombres absolument non ramifié en dehors de M. Il est défini sur K pourvu que cette extension soit choisie assez grande, *cf.* début du §2. A cause de 2.2.3, la définition précédente reste topologique. Pour avoir un problème purement algébrique, il convient de remplacer $\mathrm{H}_1(A, \mathbb{Z})$ par $\hat{T}(A) := \mathrm{H}_1^{\text{ét}}(A, \hat{\mathbb{Z}})$ ($\simeq \mathrm{H}_1(A, \mathbb{Z}) \otimes \hat{\mathbb{Z}}$ avec $\hat{\mathbb{Z}} := \prod \mathbb{Z}_l$, l nombre premier). C'est à dire qu'on regroupe des $\mathcal{O}(L)$- structures sur $\mathrm{H}_1(A, \mathbb{Z})$ qui sont dans le même genre en considérant des isomorphismes symplectiques

2.2.4 $$\hat{\alpha} : \hat{\mathcal{L}} := \mathcal{L} \otimes \hat{\mathbb{Z}} \xrightarrow{\sim} \hat{T}(A) .$$

En notant que la structure de $\mathrm{H}_1(A, \mathbb{Z})$ est controlée aux places à l'infini par 2.2.1, on déduit du principe de Hasse (*cf.* [D1] version 0, §5) que si $\mathrm{H}_1(A, \mathbb{Z})$ est relié à $\hat{\mathcal{L}}$ par un $\hat{\alpha}$, il existe un réseau de V qui lui est isomorphe comme $\mathcal{O}(L)$-module symplectique De tels $\hat{\alpha}$ définissent par passage au quotient un k comme plus haut: k peut être vu comme un isomorphisme symplectique

2.2.5 $$k : \mathcal{L}/M\,\mathcal{L} \xrightarrow{\sim} A[M] ,$$

où $A[M]$ est le noyau de la multiplication par M dans $A(\mathbb{C})$. Notons que tout quadruplet (A, θ, λ, k), k comme dans 2.2.5, provient de quadruplets $(A, \theta, \lambda, \hat{\alpha})$ pourvu que M soit assez grand; on peut choisir un tel M premier à p (*cf.* par exemple [J] th. 7.1).

Soit $\mathbb{A}$ le groupe des adèles sur $\mathbb{Q}$; $\mathbf{G}(\mathbb{A})$ opère sur la situation et donne la variété de Shimura correspondante

2.2.6 $$Sh_{\mathbb{C}} := \mathbf{G}(\mathbb{Q}) \setminus \mathbf{G}(\mathbb{A})/\mathbf{K} ,$$

où $\mathbf{K} = \mathbf{K}_\infty . \mathbf{K}_f$ est le sous-groupe compact de $\mathbf{G}(\mathbb{A})$ correspondant au niveau M. On sait (*cf.* [D1] 5.8) que $Sh_{\mathbb{C}}$ provient d'une variété algébrique Sh_E définie sur le corps de nombres $E \subset \overline{\mathbb{Q}} \subset \mathbb{C}$ "reflet" de (L, Φ). En fait, S_K est une composante connexe de la variété Sh_K déduite de Sh_E par extension des scalaires. Au §2.3, on va montrer que Sh_K a bonne réduction en v en prolongeant Sh en un schéma sur R.

On a vu que sur $\mathbb{C}$, $Sh_{\mathbb{C}}$ classifie les quadruplets $(A, \theta, \lambda, \hat{\alpha})$. Notons $\mathbb{A}_f$ les adèles finis de $\mathbb{Q}$ et $\mathbf{K}_f$ le sous-groupe compact de $\mathbf{G}(\mathbb{A}_f)$ défini par

$$\mathbf{K}_f = \{g \in \mathbf{G}(\mathbb{A}_f) | \quad g\hat{\mathcal{L}} = \hat{\mathcal{L}} \text{ et } (g-1)\hat{\mathcal{L}} \subseteq M\hat{\mathcal{L}}\} .$$

Dans un quadruplet comme ci-dessus, $\hat{\alpha}$ n'intervient que par sa classe $\hat{\alpha}.\mathbf{K}_f$. Pour X un R-schéma et A un schéma abélien sur X, il n'existe pas d'isomorphisme du type 2.2.4 à cause de la situation en caractéristique p pour les points de torsion

p-primaire. Cependant, il est possible d'adapter 2.2.4 simplement en *oubliant* sa p-partie! Ceci va permettre de construire un R-schéma Sh.

2.3. Le foncteur. — Soit $\tilde{\mathbb{Z}} = \prod_{l \neq p} \mathbb{Z}_l$ la partie première à p dans $\hat{\mathbb{Z}}$ et posons $\tilde{T}(A) := \mathrm{H}_1^{\text{ét}}(A, \tilde{\mathbb{Z}})$ ($\simeq \mathrm{H}_1(A, \mathbb{Z}) \otimes \tilde{\mathbb{Z}}$). On note $\tilde{\alpha}$ un isomorphisme symplectique:

2.3.1 $$\tilde{\alpha} : \tilde{\mathcal{L}} := \mathcal{L} \otimes \tilde{\mathbb{Z}} \xrightarrow{\sim} \tilde{T}(A) .$$

2.3.2. PROPOSITION. — *Si $A/\mathbb{C}$ est une variété abélienne munie d'une structure $\tilde{\alpha}$, il existe une structure $\hat{\alpha}$ comme en 2.2.4 la prolongeant.*

Démonstration. — Considérons deux $\mathcal{O}(L)$-modules sans torsion $\mathcal{T}$ et $\mathcal{L}$ de rang m munis de formes ρ-alternées (*cf.* avant 2.1.3), alors il existe un isomorphisme ρ-symplectique entre leurs complétés p-adiques

2.3.3 $$\alpha_{(p)} : \mathcal{L}_p \simeq \mathcal{T}_p .$$

En effet, on traduit le problème en termes de formes hermitiennes (*cf.* 2.1.3). On sait alors (*cf.* [J]) qu'il n'existe qu'une seule classe d'isomorphisme pour des $\mathcal{O}(L)$-modules hermitiens. Appliquant ceci à $T_p(A)$ et $\mathcal{L}_p$, on peut compléter $\tilde{\alpha}$ en un $\hat{\alpha}$ à l'aide d'un $\alpha_{(p)}$ comme ci-dessus.

Considérons le foncteur Sh associant à une R-algèbre B l'ensemble $Sh(B)$ des classes d'isomorphie de quadruplets $(A, \theta, \lambda, \tilde{\alpha})$ où A est un schéma abélien sur B, θ une action de $\mathcal{O}(L)$ sur A/B, $\lambda : A \to A'$ une polarisation dont le noyau est un schéma en groupes constant de structure fixée d'ordre premier à p et $\tilde{\alpha}$ un isomorphisme $\tilde{\mathcal{L}} \to \tilde{T}(A)$ à $\mathbf{K}_f$ près (*cf.* 3.0.1 et après 2.2.5); on impose les propriétés:

2.3.4) $\tilde{\alpha}$ est un isomorphisme symplectique;

2.3.5) L'involution de Rosati définie par λ envoie $\theta(a)$ sur $\theta(a^\rho)$;

2.3.6) $Lie(A/B)$ est un B-module libre et $a \in \mathcal{O}(L)$ opère dessus avec la trace $tr(a, Lie(A/B)) = Tr_\Phi(a) \in R$.

Le résultat est alors le suivant:

2.3.7. THÉORÈME. — *Si M a été choisi assez grand, Sh est représenté par un schéma Sh quasi-projectif et lisse sur R. La fibre générique de Sh est isomorphe à la variété Sh_K du §2.2*

On note $(\mathcal{A}, ...)$ le quadruplet universel sur Sh

2.4. Démonstration du théorème 2.3.7. — Comme me l' expliqué H. Carayol, *cf.* aussi [Ca] §5, la première partie se déduit assez facilement du résultat pour la famille de Siegel ([Mu1] 7.9). Quant à la lissité, on la démontre en utilisant [Me] V 1.6: il suffit de remonter la filtration de Hodge sur le $\mathrm{H}^1_{\mathrm{DR}}$ ainsi que la flèche induite par la polarisation. Ceci ne fait pas problème, le Fil^1 étant isotrope pour la forme bilinéaire alternée. on peut (et on doit) le remonter en un isotrope, l'application induite par λ envoie alors le Fil^1 pour A' sur celui pour A.

Remarque. — On vérifie (*cf.* [Gi]) que si Φ ne contient pas la norme (*i.e.* un au moins des r_μ ou s_μ est nul), Sh est compacte. Pour ce faire, on raisonne comme dans [Ca] 5.5 en appliquant le critère valuatif de propreté et le théorème de réduction semi-stable. Le point clef est que l'espace tangent d'une mauvaise réduction d'une variété abélienne classifiée par Sh ne peut contenir celui d'un tore avec action de $\mathcal{O}(L)$ sous l'hypothèse ci-dessus.

3. Uniformisation des variétés abéliennes.

3.1. Le 1-motif. — Soit maintenant R un anneau local complet et K son corps de fractions. On part d'une variété abélienne A sur K. Quitte à faire une extension de K, on peut supposer que A a une réduction semi-stable: en prenant la partie connexe du modèle de Néron de A, on obtient un modèle $\mathcal{A}$ sur R dont la réduction est extension d'une variété abélienne par un tore (qu'on suppose déployé). On sait que l'on peut remonter cette extension sur R obtenant un schéma semi-abélien G/R:

3.1.1 $$0 \to T \to G \to B \to 0 ,$$

extension d'un schéma abélien B sur R par un tore T, *cf.* [SGA 7] IX 7.1.5. Procédant de même avec la variété duale A' de A, on obtient G'/R:

3.1.2 $$0 \to T' \to G' \to B' \to 0 .$$

En fait B' s'identifie à la variété duale de B, *cf.* ci-dessous. Désignons par X (resp. X') le groupe des caractères de T' (resp. T) , sic! et $\jmath$ (resp. $\jmath'$) l'application $X \to B$ (resp.$X' \to B'$) définie par 3.1.1. Si R est de dimension 1, la théorie de Raynaud [Rn] reprise dans [BL] présente A comme quotient rigide-analytique de G par X. C'est à dire qu'en notant V^{rig} la variété rigide analytique associée à une variété algébrique V, on a une suite exacte:

3.1.3 $$0 \to X \overset{i^{\mathrm{rig}}}{\to} G^{\mathrm{rig}} \overset{e}{\to} A^{\mathrm{rig}} \to 0 ,$$

dépendant fonctoriellement de A et identifiant G^{rig} au “revêtement universel” de A^{rig}. De plus, i^{rig} provient d'un morphisme algébrique défini sur K, $i_K : X \to G_K$, et relevant j,*cf.* [SGA 7] 14.1.7. Comme par ailleurs j possède un relèvement canonique i_{can} sur R, en considérant $i_K - i_{\mathrm{can}}$, on trouve une application $X \to T$ d'où en prenant les coordonnées, une application bilinéaire

3.1.4 $$b : X \times X' \to K^* .$$

De i_K, on déduit une application bilinéaire Ψ, définie sur K, de $X \times X'$ dans la biextension de Poincaré, P, de $B \times B'$ par $\mathbb{G}_m$. On devrait pouvoir supprimer la condition sur R en adaptant convenablement [F] (sans doute en s'inspirant plus de [Mu2]).

3.2. Donnée d'une polarisation de A. — Donnons nous en plus une polarisation $\lambda : A \to A'$. Alors, λ induit une polarisation du 1-motif $X \to G$ sur K au sens de [D2] 10.2, cf aussi [Ch] p. 92. En effet on récupère une isogénie du complexe $X \to G$ vers $X' \to G'$ donc en particulier un homomorphisme (injectif) de groupes $\Phi : X \to X'$ et une polarisation $\overline{\lambda} : B \to B'$ avec une condition de compatibilité comme dans [Ch] II 3.2: notons P_λ la biextension sur $B \times B$ image réciproque de P par $Id \times \overline{\lambda}$ et Ψ_λ l'application de $X \times X$ dans P_λ déduite de $X \times X' \to P$. On doit avoir que

3.2.1 $$\overline{\lambda} \circ j = j' \circ \Phi ,$$

et que

3.2.2 $$\Psi_\lambda \text{ est symétrique } .$$

3.3. Conclusion. — Partant de $(A, \lambda)_{/R}$ on a obtenu $(X, G, i, \Phi, \overline{\lambda})$ avec les conditions 3.2.1 et 3.2.2. On a de plus la condition (*cf.* [SGA 7] IX 10.4) que

3.3.1 $X \to \mathbb{Z} : x \to v \circ \Psi_\lambda(x, x)$ est une forme quadratique définie positive ,

pour v valuation de R.

3.4. Construction de Mumford. — Généralisons la situation de 3.3 de la façon suivante. Soit R un anneau intègre normal noetherien et excellent, complet pour la topologie définie par un idéal I égal à sa racine. Si K désigne le corps des fractions de R, on pose $S = \mathrm{Spec}(R)$, $s = \mathrm{Spec}(R/I)$ et $\eta = \mathrm{Spec}(K)$. On se donne un quintuplet $(X, G, i, \Phi, \overline{\lambda})$ comme plus haut. On suppose vérifiées les conditions 3.2.1 et 3.2.2. La condition 3.3.1 devient ici:

3.4.1 Ψ se prolonge en un S-morphisme se réduisant en 0 modulo I .

La méthode de Mumford généralisée par Chai (*cf.* aussi [Br] et [F]) permet de construire un schéma semi-abélien $\mathcal{A}/S$ dont la fibre sur K est une variété abélienne; G et $\mathcal{A}$ ont même complétion I-adique. De plus $\mathcal{A}$ est muni d'une polarisation (dépendant de la donnée de $(i, \Phi, \overline{\lambda})$), *cf.* [Ch] p. 125. Enfin si R est un anneau local et I son idéal maximal, comme plus haut, les constructions 3.3 et 3.4 sont réciproques l'une de l'autre comme on le voit en passant aux schémas formels associés.

3.5. Action de $\mathcal{O}(L)$. — Comme tout ce qui précède est fonctoriel, on peut rajouter l'action de $\mathcal{O}(L)$. Ainsi, si A est munie dans 3.1 d'une action de $\mathcal{O}(L)$, il en est de même pour T, G et B; les groupes des caractères X et X' sont des $\mathcal{O}(L)$-modules sans torsion. Dans 3.2, il faut noter que λ est antilinéaire. Il en est de même pour Φ. Dans 3.4, la réciproque respecte l'action de $\mathcal{O}(L)$: si dans $(X, G, i, \Phi, \overline{\lambda})$, X et G sont munis d' actions de $\mathcal{O}(L)$ telles que i soit linéaire et Φ et $\overline{\lambda}$ soient antilinéaires, alors, par [Ch] 6.5, $\mathcal{A}/S$ est un $\mathcal{O}(L)$-schéma semi-abélien polarisé dont la fibre générique est une $\mathcal{O}(L)$-variété abélienne polarisée.

4. Construction d'une compactification de Sh.

On veut construire un espace algébrique lisse et propre $\overline{Sh}$ prolongeant Sh. La construction calquée sur [F] et esquissée en 4.2 et 4.3 utilise une longue liste d'objets combinatoires introduits en 4.1. Leur description s'inspire de la situation sur $\mathbb{C}$ ([AMRT]) ou mieux sur $\mathbb{Q}$ ([Br]), en partant de sous-groupes paraboliques maximaux de $\mathbf{G}$ (ou ce qui revient au même d'une filtration $W \subset W^{\perp}$ comme ci-dessous).

4.1. Combinatoire. — Si W désigne un sous L-espace vectoriel totalement isotrope de V, on introduit l'orthogonal $W^{\perp}$ de W pour E et $Bil(V/W^{\perp})$ l'ensemble des formes bilinéaires symétriques sur $V/W^{\perp}$ à valeurs dans $\mathbb{Z}$, $\mathcal{O}(L)$-linéaires sur la première variable $\mathcal{O}(L)$-antilinéaires sur la deuxième. On choisit une décomposition de la clôture rationnelle de l'ensemble des formes réelles définies positives $Bil(V/W^{\perp})^{+}_{\mathbb{R}}$ en cônes simpliciaux

$$[Bil(V/W^{\perp})_{\mathbb{R}}]^{\mathrm{rat}} = \bigcup_{\Sigma} \sigma$$

vérifiant les propriétes usuelles (*cf.* [Br] 4.2.5.1).

Prenons $\sigma \in \Sigma$. Les formes dans σ se factorisent par un quotient de V qu'on peut toujours écrire sous la forme $V/W_{\sigma}^{\perp}$, W_{σ} désignant un sous espace vectoriel de

W. Choisissons un réseau admissible $\mathcal{L}$ dans V (*i.e.* le réseau de départ, *cf.* après 2.1.2, ou à défaut un réseau dans le même genre, *cf.* après 2.2.4) On introduit $X_\sigma = \mathcal{L}/\mathcal{L}\cap W_\sigma^\perp$ et X'_σ le $\mathbb{Z}$ dual de $\mathcal{L}\cap W_\sigma$. Ainsi V induit un $\mathcal{O}(L)$-morphisme antilinéaire $X_\sigma \to X'_\sigma$. On note r_σ le rang de X_σ sur $\mathcal{O}(L)$. On désigne par $Bil(X_\sigma)$ le $\mathbb{Z}$-module des formes bilinéaires $X_\sigma \times X'_\sigma \to \mathbb{Z}$, $\mathcal{O}(L)$-linéaires sur la première variable, $\mathcal{O}(L)$-antilinéaires sur la deuxième et symétrique sur le sous-groupe $X_\sigma \times X_\sigma$ de $X_\sigma \times X'_\sigma$ (via Φ) et $Sym(X_\sigma)$ le $\mathbb{Z}$-dual de $Bil(X_\sigma)$: c'est le quotient de $X_\sigma \otimes X'_\sigma$ par les relations $x\otimes\Phi(y) = \Phi(y)\otimes x$ et $a\,x\otimes y = x\otimes\rho(a)\,y$. L'image de $x\otimes y$ est notée $x\odot y$. Soit $\check\sigma = \{\sum x_i\odot y_i | \forall b\in\sigma\,, \sum b(x_i,y_i)\geq 0\}$, le cône dual de σ: si $\mu\in X_\sigma$, alors $\mu\odot\mu\in\check\sigma$.

On considère aussi $S(\sigma)$ le tore de groupe de caractères $Bil(X_\sigma)$: c'est le spectre de l' anneau $\mathbb{Z}[Sym(X_\sigma)]$ de corps de fractions K_σ. Soit $S(\sigma)\to S(\sigma)_\sigma$ le plongement torique défini à l'aide de $Spec(\mathcal{R}_\sigma)$ où $\mathcal{R}_\sigma = \mathbb{Z}[Sym(X_\sigma)\cap\check\sigma]$. On a une forme bilinéaire universelle :

$$b_\sigma : X_\sigma\times X_\sigma \to S^2(X_\sigma)\hookrightarrow K_\sigma^*$$

définie par $(x,y)\to x\odot y\to q^{x\odot y}$, $q^{x\odot y}$ désignant la fonction sur $S(\sigma)$ définie par $x\odot y$ vu comme caractère. On constate que si μ est dans X_σ, $b_\sigma(\mu,\mu)$ est dans $\mathcal{R}_\sigma$ et s'annule à l'origine.

Plus généralement, pour toute face τ de σ, on a une orbite $\mathcal{O}_\tau$ dans $S(\sigma)_\sigma$ qu'on peut expliciter de la façon suivante. Ecrivons $Sym(X_\sigma)\cap\check\sigma$ sous la forme $\oplus\mathbb{N}\,s_j$ pour $j\in J$; notons q_j la fonction sur $S(\sigma)$ correspondant à s_j. Ainsi $S(\sigma)_\sigma$ s'identifie à l'espace affine sur $\mathbb{Z}$ de dimension $r_\sigma[F:\mathbb{Q}]$. La face τ est définie par des équations $s_j = 0$ pour j dans un sous-ensemble $Eq(\tau)$ de J; soit $Par(\tau)$ l'ensemble des indices des paramètres de τ: c'est le complémentaire de $Eq(\tau)$ dans J. Alors $\mathcal{O}_\tau$ est l'intersection de sa fermeture $\overline{\mathcal{O}}_\tau$ définie par les équations $q_j = 0$ pour j dans $Par(\tau)$ et de l'ouvert $S(\sigma)_\tau$ de $S(\sigma)_\sigma$ où les q_j, j dans $Eq(\tau)$, sont inversibles. L'orbite ouverte (égale au tore $S(\sigma)$) correspond à $\tau = 0$ et l'orbite fermée (réduite à l'origine) à $\tau = \sigma$. Avec les orbites, on définit une stratification de $S(\sigma)_\sigma$.

4.2. Théorie formelle locale. — Partant de $\mathcal{L}$, les données de 3.1 fournissent pour notre σ le réseau $\mathcal{L}(\sigma)$ de $\mathcal{O}(L)$-rang $m-2\,r_\sigma$ et la représentation sur le tensorisé par $\mathbb{C}$ est donnée, en regardant Φ comme une somme formelle de plongements de L dans $\mathbb{C}$, par $\Phi_\sigma = \Phi - r_\sigma Tr_{L/\mathbb{Q}}$. Ceci permet donc d'introduire un R-schéma Sh_σ, de façon analogue à Sh. Sur Sh_σ , on a un $\mathcal{O}(L)$-schéma abélien polarisé universel $\mathcal{B}_\sigma$. Les extensions G de $\mathcal{B}_\sigma$ par le tore déployé de groupe de caractères X_σ sont paramétrées par $\mathrm{Hom}_{\mathcal{O}(L)}(X_\sigma, \mathcal{B}'_\sigma)$, schéma abélien isogène au produit de r_σ copies de la duale $\mathcal{B}'_\sigma$ de $\mathcal{B}_\sigma$. Ayant un schéma classifiant les G, en considérant les couples (G, i), on en construit un revêtement principal, $Z(\sigma)$, sous le tore $S(\sigma)$. On le prolonge en Z_σ en prolongeant les fibres grâce à $S(\sigma) \hookrightarrow S(\sigma)_\sigma$: ainsi Z_σ est stratifié. Après cette construction algébrique développons l'analogue pour le complété d'un point. On choisit un point fermé dans l'orbite fermée de $S(\sigma)_\sigma$ et une variété abélienne polarisée B de dimension $d(m - 2r_\sigma)$ définie par un point fermé s de Sh_σ. Celle-ci possède une déformation universelle $\mathcal{B}_\sigma$. Les extensions de cette déformation par le tore déployé de caractères X_σ sont paramétrées par $\mathrm{Hom}_{\mathcal{O}(L)}(X_\sigma, \mathcal{B}'_\sigma)$ schéma abélien isogène au produit de r_σ copies de la duale $\mathcal{B}'_\sigma$ de $\mathcal{B}_\sigma$. Soient $\mathcal{R}_0$ l'hensélisé strict en un point fermé de $\mathrm{Hom}_{\mathcal{O}(L)}(\mathcal{L}'_\sigma, \mathcal{B}'_\sigma)$, $\mathcal{R}^{hs}$ l'hensélisé strict de $\mathcal{R}_0 \hat{\otimes} \mathcal{R}_\sigma$ au point au dessus de s, et $\mathfrak{m}$ son idéal maximal. Ainsi, $\mathcal{R}^{hs}$ apparait comme un anneau strictement local en un point fermé z de la σ-strate de Z_σ. Sur $\mathcal{R}^{hs}$, choisissons une application bilinéaire Ψ_0 de $\mathcal{L}_\sigma \times \mathcal{L}'_\sigma$ dans la biextension de Poincaré sur $\mathcal{B}_\sigma \times \mathcal{B}'_\sigma$: en déformant Ψ_0 à l'aide de la forme universelle b_σ, on construit un Ψ à valeurs dans K_σ comme en 4.1. Ainsi sur $\mathcal{R}^{hs}$, on dispose d'un quintuplet comme en 3.4: on récupère donc un $\mathcal{O}(L)$-schéma semi-abélien polarisé $\mathcal{A}$ sur le complété $\mathfrak{m}$-adique $\mathcal{R}^{hs}_{\mathfrak{m}}$ de $\mathcal{R}^{hs}$.

4.3. Théorie algébrique locale. — Si on en croit [F] §4 c, le $\mathcal{O}(L)$-schéma polarisé $\mathcal{A}$ se descend sur $\mathcal{R}^{hs}$ (il faut utiliser la théorie d'Artin [A1]).

LEMME 4.3.1. — *Pour chaque face τ, sur la τ-strate de* $\mathrm{Spec}\,\mathcal{R}^{hs}$, *la partie torique de $\mathcal{A}$ a pour groupe de caractères X_τ et la forme bilinéaire correspondante ne diffère de b_τ que par une unité. En particulier, $\mathcal{A}$ est un schéma abélien sur la 0-strate.*

Pour vérifier cela, on factorise la polarisation λ par l'isomorphisme allant d'une variété abélienne principalement polarisée $\mathcal{A}_1$ dans sa duale: on applique les résultats de [F] §4.d à $\mathcal{A}_1$: ceci permet d'identifier les $\mathbb{Q}$-espaces vectoriels engendrés et on déduit le résultat du fait que X_τ est un sous module de $\mathcal{L}$ sans cotorsion.

Le $\mathcal{O}(L)$-schéma polarisé $\mathcal{A}$ est en fait défini sur un voisinage étale affine de z:

4.3.2 $$\mathrm{Spec}(\mathcal{R}) \to Z_\sigma\ ;$$

$\mathrm{Spec}(\mathcal{R})$ est stratifié et l'analogue de 4.3.1 est encore vérifié. Par construction, si on fait le changement de base à $\mathcal{R}^{hs}_{\mathfrak{m}}$, la théorie de 3.1 et 3.2 fournit un 1-motif polarisé d'où une flèche de la σ-state de Spec($\mathcal{R}_{\mathfrak{m}}^{hs}$) vers celle de Z_σ qui devrait coïncider avec 4.3.2.

4.4. Construction de $\overline{Sh}$. — Pour chaque réseau admissible $\mathcal{L}$, pour chaque σ et chaque point de Z_σ, on vient d'obtenir un voisinage étale. On définit U comme une réunion disjointe finie de tels R-schémas affines. On choisit la réunion de façon que, pour chaque σ, la partie correspondante de **U** recouvre Z_σ (au moins sur les σ-strates). Sur la strate ouverte dense $\mathbf{U}_0$, $\mathcal{A}$ est un $\mathcal{O}(L)$-schéma abélien polarisé. Sur le produit $\mathbf{U}_0 \times \mathbf{U}_0$ au dessus de Spec(R), on considère le schéma $\mathbf{R}_0$ représentant le foncteur des isomorphismes entre $pr_1^* \mathcal{A}$ et $pr_2^* \mathcal{A}$ et **R** son normalisé dans $\mathbf{U} \times \mathbf{U}$. Le champ $\overline{Sh}$ est par définition le quotient de **U** par la relation d'équivalence définie par **R**.

THÉORÈME ATTENDU 4.4.1. — *$\overline{Sh}$ est un espace algébrique propre et lisse sur* Spec(R)*; il est réunion de strates. Le schéma abélien universel sur Sh s'étend en un schéma semi-abélien $\mathcal{A}/\overline{Sh}$. Sur la τ-strate, la partie torique de $\mathcal{A}$ a pour groupe de caractères X_τ et la forme bilinéaire correspondante ne diffère de b_τ que par une unité.*

Il devrait suffire de recopier la démonstration de [F] §4 f (en la complétant convenablement). Le fait que $\overline{Sh}$ soit un espace algébrique résulte de l'absence d'automorphismes des objets classifiés (*cf.* [DM] 4.9). Les lissités proviennent du choix convenable des σ, *cf.* 4.1, compte tenu de l'absence de torsion dans Γ_M. La propreté se démontre à l'aide du critère valuatif de [DM] en utilisant la théorie du modèle de Néron et en passant des variétés abéliennes au schémas semi-abéliens grâce à la forme bilinéaire b. Partant d'une application Spec $K \to Sh$, avec K corps des fractions d'un anneau de valuation discrète (notée v) R, on est amené à considérer un réseau admissible $\mathcal{L}$ comme en 4.1; avec [Ch] II 6.6, on définit un quotient X de $\mathcal{L}$ et $V/W^\perp$ de V. Soit σ un cône (*cf.* 4.1.1) contenant $v \circ b$ et minimal. On peut alors en principe * en déduire un morphisme Spec $R \to \mathbf{U}$ donc de vérifier le critère valuatif (c'est d'ailleurs ici que sert la panoplie de 4.1). En choisissant soigneusement Σ, on devrait pouvoir assurer que Sh est un schéma projectif (conditions de Tai, *cf.* [AMRT]).

COROLLAIRE 4.4.2. — *La fibre spéciale de Sh^0 est connexe.*

---

* malheureusement, ici, comme dans [F], l'utilisation de la théorie d'Artin n'est pas suffisamment développée.

Démonstration. — On applique le théorème principal de Zariski [DM] 4.17 ii) à la clôture de Sh^0. On revient à Sh^0 grâce à [DM] 4.15 et 4.16.

Bibliographie

[A1] ARTIN M. — *Algebraic approximation of structures over complete local rings*, Pub. Math. Ihes, **36** (1969), 23-58.

[A2] ARTINM. — *Algebraization of formal moduli I*, Global Analysis, papers in honor of K. Kodaira,21-41, Princeton University press , 1969.

[AMRT] ASH A., MUMFORD D., RAPOPORT M. , TAI Y. — *Smooth compactifications of locally symmetric varieties*, Math. Sc. Press Brookline, 1975.

[B] BAYER E. — *Unimodular hermitian and skew-hermitian forms*, J. of Algebra, **74** (1982), 341-373.

[BL] BOSCH S. ET LÜTKEBOHMERT W.. — *Stable reduction and uniformization of abelian varieties II*, Invent. Math., **78** (1984), 257-297.

[Br] BRYLINSKI J.-L. — *1-motifs et formes automorphes*, journeées automorphes, Pub. Math. de l' univ. Paris 7, n. 15, 1981.

[Ca] CARAYOL H. — *Sur la mauvaise réduction des courbes de Shimura*, compositio Math., **59** (1986), 151-230.

[Ch] CHAI C.-L. — *Compactification of Siegel moduli schemes*, London Math. Soc. Lecture Notes Series 107, Cambridge University press, 1985.

[D1] DELIGNE P. — *Travaux de Shimura*, Sém. Bourbaki, 389, 1971.

[D2] DELIGNE P. — *Théorie de Hodge III*, Pub. Math. I.H.E.S. , **41** (1975), 6-77.

[DM] DELIGNE P ET MUMFORD D. — *The irreducibility of the space of curves of given genus*, Pub. Math. I.H.E.S. , **36** (1969), 75-109.

[DR] DELIGNE P. RAPOPORT M. — *Les Schémas de modules des courbes elliptiques*, Modular functions of one variable II, Lecture Notes in Math.349, Springer-Verlag, New-York, 1973.

[F] FALTINGS G. — *Arithmetische Kompaktifizierung des Modulsraums des abelschen varietaten*, Arbeistagung Bonn 1984,p. 318-383, Lecture Notes in Math.1111 Springer-Verlag, New-York, 1985.

[Gi] GILLARD R. — Relations monomiales entre périodes p-adiques, à paraître,1987.

[Gr] GROSS B. — *On the periods of abelian integrals and a formula of Chowla and Selberg* , Inv. Math., **45** (1978), 193-211.

[J] JACOBOWITZ R. — *Hermitian forms over local fields*, Amer. J. Math., **84** (1962), 441-465.

[Me] MESSING W. — *The crystals associated to Barsotti-Tate groups: with applications to abelian schemes*, Lecture Notes in Math.264 Springer-Verlag, New-York, 1972.

[Mu1] MUMFORD D. — *Geometric invariant theory*, Ergeb. Math. 34 Springer-Verlag, New-York, 1965.

[Mu2] MUMFORD D. — *On equations defining abelian varieties I*, Invent. Math., **1** (1966), 287-354.

[Rn] RAYNAUD M. — Variétés abéliennes et géométrie rigide. Actes du congrès intern. Math. t. 1 p. 473-477,1970.

[Rp] RAPOPORT M. — *Compactifications de l'espace de modules de Hilbert-Blumenthal*, Compositio Math, **36** (1978), 255-335.

[SGA 7] GROTHENDIECK A. — *Groupes de monodromie en géométrie algébrique* , Lecture Notes in Math.288 Springer-Verlag, New-York, 1971.

[Sh] SHIMURA G. — *Automorphic forms and the periods of abelian varieties*, J. math. Soc. Japan, **31** (1979), 561-592.

WEYL'S INEQUALITY AND HUA'S INEQUALITY

D.R. Heath-Brown
Magdalen College, Oxford OX1 4AU

One central aspect of Waring's problem is the proof of an asymptotic formula for the quantity

$$R_{s,k}(N) = \#\{(n_1,\ldots,n_s)\in\mathbb{N}^s:\ \sum_1^s n_j{}^k = N\}.$$

In 1922, Hardy and Littlewood [2] showed that

$$R_{s,k}(N) = \frac{\Gamma(1+\frac{1}{k})^s}{\Gamma(\frac{s}{k})}\mathfrak{S}(N)N^{\frac{s}{k}-1} + O(N^{\frac{s}{k}-1-\delta}) \tag{1}$$

for

$$s \geqslant 5 + (k-2)2^{k-1}. \tag{2}$$

Here $\mathfrak{S}(N)$ is the usual "singular series", and δ is a small positive constant. A modern version of the proof would use the sum

$$S(\alpha) = \sum_{n=1}^{P} e(\alpha n^k),$$

where $P = [N^{1/k}]$. One then has

$$R_{s,k}(N) = \int_0^1 S(\alpha)^s e(-\alpha N)d\alpha. \tag{3}$$

Since $|S(\alpha)| \leqslant P$, there is a trivial bound

$$\int_0^1 |S(\alpha)^s e(-\alpha N)|d\alpha \leqslant P^s. \tag{4}$$

Thus one needs to save a factor $P^{k+\delta}$ in order to obtain the error term in (1). The most straightforward way of doing this uses Weyl's inequality.

Weyl's Inequality. *Let* $|\alpha - \frac{a}{q}| \leqslant q^{-2}$, *with* $(a,q) = 1$. *Then*

$$S(\alpha) \ll_\varepsilon P^{1+\varepsilon}(q^{-1} + P^{-1} + qP^{-k})^{2^{1-k}}$$

for any $\varepsilon > 0$.

It follows that

$$S(\alpha) \ll_{\varepsilon} P^{1-2^{1-k}+\varepsilon} \tag{5}$$

if $P \leqslant q \leqslant P^{k-1}$. Such α therefore contribute

$$\ll P^{s(1-2^{1-k}+\varepsilon)} \ll N^{s/k-1-\delta}$$

to (3), if $s \geqslant 1+k2^k$. Other values of α are dealt with by rather different methods, producing the main term of (1). Hardy and Littlewood obtained their bound (2) by a minor sharpening of the above argument.

Hua [5], in 1938, obtained a significantly better result by using an average bound.

Hua's Inequality. _For any_ $\varepsilon > 0$, _and any integer_ l _in the range_ $0 < l \leqslant k$, _one has_

$$\int_0^1 |S(\alpha)|^{2^l} d\alpha \ll_{\varepsilon} P^{2^l-l+\varepsilon}. \tag{6}$$

The most important case is $l=k$, when one saves a factor $P^{k-\varepsilon}$ relative to the trivial bound (4). Combining this with Weyl's inequality (5) one sees that those values of α corresponding to $P \leqslant q \leqslant P^{k-1}$ make a satisfactory contribution $O(P^{s-k-\delta})$ to (3), if

$$s \geqslant 1+2^k. \tag{7}$$

The remaining values of α again require different methods, and one obtains the same asymptotic formula (1), but for the larger range (7).

Recently, Vaughan [7],[8] has gone one step further, by proving the asymptotic formula for $s = 2^k$. However, in his method only a power of $\log N$ is saved, rather than N.

For large values of k one can do better, by methods due to Vinogradov (see Vaughan [6; Chapter 5]). For a suitable range of q, one may replace the exponent 2^{1-k} in (5) by a rather complicated expression σ_k, satisfying

$$\sigma_k \sim (4k^2 \log k)^{-1} \qquad (k \to \infty)$$

and $\sigma_k > 2^{1-k}$ for $k \geqslant 12$. Similarly one can prove (1) in a range $s \geqslant s(k)$, where

$$s(k) \sim 4k^2 \log k \qquad (k \to \infty)$$

and $s(k) < 1+2^k$ for $k \geqslant 11$.

Very recently (Heath-Brown [3]) Weyl's inequality has been sharpened as follows.

THEOREM 1. *Let* $\left|\alpha - \frac{a}{q}\right| \leqslant q^{-2}$ *with* $(a,q) = 1$. *Then*

$$S(\alpha) \ll_{\varepsilon} P^{1+\varepsilon}(Pq^{-1} + P^{-2} + qP^{1-k})^{\frac{4}{3}2^{-k}}$$

for any $\varepsilon > 0$.

Thus

$$S(\alpha) \ll P^{1 - \frac{8}{3}2^{-k} + \varepsilon} \qquad (8)$$

if

$$P^3 \leqslant q \leqslant P^{k-3}. \qquad (9)$$

This is better than Weyl's inequality, but the condition on q requires $k \geqslant 6$. In general, for $k \geqslant 6$, Theorem 1 gives a sharper bound than Weyl's inequality, but for a shorter range of q.

In the same circle of ideas one has (Heath-Brown [4]) the following.

THEOREM 2. *Let* $\alpha \in \mathbb{R}$ *and* $\varepsilon > 0$ *be given. For any integer* $k \geqslant 6$ *there are infinitely many* $n \in \mathbb{N}$ *with*

$$\|\alpha n^k\| \leqslant n^{\varepsilon - \frac{8}{3}2^{-k}}.$$

This should be compared with the result of Danicic [1].

Let $\varepsilon > 0$ *and* $k \in \mathbb{N}$ *be given. Then there exists* $N(\varepsilon, k)$ *such that, for any* $\alpha \in \mathbb{R}$ *and any* $N \geqslant N(\varepsilon, k)$, *one can find* $n \leqslant N$ *with*

$$\|\alpha n^k\| \leqslant N^{\varepsilon - 2^{1-k}}.$$

It follows from Danicic's result that

$$\|\alpha n^k\| \leqslant n^{\varepsilon - 2^{1-k}}$$

for infinitely many n. We therefore have a better exponent in Theorem 2, at least for $k \geqslant 6$. However we have lost the "localization" property $n \leqslant N$. This is because the bound (8) holds for only a rather short range (9).

Developing the methods used to prove Theorem 1, one may obtain a sharpening on Hua's inequality (see Heath-Brown [3]).

THEOREM 3. *Let* $k \geqslant 6$ *and* $\varepsilon > 0$. *Then*

$$\int_0^1 |S(\alpha)|^{\frac{7}{8}2^k} d\alpha \ll_{\varepsilon} P^{\frac{7}{8}2^k - k + \varepsilon}.$$

Thus one can save $P^{k-\varepsilon}$ as before, but using only $\frac{7}{8}2^k$ k-th powers. As a corollary one finds:

COROLLARY *The asymptotic formula* (1) *holds for* $k \geqslant 6$ *and*

$$s \geqslant 1 + \frac{7}{8}2^k.$$

The expression $\frac{7}{8}2^k$ arises as follows. Using Dirichlet's approximation theorem we write

$$\left|\alpha-\frac{a}{q}\right| \leqslant \frac{1}{qP^{k-3}} \quad , \quad q \leqslant P^{k-3} \quad , \quad (a,q)=1 \quad ,$$

and we take $\mathfrak{m}$ (the minor arcs) to consist of those values of α for which the corresponding q satisfies $q \geqslant P^3$. On using the estimate (8) and the case $l=k-1$ of Hua's inequality (6) we then find

$$\begin{aligned}\int_{\mathfrak{m}} |S(\alpha)|^{\frac{7}{8}2^k} d\alpha &\ll (P^{1-\frac{8}{3}2^{-k}+\varepsilon})^{\frac{3}{8}2^k} \int_{\mathfrak{m}} |S(\alpha)|^{2^{k-1}} d\alpha \\ &\ll P^{\frac{3}{8}2^k-1+\varepsilon'} \int_0^1 |S(\alpha)|^{2^{k-1}} d\alpha \\ &\ll P^{\frac{3}{8}2^k-1+\varepsilon'} . P^{2^{k-1}-(k-1)+\varepsilon} \\ &\ll P^{\frac{7}{8}2^k-k+\varepsilon''}\end{aligned}$$

(for arbitrary small $\varepsilon', \varepsilon''$). The remaining α, in the "major arcs" require a more complicated argument, but it is the above estimation which limits the method to $s > \frac{7}{8}2^k$.

Let us now look at the proof of Theorem 1. We first examine Weyl's method. Let ∇_h denote the forward difference operator, so that

$$(\nabla_h f)(x) = f(x+h) - f(x).$$

One then proves, by induction, a bound

$$|S(\alpha)|^{2^j} \ll P^{2^j-j-1} \sum_{|h_1|,\ldots,|h_j|<P} \left| \sum_{n\in I(h_1,\ldots,h_j)} F(n) \right| \tag{10}$$

where

$$F(n) = e(\alpha\nabla_{h_1}\cdots\nabla_{h_j}(n^k))$$

and $I(h_1,\ldots,h_j)$ is a subinterval of $[1,P]$. The function

$$\nabla_{h_1}\cdots\nabla_{h_j}(n^k)$$

is a polynomial in n, with degree k-j. Thus if one takes j=k-1, the innermost summation in (10) may readily be estimated as a sum of a finite geometric progression. This leads to Weyl's inequality, as given earlier.

If instead one chooses j=k-3, there will be a cubic sum

$$\Sigma = \sum_n e(An^3+Bn^2+Cn+D) \tag{11}$$

say, to handle. We can estimate Σ itself by an appropriate generalization of Weyl's inequality. For suitable values of α this will give

$$\Sigma \ll P^{\frac{3}{4}+\varepsilon}. \tag{12}$$

If one were to insert this into (10) one would recover the bound (5). Thus nothing has yet been lost. However, we can improve on the above process, by observing that we have many different sums Σ, corresponding to different sets of h's. One expects the average size of Σ to be $O(P^{\frac{1}{2}+\varepsilon})$, which is distinctly better than (12). It should be clear now why one does not choose $j=k-2$. This would lead to a quadratic sum Σ, for which the Weyl bound (taking the form $O(P^{\frac{1}{2}+\varepsilon})$) is already best possible. Thus nothing would be gained by averaging.

In order to put the above idea into practice one must simplify Σ, since it is not feasible to average over all four variables A,B,C and D. However, by employing the symmetric difference

$$(\delta_h f)(x) = f(x+h) - f(x-h)$$

in place of the forward difference ∇_h, one obtains an inequality of the form (10), in which the innermost sum takes the same shape (11) as before, but with $B=D=0$. We then use the following mean-value estimate.

LEMMA <u>Let</u> $\varepsilon>0$. <u>Then</u>

$$\int_0^1\int_0^1 \Big|\sum_{n=1}^{P} e(An^3+Cn)\Big|^6 dA\,dC \ll_\varepsilon P^{3+\varepsilon}.$$

This easy result is proved by observing that the double integral counts solutions of the simultaneous equations

$$\left.\begin{aligned} n_1^3+n_2^3+n_3^3 &= n_4^3+n_5^3+n_6^3 \\ n_1+n_2+n_3 &= n_4+n_5+n_6 \end{aligned}\right\} \qquad 1\le n_i\le P. \tag{13}$$

Since

$$(n_1+n_2+n_3)^3 - (n_1^3+n_2^3+n_3^3) = 3(n_1+n_2)(n_2+n_3)(n_3+n_1),$$

the conditions (13) yield

$$(n_1+n_2)(n_2+n_3)(n_3+n_1) = (n_4+n_5)(n_5+n_6)(n_6+n_4).$$

Thus each triple n_1,n_2,n_3 determines $O(P^{\varepsilon})$ divisors $n_4+n_5, n_5+n_6, n_6+n_4$, which then fix n_4,n_5,n_6. It follows that (13) has $O(P^{3+\varepsilon})$ solutions.

The lemma shows that our modified Σ has mean value $O(P^{\frac{1}{2}+\varepsilon})$. In fact however there is a factor P lost in passing from the sum over the h's, which occurs in (10), to the integral over A and C. Thus Σ has, in effect, average size $O(P^{2/3+\varepsilon})$. This is still better than the Weyl estimate (12), and suffices for Theorem 1.

References

1. I. Danicic, Contributions to number theory, Ph.D. Thesis, London, 1957.

2. G.H. Hardy and J.E. Littlewood, Some problems of "Partitio Numerorum"; IV, Math. Zeit. 12 (1922), 161-188.

3. D.R. Heath-Brown, Weyl's inequality, Hua's inequality, and Waring's Problem, J. London Math. Soc. to appear.

4. D.R. Heath-Brown, The fractional part of αn^k, Mathematatika, to appear.

5. L.-K. Hua, On Waring's problem, Quart. J. Math. Oxf. Ser. 9 (1938), 199-202.

6. R.C. Vaughan, The Hardy-Littlewood method (Cambridge University Press, London, 1981)

7. R.C. Vaughan, On Waring's problem for cubes, J. Reine Angew. Math. 365 (1986), 122-170.

8. R.C. Vaughan, On Waring's problem for smaller exponents. II, Mathematika, 33 (1986), 6-22.

POSITIVE DEFINITE BINARY QUADRATIC FORMS OVER k[X]

Yves HELLEGOUARCH

The main theme of this paper is the extension to k[X] of Gauss' theory of binary quadratic forms in the positive definite case.

The ground field k is always supposed to be of caracteristic $\neq 2$ and when it is finite (or more generally when it is "suitable") the new theory is closely similar to the old one provided -1 is not a square in k.

Alongside this main theme, two secondary themes are introduced.

The first one is a g-adic method of computation of products of powers of certain ideals in a quadratic field ("pure" ideals) or in the more general situation of paragraph 6.

The second one is a method of computation of the sum of points on the Jacobian of an elliptic or hyperelliptic curve defined over k. This method is deduced from the preceding one via the theory of reduction of quadratic forms.

1. DEFINITION OF A "NEAR-ORDERING" ON k[K]

Let us consider the object :

$$A = (k[X], k(X))$$

constituted by a polynomial ring k[X] over a field k such that $1+1 \neq 0$, and by the field of fractions k(X) of k[X].

Lüroth's theorem implies that the group of k-isomorphisms of this object is the group GA(k) of affine transforms :

$$x \longmapsto \lambda X + \mu, \quad \lambda \in k^*, \quad \mu \in k$$

When k is real closed, we know that k(X) is formally real and, so, it can be endowed with different orderings ([9] p. 227). Among those the only ones which are invariant under the translations $X \longmapsto X+\mu$ are induced by the orderings of $k((\frac{1}{X}))$ in which X is infinitesimally great positive (resp. infinitesimally great negative).

In fact those two orderings are permuted under GA(k), as shown by the transformation $X \longmapsto \lambda X$, with $\lambda \in k^* \setminus k^{*2}$, and it is enough to consider the first one.

Going back to the general case, where k is arbitrary, we can "orientate" the object A by considering the subgroup :

$$GA^+(k) = \{x \longmapsto \lambda^2 X+\mu \ ; \ \lambda \in k^*, \ \mu \in k\}$$

and we will say that $(A, GA^+(k))$ is an orientated arithmetic rational field over k.

DEFINITION 1.- Let be given an orientated arithmetic rational field A over k, and let $\alpha \in k(X)$. Modulo the imbedding $k(X) \longrightarrow k((\frac{1}{X}))$, α can be written as a formal power series in $\frac{1}{X}$:

$$\alpha = a_n X^n + a_{n-1} X^{n-1} + \ldots \quad \text{with} \quad a_m \in k^*.$$

Then we will say that α is positive, and we will write $\alpha > 0$ iff $a_n \in k^{*2}$.

Remark :

1) a_n will be referred to as the signature of α, and we will write :

$$a_n = \sigma(\alpha).$$

2) σ is a homomorphism :

$$k((\frac{1}{X}))^* \longrightarrow k^*/_{k^{*2}}$$

so :

$$a > 0 \text{ and } \beta > 0 \Rightarrow \alpha\beta > 0$$

3) If we define $\alpha > \beta$ by $\alpha - \beta > 0$, we do not have the transitivity property. Indeed, taking $k = \mathbb{F}_3$ we have :

$$-1 \notin k^{*2}$$

and if the transitivity property did hold, we would have an ordering on $k((\frac{1}{X}))$, which is impossible since the characteristic of k is not zero.

4) But if k is real closed, we do have an ordering on $k((\frac{1}{X}))$.

2. "POSITIVE DEFINITE" BINARY QUADRATIC FORMS OVER $k[X]$

2.1. Notations

From now on, we will consider an orientated arithmetic rational field $A = (k[X],k(X))$ and denote $k[X]$ by Z and $k(X)$ by Q.

By abuse of language we will define a "positive definite" quadratic form on Q^n as follows.

DEFINITION 2.- Let $q : Q^n \longrightarrow Q$ be a quadratic form. Then we will say that q is "positive definite" iff :

$$x \in Q^n \setminus \{0\} \Rightarrow q(x) > 0.$$

EXAMPLE.- If k is real-closed, then this is the classical definition of a positive definite quadratic form (without inverted commas).

2.2. Our main example

Let D be a square free polynomial in $k[X]$ of odd degree, and suppose that :

$$-D > 0.$$

The typical choice of D will be :

$$D = -X^{2g+1} + a_1X^{2g} + \ldots + a_{2g+1}$$

where g is an integer ≥ 0.

Then we will consider the quadratic extension $(Z[\sqrt{D}],Q(\sqrt{D}))$ of A, and denote $Z[\sqrt{D}]$ by R, and $Q(\sqrt{D})$ by K.

It is easy to show that R is the integral closure of Z in K.

Now K can be considered as a Q-vector space of dimension 2, and the norm form K to Q is a canonical binary quadratic form on K.

LEMMA 1.- The norm : $x \overset{N}{\longmapsto} \mathrm{Norm}_{K/Q}(x)$ is a positive definite binary quadratic form on K.

Proof :

1) It is sufficient to prove the assertion for an integral x.

2) Suppose $x \in R$, then we can write :

$$x = U + V\sqrt{D}, \ U \text{ and } V \in Z$$

Thus we have :

$$\mathrm{Norm}_{K/Q}(x) = U^2 - V^2D$$

and the hypothesis on D easily implies that :

$$\sigma(U^2 - V^2D) \in k^{*2} \qquad \square$$

Denoting by N this quadratic form, the couple (K,N) is then analogous to a two-dimensional euclidean space and we can also define an orientation on K.

DEFINITION 3.- Let (α_1,α_2) be a basis of K over Q. We will say that this basis is direct (or positive) iff :

$$\begin{vmatrix} \alpha_1 & \alpha'_1 \\ \alpha_2 & \alpha'_2 \end{vmatrix} > 0$$

where α'_i denotes the conjugate of α_i over Q.

Remark : If $-1 \in k^{*2}$ then :

$$(\alpha_1,\alpha_2) \text{ direct} \Rightarrow (\alpha_2,\alpha_1) \text{ direct.}$$

LEMMA 2.- Suppose (α_1,α_2) is direct.

i) If $\begin{pmatrix} a & b \\ c & d \end{pmatrix}$ is an inversible matrix with entries in Z such that :

$$\begin{vmatrix} a & b \\ c & d \end{vmatrix} > 0$$

then the basis $(a\alpha_1 + b\alpha_2, c\alpha_1 + d\alpha_2)$ is direct.

ii) If $\gamma \in K^*$, then the basis $(\gamma\alpha_1,\gamma\alpha_2)$ is also direct.

Proof : In both cases we use the second remark of paragraph 1 and in the second case we use lemme 1. $\square$

DEFINITION 4.- Two binary quadratic forms q_1 and q_2 over Z^2 will be said to be equivalent iff there is a matrix $\begin{pmatrix} a & b \\ c & d \end{pmatrix} \in \mathcal{M}_2(Z)$ such that :

i) $$q_2(Z_1,Z_2) = q_1(aZ_1+bZ_2,cZ_1+dZ_2)$$

ii) $$\begin{vmatrix} a & b \\ c & d \end{vmatrix} \in k^* \quad (\text{resp. } \begin{vmatrix} a & b \\ c & d \end{vmatrix} \in k^{*2})$$

Following the classical theory of binary quadratic forms ([4], p. 192) we now want to associate with each ideal class of R a strict class of forms, so we need to define the norm of an ideal.

DEFINITION 5.- Let $I = Z\alpha_1 + Z\alpha_2$ be an ideal of K, $I \neq 0$, we define :

$$\text{Norm}(I) := \frac{\det_{\mathfrak{B}}(\alpha_1,\alpha_2)}{\sigma[\det_{\mathfrak{B}}(\alpha_1,\alpha_2)]}$$

where $\mathfrak{B} = (1,\sqrt{D})$ is the canonical basis of K.

Example : If :

$$I = Zc + Z(b - \sqrt{D}) = (c, b - \sqrt{D})$$

where c is a monic polynomial, then $N(I) = c$.

DEFINITION 6.- Let (α_1,α_2) be a direct base of the ideal $I \neq 0$. We define a positive definite binary quadratic form q_{α_1,α_2} on Q^2 by :

$$q_{\alpha_1,\alpha_2}(Z_1,Z_2) := \frac{N(\alpha_1 Z_1 + \alpha_2 Z_2)}{\text{Norm}(M)}$$

Remark : Each ideal has a direct base, because :

$$\left\{ \sigma\left(\begin{vmatrix} \alpha_1 & \alpha'_1 \\ \alpha_2 & \alpha'_2 \end{vmatrix}\right)\right\} = \lambda \Rightarrow \left\{ \sigma\left(\begin{vmatrix} \lambda\alpha_1 & \lambda\alpha'_1 \\ \alpha_2 & \alpha'_2 \end{vmatrix}\right) = \lambda^2 \right\}$$

THEOREM 1.-

i) If $I = Z\alpha_1 + Z\alpha_2$ is an ideal of K, $I \neq 0$, then q_{α_1,α_2} is a primitive positive definite binary quadratic form of discriminant $-D$.

ii) If $-1 \notin k^{*2}$, the application :

$$I \overset{\varphi}{\longmapsto} \text{strict class of } q_{\alpha_1,\alpha_2}$$

induces a bijection between classes of ideals of K and strict classes of above forms.

iii) If $-1 \in k^{*2}$, this application induces a bijection between the orbits of ideal classes under Galois (K/Q) and the strict classes of above forms.

Proof :

1) It is easy to see that :

$$q_{\gamma\alpha_1,\gamma\alpha_2} = \lambda^2 q_{\alpha_1,\alpha_2}, \quad \text{with} \quad \lambda \in k^*$$

so $q_{\gamma\alpha_1,\gamma\alpha_2}$ is strictly equivalent to q_{α_1,α_2}.

2) It is well known ([1], p. 164) that each ideal I has a canonical Z-base of the type :

$$r(c, b - \sqrt{D})$$

with $r \in Q^*$ and c and $b \in Z$ with :

$$b^2 - D = ac, \quad a \in Z.$$

Then there is a $\lambda \in k^*$ such that $(\lambda c, b - \sqrt{D})$ is direct, and :

$$q_{r\lambda c, r(b-\sqrt{D})} \sim q_{\lambda c, b-\sqrt{D}}$$

Finally :

$$q_{\lambda c, b-\sqrt{D}}(Z_1, Z_2) = \lambda^2 cZ_1^2 + 2\lambda bZ_1Z_2 + aZ_2^2$$

which shows that we have a primitive form of discriminant $-\lambda^2 D \sim -D$.

3) Suppose $\begin{pmatrix} a_{11} & a_{12} \\ a_{21} & a_{22} \end{pmatrix} \in \mathcal{M}_2(Z)$ with $\begin{vmatrix} a_{11} & a_{12} \\ a_{21} & a_{22} \end{vmatrix} = \lambda^2 \in k^{*2}$, then we have :

$$q_{\alpha_1,\alpha_2}(a_{11}Z_1 + a_{21}Z_2, a_{12}Z_1 + a_{22}Z_2) = q_{\beta_1,\beta_2}(Z_1, Z_2)$$

So the polynomials :

$$N(\beta_1 T + \beta_2) \quad \text{and} \quad N((a_{11}T + a_{21})\alpha_1 + (a_{12}T + a_{22})\alpha_2)$$

must have the same roots which are, respectively :

$$\left\{ -\frac{\beta_2}{\beta_1}, -\frac{\beta_2'}{\beta_1'} \right\}, \quad \left\{ -\frac{a_{21}\alpha_1 + a_{22}\alpha_2}{a_{11}\alpha_1 + a_{12}\alpha_2}, -\frac{a_{21}\alpha_1' + a_{22}\alpha_2'}{a_{11}\alpha_1' + a_{12}\alpha_2'} \right\}$$

So there is a $\gamma \in K^*$ such that :

$$\text{(I)} \qquad \begin{cases} a_{11}\alpha_1 + a_{12}\alpha_2 = \gamma\beta_1 \\ a_{21}\alpha_1 + a_{22}\alpha_2 = \gamma\beta_2 \end{cases}$$

or

$$\text{(II)} \qquad \begin{cases} a_{11}\alpha_1 + a_{12}\alpha_2 = \gamma\beta'_1 \\ a_{21}\alpha_1 + a_{22}\alpha_2 = \gamma\beta'_2 \end{cases}$$

We deduce that in the case (I) :

$$\begin{vmatrix} a_{11} & a_{12} \\ a_{21} & a_{22} \end{vmatrix} \begin{vmatrix} \alpha_1 & \alpha'_1 \\ \alpha_2 & \alpha'_2 \end{vmatrix} = N(\gamma) \begin{vmatrix} \beta_1 & \beta'_1 \\ \beta_2 & \beta'_2 \end{vmatrix}$$

and, in the case (II) :

$$\begin{vmatrix} a_{11} & a_{12} \\ a_{21} & a_{22} \end{vmatrix} \begin{vmatrix} \alpha_1 & \alpha'_1 \\ \alpha_2 & \alpha'_2 \end{vmatrix} = - N(\gamma) \begin{vmatrix} \beta_1 & \beta'_1 \\ \beta_2 & \beta'_2 \end{vmatrix}$$

Since $-1 \notin K^{*2}$ the case (II) must be rejected, and this proves our assertion.

5) If $-1 = \lambda^2$, $\lambda \in k^*$, the case (II) cannot be rejected and we get two classes of ideals ; those of :

$$Z\beta_1 + Z\beta_2 \text{ and } Z\lambda\beta'_1 + Z\beta'_2.$$

□

LEMMA 3.- Given a primitive positive definite binary quadratic form q and a modulus $M \in Z$, there exists an element $d \in Z$, prime to M, which can be represented by q.

Proof :

1) Suppose $M = M_1 \ldots M_r$, where M_i is a power of a prime π_i, then if we can find, for each i, an element $d_i = q(u_i,v_i)$ prime to M_i, we can also, by the Chinese remainder theorem, find u and v in Z such that :

$$(u,v) \equiv (u_i,v_i), \text{ mod } M_i$$

for each $i \in \{1,\ldots,r\}$. Then :

$$q(u,v) \equiv d_i, \qquad \text{mod } M_i$$

for each $i \in \{1,\ldots,r\}$, so $d := q(u,v)$ is prime to M.

2) Accordingly it is sufficient to treat the case where M is a prime π. Suppose :

$$q(Z_1,Z_2) = cZ_1^2 + 2bZ_1Z_2 + aZ_2^2.$$

Since q is primitive π does not divides all the coefficients a, b, c. Reducing modulo π, we have a quadratic form $\overline{q}$ on the field $F = Z/_{\pi Z}$ which is non zero :

$$\overline{q}(Z_1,Z_2) = \overline{c}Z_1^2 + 2\overline{b}Z_1Z_2 + \overline{a}Z_2^2$$

then it is clear that we can find $\overline{u}$ and $\overline{v}$ in $Z/_{\pi Z}$ such that $\overline{q}(\overline{u},\overline{v}) \neq \overline{0}$. So any (u,v) above $(\overline{u},\overline{v})$ gives a $d := q(u,v)$ prime to π. □

Lemma 3 gives a means of defining a group structure on the set of strict classes of above forms by Dirichlet's method (see [2] p. 333-8). Let us recall the definition of the product.

If the classes $\mathcal{C}_1$ and $\mathcal{C}_2$ are represented by concordant forms :

$$\begin{cases} q_1 = [c_1, b, .] \\ q_2 = [c_2, b, .] \end{cases}$$

with $(c_1,c_2) = 1$ and $(c_1c_2,D) = 1$, then $\mathcal{C}_1\mathcal{C}_2$ is the class of $q_3 := [c_1c_2, b, .]$.
But if we take :

$$I_1 = (c_1, b + \sqrt{D}), \quad I_2 = (c_2, b + \sqrt{D})$$

we have :

$$I_1I_2 = (c_1c_2, b + \sqrt{D})$$

(see paragraph 3) and :

$$\varphi(I_1) = \lambda_1^2 q_1, \quad \varphi(I_2) = \lambda_2^2 q_2, \quad \varphi(I_1I_2) = \lambda_3^2 q_3$$

so φ is a homomorphism of groups.
We can state :

<u>THEOREME 2</u>.-

i) <u>The following two groups are isomorphic</u>

1) <u>the ideal class group of</u> K

2) J(k), <u>where</u> J <u>denotes the Jacobian of the curve</u> :

$$Y^2 = D(X)$$

ii) <u>When</u> $-1 \notin k^{*2}$, <u>the group of strict classes of above forms is isomorphic to</u> J(k).

iii) When $-1 \in k^{*2}$, the group of strict classes of above forms is isomorphic to $J(k)/_{[2]J(k)}$.

The proof results of the preceding remark and, for 1) $\Longleftrightarrow$ 2), of the ramification of the place at infinity of Q in K (see [7], p. 15).

3. CALCULUS OF IDEALS

This paragraph, which is of an algorithmic nature, will be generalized later.

We said above, in paragraph 2, that each ideal I of R has a canonical Z-basis of the form :

$$r(c, b - \sqrt{D})$$

with $r, c, b \in Z$ and :

$$b^2 - D = ac \quad (a \in Z).$$

DEFINITION 7.- We will say that I is a **pure ideal** iff :
$r = 1$ and c and D are relatively prime.

LEMMA 4.- Each ideal class contains a pure ideal.

Proof :

1) Let $\mathcal{C}$ be the image of the ideal class of I by the application φ of paragraph 2. By lemma 3 we know that $\mathcal{C}$ represents a polynomial $d \in Z$, prime to any given modulus M, and to D in particular.

Thus, if $[c, 2b, a] \in \mathcal{C}$, we have :

$$d = cu^2 + 2buv + av, \text{ with } (u,v) \in Z^2.$$

It is clear that we can suppose that u and v are relatively prime, so there is $(s,t) \in Z^2$ such that :

$$\begin{vmatrix} u & v \\ s & t \end{vmatrix} = 1$$

Take :

$$\begin{pmatrix} Z_1 \\ Z_2 \end{pmatrix} = \begin{pmatrix} u & v \\ s & t \end{pmatrix} \begin{pmatrix} Z_1' \\ Z_2' \end{pmatrix}$$

Then q is strictly equivalent to :

$$q'(Z'_1, Z'_2) = dZ'^2_1 + 2b'Z'_1 Z'_2 + a'Z'^2_2$$

which is associated to the ideal :

$$I'_1 = (d, b' - \sqrt{D})$$

If $-1 \notin k^{*2}$, I and I_1 belong to the same class.
If $-1 \in k^{*2}$, I and I_1 might belong to conjugate classes, but then I and I_1 are in the same class. □

Our aim is now the study of pure ideals.
So let be given the pure ideal $(c, b - \sqrt{D})$ with $c \notin k^*$ and consider the ring Z_c of c-adic integers defined by :

$$Z_c = \varprojlim_{n} Z/c^n Z .$$

Then we will associate to the ideal M a c-adic integer b by the following construction. □

PROPOSITION.- Write :

(1) $$\mathfrak{b} = b \sum_{n \geq 0} (-1)^n \binom{1/2}{n} \left(\frac{ac}{b^2}\right)^n$$

then one obtains in Z_c :

$$\begin{cases} \mathfrak{b} \equiv b & (\text{mod } c) \\ \mathfrak{b}^2 = D \end{cases}$$

Proof :

1) Since $(c,D) = 1$, we have $(c,b) = 1$ and the series is convergent in Z_c.

2) Since the series is the binomial expansion of :

$$b\left(1 - \frac{ac}{b^2}\right)^{1/2}$$

we have :

$$\mathfrak{b}^2 = b^2 - ac = D$$

□

(for example all polynomials of degree less than the degree of c). An element $\tilde{a} \in Z_c$, can be written as :

$$\tilde{a} = \sum_{\nu=0}^{\infty} \alpha_\nu c^\nu , \qquad \text{with} \quad \alpha_\nu \in \mathcal{S}$$

Then we will define the polynomial $a_n \in Z$ **truncated at rank** n, by the formula :

$$\tilde{a}_{|n} = a_n = \sum_{\nu=0}^{n-1} \alpha_\nu c^\nu .$$

THEOREM 3.- Suppose that $I = (c, b - \sqrt{D})$ is a pure ideal $\neq R$, and consider the Z_c-module

$$I_c = (c, \tilde{b} - \sqrt{D})$$

in $Z_c \otimes_Z R$.

Then for each integer $n \geq 0$, we have :

$$I_c^n = (c^n, \tilde{b} - \sqrt{D})$$

Proof : We proceed by induction on n.

1) The result is obvious for $n = 0$ or 1.
2) Suppose it is proven for n, then we have :

$$\begin{aligned} I_c^{n+1} &= (c^n, \tilde{b} - \sqrt{D})(c, \tilde{b} - \sqrt{D}) \\ &= (c^{n+1}, c(\tilde{b} - \sqrt{D}), 2\tilde{b}(\tilde{b} - \sqrt{D})) \end{aligned}$$

Since $\tilde{b} \equiv b \pmod{c}$ and $(2b,c) = 1$, one obtains the result.

□

COROLLARY 1.- A necessary and sufficient condiiton for a pure ideal to be a nth power is that its norm be a nth power.

Proof :

1) Let b_n be the polynomial deduced from $\tilde{b}$ by truncation at rank n.

2) It is clear by theorem 3 that :

$$I^n = (c^n, b_n - \sqrt{D})$$

3) Since the nth power of an impure ideal cannot be pure, the lemma is proven.

COROLLARY 2.- A necessary and sufficient condition for an ideal class to be a nth power is that it contains a pure ideal whose norm is a nth power.

DEFINITION 9.- Let be given a finite set of **pure** ideals $I_1,\dots,I_r$.
We will say that $I_1,\dots,I_r$ are **concordant ideals** if their norms are relatively prime two by two.

LEMMA 5.- Any finite set of ideal classes can be represented by concordant ideals.

The proof of this lemma is similar to the proof of lemma 3 : one proceeds inductively and just multiplies the modulus M by the norms of the preceeding ideals.

THEOREM 4.- Let $\mathcal{C}_1,\dots,\mathcal{C}_n$ be ideal classes represented by concordant ideals of norms $c_1,\dots,c_r$.
Put $c = c_1\dots c_r$ and consider the ring Z_c of c-adic integers.
Then there exists $\tilde{b} \in Z_c$ such that the class $\mathcal{C}_1^{n_1} \dots \mathcal{C}_r^{n_r}$ can be represented by :

$$\left(c_1^{n_1} \dots c_r^{n_r},\ b_n - \sqrt{D}\right)$$

where b_n means $\tilde{b}$ **truncated** at rank $n = \sup\{n_1,\dots,n_r\}$.

Proof :

1) Let $I_1,\dots,I_r$ be concordant ideals in $\mathcal{C}_1,\dots,\mathcal{C}_r$ with $\mathrm{Norm}(I_1) = c_1,\dots,\mathrm{Norm}(I_r) = c_r$. We can associate to each I_i a c_i-adic integer $\tilde{b}^i$ and since the norms $c_1,\dots,c_r$ are relatively prime we have :

$$\begin{cases} Z_{c_1} \times \dots \times Z_{c_r} \xrightarrow{\sim} Z_c \\ (\tilde{b}_1,\dots,\tilde{b}_r) \longmapsto \tilde{b} \end{cases}$$

2) Proceeding by induction we can suppose :

$$n_1 = n_2 = 1, \quad n_3 = \dots = n_r = 0$$

and tensorizing with Z_c, we must verify that :

$$(c_1, \tilde{b} - \sqrt{D})\,(c_2, \tilde{b} - \sqrt{D}) = (c_1c_2, \tilde{b} - \sqrt{D})$$

3) We have :

$$(c_1, \tilde{b} - \sqrt{D})\,(c_2, \tilde{b} - \sqrt{D}) = (c_1c_2,\ c_1(\tilde{b} - \sqrt{D}),\quad c_2(\tilde{b} - \sqrt{D}),\ 2b(\tilde{b} - \sqrt{D}))$$

Since $(c_1,c_2) = 1$, we get the result. □

4. GAUSS THEOREMS

DEFINITION 10.- Let q_1 and q_2 be two quadratic forms : $Z^n \longrightarrow Z$. We say that q_1 and q_2 belong to the same **genus** iff they are Z_v-equivalent for all valuations v of Q which are trivial on k.

We will apply this terminology to the set of positive definite binary quadratic forms of discriminant -D on Z^2, and we will say that the genus of N is the **principal genus** G :

$$G := \{q \ ; \ q \sim N \text{ everywhere}\}$$

Our aim is to give an description of G and of the group of genera in terms of the jacobian J(k) of the curve :

$$Y^2 = D(X).$$

4.1. Legendre's theorem

DEFINITION 11.- Let be given a field k. We will say that k is **"suitable"** iff any non degenerated ternary quadratic form on k^3 is isotropic.

EXAMPLES.-

1) any finite field $\mathbb{F}_q$ is "suitable".
2) any quadratically closed field ([9] p. 41) is suitable too.

DEFINITION 12.- Let D be as in 2.2).

For any $c \in Z\setminus\{0\}$ and π, prime in Z, we write :

$$\left(\frac{c,D}{\pi}\right) = \begin{cases} 1 & \text{if } cZ_1^2 + DZ_2^2 - Z_3^2 = 0 \text{ is isotropic over } Q_\pi \\ -1 & \text{otherwise.} \end{cases}$$

Remark : One can easily see that :

$$\left\{\left(\frac{c,D}{\pi}\right) = 1\right\} \Longleftrightarrow \left\{\exists x \in Q_\pi \otimes_Q K, \text{ such taht } c = N(x)\right\}$$

LEGENDRE'S THEOREM :

For a "suitable" k and any $c \in Z\setminus\{0\}$ the following confitions are equivalent :

i) c is a norm from K.

ii) for every π dividing cD we have : $\left(\frac{c,D}{\pi}\right) = 1.$

Sketch of the proof :

On can follow any textbook ([10], p. 41-7, [11], p. 218-225 or [13], p. 74-75) using the absolute value at infinity on k[X] (i.e. the degree) for the descents.

The ultimate descent gives a quadratic form xith coefficients in k* of the type :

$$\alpha_1 Z_1^2 + \alpha_2 Z_2^2 + \alpha_3 Z_3^2$$

Since k is suitable, this form admits an isotropic vector. □

COROLLARY 1.- If C is algebraically closed, and if t is transcendantal over C, then C(t) is "suitable".

Proof :

1) Take a diagonalized ternary form on k = C(t) and multiply it by a common denominator, we have the integral form :

$$A_1(t)Z_1^2 + A_2(t)Z_2^2 + A_3(t)Z_3^2, \quad \text{with} \quad A_i(t) \in C[t].$$

2) A linear transformation of the Z_i allows one to take square free $A_i(t)$ and divisibility considerations allow one to suppose that :

$$(A_2,A_3) = (A_3,A_1) = (A_1,A_2) = 1.$$

3) Multiplying by $-A_3(t)$ and changing Z_3, one has to study the quadratic form :

$$c(t)Z_1^2 + D(t)Z_2^2 - Z_3^2$$

4) Apply then Legendre's theorem with k = C (which is "suitable") and $\pi = t - \alpha$, $\alpha \in C$. Since C is algebraically closed the form in 3) admits an isotropic vector over Q_π. □

COROLLARY 2.- For a "suitable" k and any c = N(I), where I is an ideal of R $(I \neq 0)$, the following conditions are equivalent :

i) c is a norm from K

ii) for every π dividing D we have : $\left(\frac{c,D}{\pi}\right) = 1.$

Proof :

1) We know that the class of I contains an ideal of the type :

$$(c, b - \sqrt{D})$$

with $c, b \in Z$ and :

$$b^2 - D \equiv 0 \pmod{c}$$

2) Then one sees that the condition $\left(\frac{c,D}{\pi}\right) = 1$ for a prime divisor π of c is automatically verified since D is square-free.

4.2. Gauss' principal genus theorem

THEOREM 5.- Suppose that k is "suitable" and that $\mathcal{C}$ is the strict class of quadratic forms associated to the ideal $(c, b - \sqrt{D})$ of R.

Then the following conditions are equivalent :

i) $\mathcal{C}$ is contained in the principal genus G

ii) the ideal class of $(c, b - \sqrt{D})$ is divisible by two in $J(k)$.

Proof :

1) Let us show first that ii) implies i).

Corollary 2 of theorem 2 shows that $\mathcal{C}$ represents a square d^2 prime to D.

If π does not divides d, then the reduced form q over $Z/\pi Z$ is hyperbolic, so universal, and Hensel's lemma gives the result.

2) The converse is more difficult and we will adapt Cassels' proof ([2], p. 339)

2.1) Since $\mathcal{C}$ belongs to the principal genus, 1 is representable by $q_{c,\, b - \sqrt{D}}$ over each Z_π. Legendre's theorem implies then that 1 is representable by $q_{c,\, b - \sqrt{D}}$ over Q.

2.2) It remains to show that if the form $q_{c,\, b - \sqrt{D}}$ represents 1 over Q, it represents 1 over Z. Let π be a prime divisor of D in Z, and let $(u_\pi, v_\pi) \in Z_\pi^2$ be such that :

$$q_{c,\, b - \sqrt{D}}(u_\pi, v_\pi) = cu_\pi^2 + 2bu_\pi v_\pi + av_\pi^2 = 1,$$

we will show in lemma 6 that there exists $(u,v) \in Q^2$ such that :

$$\begin{cases} q_{c,\, b - \sqrt{D}}(u,v) = 1 \\ |u - u_\pi|_\pi < 1 \quad \text{for all } \pi \text{ dividing } D \end{cases}$$

2.3) Then, multiplying u and v by their least common denominator d, we would have :

$$q_{c,\, b - \sqrt{D}}(u^*, v^*) = d^2$$

with $(d,D) = 1$ and $(u^*, v^*) \in Z^2$, hence $\mathcal{C}$ would be the square of another class $\mathcal{C}_1$ by corollary 2 of theorem 3. □

To prove the assertion in 2.2) we put :

$$q(Z_1, Z_2, Z_3) = q_{c,\, b - \sqrt{D}}(Z_1, Z_2) - Z_3^2$$

and we introduce the ring Z_M of M-adic numbers, where M is the product of a finite number of primes : $M = \pi_1 \ldots \pi_h$. We know that the Chinese remainder theorem extends to :

$$Z_M \simeq Z_{\pi_1} \times \ldots \times Z_{\pi_h}.$$

LEMMA 6.- Let q be an isotropic form over Q in $n \geq 3$ variables, let $\varepsilon > 0$ be arbitrarily small and let $M = \pi_1 \ldots \pi_h$ as above.

Let $\vec{\beta} \in Z_M^n \setminus \{0\}$ be given such that $q(\vec{\beta}) = 0$, then there is a $\vec{c} \in Q^n$ with $q(\vec{c}) = 0$ such that :

$$\| \vec{c} - \vec{\beta} \|_M < \varepsilon$$

where $\| \ \|_M$ denotes an M-adic norm of Z_M^n.

Proof : Let $a \in Q^n$, $a \neq 0$ be such that $q(a) = 0$.

1) Suppose that none of the π_i-adic coordinates of $f(a,\beta)$ is zero, where f is the bilinear form belonging to q.

By the Chinese remainder theorem there is a $b \in Q^n$ which is arbitrarily close to β.

Then we have for λ and μ in Q :

$$q(\lambda a + \mu b) = 2\lambda\mu f(a,b) + \mu^2 q(b)$$

and none of the coordinates of $f(\vec{a}, \vec{b})$ is zero in $\mathbb{Q}_{\pi_1} \times \ldots \times \mathbb{Q}_{\pi_h}$.

So, taking $\mu = 1$ and $\lambda = -\dfrac{q(\vec{b})}{2f(\vec{a}, \vec{b})}$ we see that $\lambda\vec{a} + \vec{b}$ is isotropic or null.

When b tends M-adically to β, λ tends M-adically to zero and $c = \lambda a + b$ is arbitrarily close to β.

2) If a π_i-adic coordinate of $f(a,\beta)$ is null we can replace the π_i-adic component of β by an arbitrarily close similar component such that the π_i-adic coordinate of $f(a,\beta)$ is not zero and $q(\beta) = 0$ (see [2], lemma 2.8. p. 62). □

COROLLARY 1.- Suppose that k is "suitable" and suppose $(c, b - \sqrt{D})$ is a pure ideal $\neq R$ (i.e. $(c,D) = 1$ and $c \notin k^*$) then the following conditions are equivalent :

i) the class of $(c, b - \sqrt{D})$ is divisible by two in $J(k)$

ii) $q_{c,\, b - \sqrt{D}} \in G$

iii) c is a norm in Z_D

iv) c is a square in $Z/(D)$.

Proof :

1) i) $\Longleftrightarrow$ ii) by Gauss'principal genus theorem.

2) ii) $\Longleftrightarrow$ iv) by Hensel's lemma.

3) Let us show that ii) $\Rightarrow$ iii)

We must prove that if the prime π divides D we have :

$$\begin{pmatrix} (u,v) \in Q_\pi^2 \\ c = u^2 - Dv^2 \end{pmatrix} \Rightarrow (u,v) \in Z_\pi^2$$

and this follows from the fact that :

$$0 \leq v_\pi(c) = \inf\{v_\pi(u^2),\ v_\pi(Dv^2)\}$$

because $v_\pi(u^2) \neq v_\pi(Dv^2)$.

4) Let us show that iii) $\Rightarrow$ ii).

Since c is a norm in Z_D, Legendre's theorem implies that c is a norm form K which implies ii).

COROLLARY 2.- When k is "suitable" there is a cononical monomorphism :

$$J(k)/_{[2]J(k)} \longrightarrow L^*/_{L^{*2}}$$

where L denotes the ring $Z/_{(D)}$.
This injection is defined, for pure ideals, by :

$$\text{Class}(c, b - \sqrt{D}) \xmapsto{\text{Norm}} c \longmapsto \bar{c} \in L^*/_{L^{*2}}$$

Proof :

1) Corollary 1 proves the injectivity.

2) Suppose that $(c_1, b_1 - \sqrt{D})$ and $(c_2, b_2 - \sqrt{D})$ are concordant (thence pure). Then the product of those ideals is also pure, and its norm is c_1c_2 (this is well known and it results also from paragraph 3). □

4.3. Gauss'second theorem

Let us denote by $\mathfrak{G}$ the group of classes of primitive positive definite binary quadratic forms of discriminant -D, we know, by theorem 2, that if $-1 \notin k^{*2}$ it is isomorphic with J(k) and if $-1 \in k^{*2}$ it is isomorphic with $J(k)/_{[2]J(k)}$.

THEOREM 6.- If $k = \mathbb{F}_q$, then the order of $J(k)/_{[2]J(k)}$ is 2^{t-1} where t denotes the number of prime factors of D.

Proof :

1) If we translate to this case the classical proof ([4], p. 172-6) using the quadratic reciprocity law ([1], § 15 or [6]) we get the order of the group of genera :

$$\left| \mathfrak{G} / [2]\mathfrak{G} \right| = 2^{t-1}$$

2) Then the above remark gives :

$$\mathfrak{G} / [2]\mathfrak{G} \cong J(k) / [2]J(k)$$

in both cases $((-1) \notin k^{*2}$ or $-1 \in k^{*2})$. □

5. A CONTINUED FRACTION ALGORITHM

According to a general definition ([12], p. 78) we will define the notion of a reduced ideal.

DEFINITION 13.- An integral ideal $I = s(c, -b + \sqrt{D})$ is said to be **reduced** iff :

i) $s \in k^*$

ii) $|N(c)|$ is minimal in $\{ |N(\gamma)| \ ; \ \gamma \in I \setminus \{0\} \}$ where $| \ |$ denotes the absolute value at infinity on Z.

Remarks :

1) If F is any polynomial in Z, we have :

$$|F| = \rho^{\deg(F)}$$

where ρ is a fixed real number > 1.

2) Lemma 7 will give E. Artin's definition ([1], p. 178).

LEMMA 7.- Suppose $I = (c, b - \sqrt{D})$. Then the following conditions are equivalent :

i) I is reduced

ii) $|c| < |D|^{1/2}$.

Proof :

ii) ⇒ i)

Let $\gamma = uc + v(b - \sqrt{D}) = (uc - vb) + v\sqrt{D}$ be in M.

We have :

$$N\gamma = (uc - vb)^2 - v^2 D$$

so : $$|N\gamma| = \sup\{\,|uc - vb|^2,\ |v^2D|\,\}$$

If $|N\gamma| \le |c^2| < |D|$ we must have $v = 0$ and $u \in k^*$.

i) ⇒ ii)

Suppose $|N(c)|$ is minimal, then :

$$|c^2| \le |(uc-b)^2 - D|$$

for all $u \in Z$.

We can choose u such that $|uc-b| < |c|$, then we must have :

$$|c^2| \le |D|\,.$$

□

THEOREM 7.-

i) each ideal class contains one and only one reduced ideal.

ii) the reduced ideal which is equivalent to a given ideal can be computed by a continued fraction algorithm in a finite number of steps.

Proof :

1) We will show the unicity of the reduced ideal. Suppose I_1 and I_2 both reduced and such that :

$$I_2 = \alpha I_1\,,\quad \alpha \in K^*$$

Write :

$$I_i = (c_i,\, b_i - \sqrt{D}),\qquad i=1,2.$$

Then it is clear that αc_1 is the element of minimal absolute norm in αI_1, so we must have by lemma 7 :

$$\alpha c_1 = \lambda c_2,\ \text{ with } \lambda \in k^*$$

which shows that $\alpha \in Q^*$.

Writing $\alpha = \frac{n}{d}$ in lowest terms in Z, we have :

$$dI_2 = nI_1$$

looking at the coefficients of $\sqrt{D}$, one gets :

$$d = \mu n,\ \text{ with } \mu \in k^*.$$

2. Continued fraction algorithm

Remark that :

$$I \sim \left(1, \frac{b - \sqrt{D}}{c}\right)$$

Put $\alpha_0 = \dfrac{b - \sqrt{D}}{c}$.

2.1. If $|N\alpha_0| \geq 1$ then M is reduced.

2.2. If not, consider :

$$- \frac{1}{\alpha_0} = \frac{c(-b - \sqrt{D})}{ca} = \frac{-b - \sqrt{D}}{a}$$

take $c_1 = a$, $b_1 =$ remainder of $-b$ mod c_1, and :

$$a_1 := \frac{b_1^2 - D}{c_1}, \qquad \alpha_1 := \frac{b_1 - \sqrt{D}}{c_1},$$

Then :

$$I \sim (1, \alpha_1) \sim (c_1, b_1 - \sqrt{D}) := I_1$$

If $|N\alpha_1| \geq 1$, then I_1 is reduced.

2.3. If not, go on...

At each step we have :

$$|c_{n+1}| = |a_n| = |N(\alpha_n)|\,|c_n|$$

so :

$$|c_{n+1}| = |N(\alpha_n \ldots \alpha_0)|\,|c_n|$$

and since $|N(\alpha_i)| \leq \dfrac{1}{\rho}$ for each $i \in \{0, \ldots, n-1\}$ we see that :

$$|N(\alpha_n)| = \frac{|c_{n+1}|}{|c|\,|N(\alpha_{n-1} \ldots \alpha_0)|} \geq \frac{1}{\rho^{d-n}}$$

where $d :=$ degree(c).

So the algorithm must stop in $d-1$ steps at most (be it understood that 2.1) is not considered as a real step). □

COROLLARY 1.- Let k be an arbitrary field of characteristic $\neq 2$, and suppose that the class of the reduced ideal $(c, -b + \sqrt{D})$ belongs to $[2]J(k)$, then c is a square in $Z/_{DZ}$.

Proof :

By paragraph 3 we can write :

$$(c, b - \sqrt{D}) \sim (c', b' - \sqrt{D})^2 = (c'^2, b'_1 - \sqrt{D})$$

so $(c, b - \sqrt{D})$ is the reduced ideal belonging to the class of $(c'^2, b'_1 - \sqrt{D})$. But the reduction algorithm implies at each step :

$$\begin{cases} c_{n+1} = a_n \\ c_n a_n \equiv b_n^2 \pmod{D} \end{cases}$$

so :

$$c_n\, c_{n+1} \equiv \square \pmod{D}$$

Since c_0 is a square we get that all the c_n are squares (mod D) and, in particular, that c is a square (mod D). □

COROLLARY 2.- Let k be an arbitrary field of characteristic $\neq 2$, and suppose that :

$$D(X) = -X^3 + a_1 X^2 + a_2 X + a_3$$

the the following conditions are equivalent :

i) the point $(\alpha,\beta) \in [2]\, J(k)$, with $\beta \neq 0$

ii) $X-\alpha$ is a square in $(Z/DZ)^*$.

Remark.- One can compare with ([8], p. 47)

Proof :

i) ⇒ ii) results from corollary 1, but we give it again to introduce the second part of the proof.

Suppose we have :

$$(c, b - \sqrt{D}) \sim (c_1, b_1 - \sqrt{D})^2$$

Then we know by theorem 3 that :

$$(c_1, b_1 - \sqrt{D})^2 = (c_1^2, b_2 - \sqrt{D})$$

with :

$$b_2^2 - D = c_1^2\, a_2$$

Reducing this latter ideal we have :

$$(c_1^2, b_2 - \sqrt{D}) \sim (a_2, -b_2 - \sqrt{D})$$

and theorem 7 implies that : $c = a_2$.

But

$$b_2^2 \equiv c_1^2\, a_2 \pmod{D}$$

so a_2 is a square in Z/DZ.

ii) ⇒ i). We consider the ideal :

$$(X-\alpha, \beta - \sqrt{D})$$

and take $c = X-\alpha$, $b = \beta$, $b^2 - D = ac$, so that :

$$|b| = 1, \quad |c| = \rho, \quad |a| = \rho^2 .$$

Since c is a square modulo D, it cannot be the square of a polynomial of degree ≤ 1. So we write :

$$c = d^2 \bmod D, \quad \text{with} \quad |d| = \rho^2$$

Now we replace b by $b + \lambda c$, with $\lambda \in k$, in order to construct b_2 as in the first part. We get :

$$(b+\lambda c)^2 - D = ca_\lambda$$

so :

$$(bd^{-1} + \lambda d)^2 \equiv a_\lambda \mod D$$

where bd^{-1} means a polynomial of degree ≤ 2.

Since $|d| = \rho^2$, we can choose λ such that :

$$|bd^{-1} + \lambda d| \leq \rho$$

Then surely $bd^{-1} + \lambda d$ is to be identified with b_2 in the first part and a_λ with c_1^2.

In fact, since $|a_\lambda| \leq \rho^2$, the unicity of the remainder in the division by D gives :

$$(bd^{-1} + \lambda d)^2 = a_\lambda$$

Taking now :

$$\begin{cases} c_1 = \pm (bd^{-1} + \lambda d) \\ b_1 \equiv -(b + \lambda c) \qquad (\bmod c_1) \end{cases}$$

and noticing that $(c_1,b_1) = 1$, one has :

$$\begin{aligned}(c_1, b_1 - \sqrt{D})^2 &= (a_\lambda, -(b+\lambda c) - \sqrt{D}) \\ &\sim (c, b - \sqrt{D})\end{aligned}$$

Example :

Start from Pythagoras' relation :

$$\alpha^2 + \beta^2 = \gamma^2, \quad \alpha, \beta, \gamma \in k^* \quad \text{with} \quad \alpha^2 \neq \beta^2$$

and consider the elliptic curve :

$$C_{\alpha,\beta,\gamma} : Y^2 = -X(X+\alpha^2)(X+\beta^2) = D(X)$$

which admits the rational point P :

$$x = -\gamma^2 \qquad y = \alpha\beta\gamma = b$$

This point corresponds to the class of the reduced pure ideal :

$$(X+\gamma^2,\ \alpha\beta\gamma - \sqrt{D})$$

It is clear that $c = X+\gamma^2$ is a square in $(Z/_{DZ})^*$ since :

$$\begin{cases} c \equiv \gamma^2 & \text{mod } X \\ c \equiv \beta^2 & \text{mod } (X+\alpha^2) \\ c \equiv \alpha^2 & \text{mod } (X+\beta^2) \end{cases}$$

so it is divisible by 2 in $J(k)$.

In order to find all points $(u,v) \in J(k)$ such that :

$$[2](u,v) = (x,y)$$

we will use the method in the proof of Corollary 2. Choosing signs for α, β, γ in such a way that $\alpha\beta\gamma = y$, we have (by Lagrange's interpolation formula) :

$$\begin{aligned} d(X) &= \gamma\frac{(X+\alpha^2)(X+\beta^2)}{\alpha^2\beta^2} + \beta\frac{X(X+\beta^2)}{\alpha^2(\alpha^2-\beta^2)} - \alpha\frac{X(X+\alpha^2)}{\beta^2(\alpha^2-\beta^2)} \\ &= -\frac{(\gamma-\alpha)(\gamma-\beta)}{\alpha^2\beta^2(\alpha+\beta)}X^2 + \left(\frac{\gamma^3-\alpha^3-\beta^3}{\alpha^2\beta^2} - \frac{1}{\alpha+\beta}\right)X + \gamma \end{aligned}$$

and :

$$\begin{aligned} bd^{-1}(X) &= \alpha\beta\frac{(X+\alpha^2)(X+\beta^2)}{\alpha^2\beta^2} + \alpha\gamma\frac{X(X+\beta^2)}{\alpha^2(\alpha^2-\beta^2)} - \beta\gamma\frac{X(X+\alpha^2)}{\beta^2(\alpha^2-\beta^2)} \\ &= \frac{\alpha+\beta-\gamma}{\alpha\beta(\alpha+\beta)}X^2 - \frac{\gamma(\gamma-\alpha)(\gamma-\beta)}{\alpha\beta(\alpha+\beta)}X + \alpha\beta \end{aligned}$$

The condition :

$$|bd^{-1}(X) + \lambda d(X)| \leq \rho$$

gives :

$$\lambda = \frac{\alpha\beta(\alpha+\beta-\gamma)}{(\gamma-\alpha)(\gamma-\beta)}$$

and the resulting polynomial c_1 is :

$$c_1(X) = X + (\gamma+\alpha)(\gamma+\beta)$$

So we get a new point $Q = (u,v)$ with $u = -(\gamma+\alpha)(\gamma+\beta)$ and $v = b_1 = (\gamma+\alpha)(\gamma+\beta)(\alpha+\beta)$ determined by :

$$b_1 \equiv -(b+\lambda c) \pmod{c_1}$$

All in all we have 20 points whose abscissas are :

$$\infty,\ 0,\ -\alpha^2,\ -\beta^2,\ -\gamma^2,\ \frac{-\alpha^2\gamma^2}{\beta^2},\ -\frac{\beta^2\gamma^2}{\alpha^2},\ -\frac{\alpha^2\beta^2}{\gamma^2},\ -(\gamma+\alpha)(\gamma+\beta),\ -(\gamma-\alpha)(\gamma-\beta),$$

$$(\gamma+\alpha)(\beta-\gamma),\quad (\gamma+\beta)(\alpha-\gamma)$$

Since no elliptic curve on $\mathbb{Q}$ can have more than 16 rational points of finite order (Mazur), one of the points Q must be of infinite order.

Remark.- One must take care to distinguish between this curve and the curve :

$$E_{\alpha,\beta,\gamma} : Y^2 = X(X-\alpha^p)(X+\beta^p)$$

associated to Fermat's equation :

$$\alpha^p + \beta^p + \gamma^p = 0$$

which was introduced in [5], p. 5.2.1-6 and studied there and in [3].

6. TOWARDS A GENERALIZATION

We will now consider the general situation of Kummer's theorem on the extension of ideals.

Let R be a Dedekind ring, K its field of fractions, $L = K(\alpha)$ a finite separable extension of K of degree n, generated by an integral element α (over R), f(X) the monic irreductible polynomial of α over K and S the integral closure of R in L : we suppose that the R-module S is free.

Let us recall Kummer's classical result.

KUMMER'S THEOREM.- Let $\mathfrak{p}$ be a prime ideal of R which does not divide the determinant :

$$\det_{\mathfrak{B}} (1,\alpha,\ldots,\alpha^{n-1}),\quad \mathfrak{B} = \text{R-base of } S$$

Then suppose that :

$$\bar{f}(X) = \bar{f}_1^{e_1}(X) \ldots \bar{f}_s^{e_s}(X)$$

is the prime decomposition of $\bar{f}(X)$ in $R/\mathfrak{p}\,[X]$.
Then the prime decomposition of $\mathfrak{p}S$ in S is :

$$\mathfrak{p}S = P_1^{e_1} \ldots P_s^{e_s}$$

where:

$$P_i = \mathfrak{p}S + f_i(\alpha)S$$

for any choice of $F_i(X)$ in R[X] such that $f_i(X) \to \bar{f}_i(X)$.

Now we will use Hensel's lemma to give a decomposition of f(X) in $R_{\mathfrak{p}}[X]$ where $R_{\mathfrak{p}}$ is the ring of $\mathfrak{p}$-adic integers :

$$f(X) = \tilde{g}_1(X) \dots \tilde{g}_s(X), \quad \tilde{g}_i(X) \in R_{\mathfrak{p}}[X]$$

in which $\tilde{g}_i(X)$ reduces to $\bar{f}_i^{e_i}(X)$ modulo $\mathfrak{p}$.

THEOREM 8.- For each positive integer n we have :

$$P_i^{e_i n} = \mathfrak{p}^{n_s} + \tilde{g}_i|_n(\alpha)\, S$$

where $\tilde{g}_i|_n(X)$ is the polynomial of R[X] deduced from $\tilde{g}_i(x)$ by truncation at rank n.

Proof :

It is based on the following observations (in which S_P means the ring of P-adic integers).

$$\begin{cases} \tilde{g}_1(\alpha) \dots \tilde{g}_s(\alpha) = 0 & \text{in } R_{\mathfrak{p}}[\alpha] \\ \tilde{g}_i(\alpha) \text{ is a unit in } S_{P_j} & \text{if } i \neq j \\ \tilde{g}_i(\alpha) = 0 & \text{in } S_{P_i} \end{cases}$$

from which we deduce that :

$$v_{P_j}\left(\tilde{g}_i|_n(\alpha)\right) \quad \begin{cases} \geq n e_i & \text{if } i = j \\ 0 & \text{if } i \neq j \end{cases}$$

□

We must now define a notion similar ro the notion of "pure" ideal of paragraph 3.

DEFINITION 11.- We will say that an ideal $I \neq 0$ of S is a **candid ideal** if it is unramified and if in its decomposition :

$$I = P_1^{v_1} \dots P_r^{v_r}$$

the norms $N(P_1), \dots, N(P_r)$ are prime to each other two by two.

We will say that the extension S/R is a candid extension if in each ideal class of S, there is a candid ideal prime to a given modulus M.

Examples :

1) Each pure ideal is candid, so R/Z in paragraph 2 is candid since each ideal class of R contains a pure ideal (lemma 4) prime to a given modulus.

2) If R is a maximal order in a number field, S/Z is candid since we can find infinitely many prime ideals in any ideal class.

We can now generalize theorem 4.

THEOREM 9.- Let $\mathcal{C}_1,\ldots,\mathcal{C}_r$ be r ideal classes in a candid extension S/R. Then :

i) we can represent $\mathcal{C}_1,\ldots,\mathcal{C}_r$ by candid ideals $I_1,\ldots,I_r$ whose norms $c_1,\ldots,c_p$ are prime to each other two by two.

ii) If we put $c = c_1,\ldots,c_r$ there is a c-adic polynomial $g(X)$ such that for $i = 1,\ldots,r$ we have :

$$I_i = (c_i, \tilde{g}|_{n_i}(\alpha)) := c_iS + \tilde{g}|_{n_i}(\alpha)S$$

iii) the class $\mathcal{C}_1^{n_1}\ldots\mathcal{C}_r^{n_r}$ can be representel by

$$I_1^{n_1}\ldots I_r^{n_r} = \left(c_1^{n_1}\ldots c_r^{n_r},\ \tilde{g}|_n(\alpha)\right)$$

where

$$n = \sup(n_1,\ldots,n_r).$$

Proof :

This follows from theorem 8 in noticing that, by the Chinese remainder theorem :

$$R_c \cong R_{c_1} \times\ldots\times R_{c_r}$$

So :

$$R_c[X] \cong R_{c_1}[X] \times\ldots\times R_{c_r}[X]$$

and we take, for $\tilde{g}(X)$, the polynomial which corresponds to $(\tilde{g}_1(X),\ldots,\tilde{g}_r(X))$. □

REFERENCES

[1] E. ARTIN.- Quadratischer Körper im Gebiet der höheren Kongruenzen I. Math. Zeitsch. 19 (1924), 153-206 (= Coll. p. 1-54).

[2] J.W.S. CASSELS.- Rational quadratic forms. Ac. Press 1978.

[3] G. FREY.- Rationale Punkte auf Fermatkurven und gewiwteten Modulkurven. J. Crelle 331 (1982).

[4] E. HECKE.- Lectures on the theory of algebraic numbers. Springer 1981.

[5] Y. HELLEGOUARCH.- Courbes elliptiques et équation de Fermat. Thèse Besançon 1972.

[6] Y. HELLEGOUARCH.- Loi de réciprocité, critère de primalité dans $\mathbb{F}_q[t]$. C.R. Ac. Sci. Canada, vol VIII, n° 5, oct. 1986, 291-296.

[7] Y. HELLEGOUARCH, D.L. McQUILLAN, R. PAYSANT-LE ROUX.- Unités de certains sous-anneaux des corps de fonctions algébriques. Acta Arithmetica XLVIII 1987, 9-47.

[8] N. KOBLITZ.- Introduction to elliptic curves and modular forms. Springer 1984.

[9] T.Y. LAM.- The algebraic theory of quadratic forms. Benjamin 1973.

[10] A.M. LEGENDRE.- Théorie des Nombres. T. I, Blanchard 1955.

[11] T. NAGELL.- Number Theory. Chelsea 1964.

[12] R. PAYSANT-LE ROUX.- Calibre d'un corps arithmétique. Unités. Thèse, Caen, 1987.

[13] J.P. SERRE.- Cours d'arithmétique. P.U.F., 1970.

Address:
Département de Mathématiques et de Mécaniques
F-14032 Caen Cedex

MEAN VALUE ESTIMATES FOR EXPONENTIAL SUMS

M. Jutila

1. INTRODUCTION

1.1. A general mean value problem. Exponential sums of the type

$$(1.1)\qquad S(M,M';t) = \sum_{M}^{M'} d(m)g(m,t)e(f(m,t)),$$

where $d(m)$ is the usual divisor function, $e(x) = e^{2\pi ix}$, f and g are certain functions, and t is a real parameter, occur for instance in the theory of Riemann's zeta-function and in Dirichlet's divisor problem. A good example and model of a sum like this is the Dirichlet polynomial

$$(1.2)\qquad \sum_{M}^{M'} d(m)m^{-1/2-it}.$$

We are interested in the mean square of the sum (1.1) over a short interval, or over a system of non-overlapping short intervals. Still more generally, we shall consider, for sums of the type

$$(1.3)\qquad S(M,M';v,y) = \sum_{M}^{M'} d(m)g(m,v,y)e(f(m,v,y)),$$

the mean value

$$(1.4)\qquad \sum_{r=1}^{R} \int_{0}^{V} |S(M,M';v,y_r)|^2 dv,$$

where $0 < Y \leqq y_1 < \dots < y_R \leqq 2Y$, and $y_{r+1} - y_r \geqq Y_0 > 0$. A particular case of (1.4) is the above mentioned mean square of the sum (1.1) over a system of R non-overlapping intervals $[t_t, t_r+T_0] \subseteq [T,2T]$. Indeed, defining $f(x,v,y) = f(x,v+y)$, $g(x,v,y) = g(x,v+y)$, we have

$$(1.5)\qquad \sum_{r=1}^{R} \int_{t_r}^{t_r+T_0} |S(M,M';t)|^2 dt = \sum_{r=1}^{R} \int_{0}^{T_0} |S(M,M';v,t_r)|^2 dv.$$

In [7] we studied the mean square of the Dirichlet polynomial (1.2) over a single interval $[T,T+T_0]$ with $T^{1/2+\varepsilon} \ll T_0 \ll T^{2/3}$, and similar arguments apply to the mean value (1.4) as well. A more serious difficulty comes at the very end of the proof, where a generalization of a theorem of E. Bombieri and H. Iwaniec [1] is required; such a result is given in Lemma 2 below.

Methods and results pertaining to the above sums can be carried over to similar sums involving Fourier coefficients of cusp forms for the modular group. The reason for this analogy can be found in common functional properties of the associated Dirichlet series (see [8]).

1.2. The assumptions on the functions f and g. Define

$$D(a,b;\mu) = \{z \in \mathbb{C} : |z-x| < \mu \text{ for some } x \in [a,b]\},$$

the "μ-hull" of the interval $[a,b]$. Let M and M' be large positive numbers with $M < M' \leqq 2M$, and write $D_1 = D(M,M';c_1M)$, $D_2 = D(0,V;c_2V)$, where $V > 0$ and the c_i are positive constants. Let further $M_1 = (1-c_1)M$, $M_2 = M' + c_1M$, $V_1 = -c_2V$, $V_2 = (1+c_2)V$, $Y > 0$, $E_1 = (M_1,M_2)\times(V_1,V_2)\times[Y,2Y]$, $E_2 = [M,M']\times[0,V]$, $E_3 = (M_1,M_2)\times[Y,2Y]$, and $E_4 = (V_1,V_2)\times[Y,2Y]$. The relation $A \asymp B$ means that $A \ll B \ll A$.

Throughout this paper, the functions $f(x,v,y)$ and $g(x,v,y)$ are supposed to satisfy the following conditions.

(i) $f(x,v,y)$ is real for $(x,v,y) \in E_1$.

(ii) For fixed $y \in [Y,2Y]$, f is a holomorphic function of (x,v) in $D_1\times D_2$. Also, f_x is continuously differentiable (as a function of (x,y)) in E_3 for fixed $v \in (V_1,V_2)$.

(iii) There are positive numbers F and T such that

(1.6) $\quad f_x \ll FM^{-1}$ in $D_1\times E_4$,

(1.7) $\quad f_{vx} \ll FM^{-1}T^{-1}$ for $(x,y)\in E_3$, $v \in D_2$,

and in the set E_1 we have

(1.8) $\quad |f_{xx}| \asymp FM^{-2}$,

(1.9) $\quad |f_{vx}| \asymp FM^{-1}T^{-1}$,

(1.10) $\quad |f_{xy}| \asymp FM^{-1}Y^{-1}$.

(iv) The function g, defined for $(x,v,y) \in [M,M']\times[0,V]\times[Y,2Y]$, is continuous in E_2 for fixed $y \in [Y,2Y]$, and g_x is a continuous function of x for fixed v and y. Also, $g \ll G$ and $g_x \ll G'$.

1.3. The main theorem. We may now formulate our result on the mean value (1.4).The case $R = 1$ of the following theorem should be understood in the sense that the sum S is a function of v only (the appearance of y_1 is irrelevant), and the conditions involving y_r, Y, and Y_0 are to be omitted. As usual, ε will stand generally for a small positive constant, not necessarily the same everywhere.

THEOREM 1. *Suppose that the functions f and g satisfy the conditions (i) - (iv). Suppose further that* $M^{-1/2+\varepsilon}T \ll V \ll M^{-1/3}T$, $Y \leqq y_1 < \dots < y_R \leqq 2Y$, $y_{r+1} - y_r \geqq Y_0 > 0$ *for* $r = 1, \dots, R-1$, *and that*

(1.11) $Y_0 \gg T^{-1}VY,$

(1.12) $F^{2/3+\varepsilon} \ll M \ll F.$

Then

(1.13) $$\sum_{r=1}^{R} \int_{0}^{V} |S(M,M';v,y_r)|^2 dv \ll (G + MG')^2 MF^{\varepsilon}\{RV + R^{1/2}F^{-1/2}T^{3/2}V^{-1/2}\min(R^{1/2}, (F/M)^{1/2})\}.$$

The same estimate holds if the coefficients $d(m)$ are replaced by $a(m)m^{-(\kappa-1)/2}$, where the $a(m)$ are the Fourier coefficients of a cusp form of weight κ for the modular group.

Remark 1. The theorem gives an upper bound for the left hand side of (1.5) if the functions $f(x,t)$ and $g(x,t)$ satisfy suitable conditions, corresponding to (i) - (iv). We may understand x (resp. t) either as a complex variable in the domain $D(M,M';c_1M)$ (resp. $D(T,2T;c_3T)$), or as a real variable in the interval (M_1,M_2) (resp. $((1-c_3)T,(2+c_3)T)$). The modified conditions (i) - (ii) are: (i) f is real for real x and t, (ii) f is a holomorphic function of the complex variables x and t. In (iii), f_{vx} is replaced by f_{tx}, and the condition (1.10) is omitted. Finally, in (iv), g is a continuous function of the real variables x and t, g_x is a continuous function of x for fixed t, and the bounds for g and g_x are as before. Then the functions $f(x,v,y)$ and $g(x,v,y)$, defined before (1.5), satisfy the conditions of the theorem. With $y_r = t_r$, $Y = T$, $V = Y_0 = T_0$, and supposing that $M^{-1/2+\varepsilon}T \ll T_0 \ll M^{-1/3}T$, $F^{2/3+\varepsilon} \ll M \ll F$, we obtain

(1.14) $$\sum_{r=1}^{R} \int_{t_r}^{t_r+T_0} |S(M,M';t)|^2 dt \ll (G + MG')^2 MF^{\varepsilon}\{RT_0 + R^{1/2}F^{-1/2}T^{3/2}T_0^{-1/2}\min(R^{1/2},(F/M)^{1/2})\}.$$

Remark 2. If the variable v is replaced by Av, then V and T are replaced by V/A and T/A. Hence the value of T is at our disposal, and we may suppose that $T = F$, to emphasize the analogy between Dirichlet polynomials and more general sums.

Remark 3. It suffices to prove the theorem in the case g = constant, for the general case then follows by partial summation. To have a perfect analogy between the sum (1.3) and the Dirichlet polynomial (1.2), we may specify $g = G = M^{-1/2}$.

Remark 3. The right hand side of (1.14) can be replaced by

(1.15) $$\ll (G + MG')^2 MF^{\varepsilon}\{RT_0 + R^{2/3}F^{-1/3}T \min(R^{1/3}, (F/M)^{1/3})\}.$$

This is clear if RT_0 dominates in $\{\ldots\}$ in (1.14). Otherwise we replace T_0 by $T_0' = (RF)^{-1/3}T(\min(\ldots))^{2/3}$ ($\geq T_0$) to obtain (1.15). Note that the intervals $[t_r, t_r+T_0']$ may overlap, but then we may cover the union of the intervals $[t_r, t_r+T_0]$ by R' ($\leq R$) non-overlapping intervals $[t_r', t_r'+T_0]$ and apply (1.14) to this system.

1.4. Applications to Dirichlet series. Specialized to a Dirichlet polynomial, (1.14) and (1.15) yield for $M \asymp T$

$$\sum_{r=1}^{R} \int_{t_r}^{t_r+T_0} \left|\sum_{M}^{M'} d(m)m^{-1/2-it}\right|^2 dt \ll (RT_0 + (RT)^{2/3})T^{\varepsilon}.$$

This implies the mean value theorem

$$\sum_{r=1}^{R} \int_{t_r}^{t_r+T_0} |\zeta(1/2+it)|^4 dt \ll (RT_0 + (RT)^{2/3})T^{\varepsilon} \tag{1.16}$$

due to H. Iwaniec [5]. Actually Iwaniec's estimate was $\ll (RT_0+R^{1/2}T_0^{-1/2}T)T^{\varepsilon}$, but as we saw above, this implies (1.16).

Consider next the cusp form L-function

$$\varphi(s) = \sum_{n=1}^{\infty} a(n)n^{-s}.$$

Applying (1.14) with $F = T$, $R = 1$, and $T_0 = T^{2/3}$ to the corresponding Dirichlet polynomial in which n runs over an interval $[M,M']$ with $M \ll T$, we reprove the estimate

$$\int_{T}^{T+T^{2/3}} |\varphi(\kappa/2+it)|^2 dt \ll T^{2/3+\varepsilon} \tag{1.17}$$

due to A. Good [2] (see also [7]).

Though the results (1.16) and (1.17) are not new, it is anyway of methodical interest that these can be proved in a unified way and more elementarily than previously, for the heavy machinery of "Kloostermania" is not needed.

1.5. Applications to $\Delta(x)$ and $E(T)$. These are the error terms in the asymptotic formulae

$$\sum_{n\leq x}{}' d(n) = x(\log x + 2\gamma - 1) + 1/4 + \Delta(x),$$

$$\int_0^T |\zeta(1/2+it)|^2 dt = T(\log(T/2\pi) + 2\gamma - 1) + E(T),$$

where γ is Euler's constant and the summation convention $\sum'$ is defined in the usual way (see [4]). We are interested in the mean square of $\Delta(x+U) - \Delta(x)$ and $E(t+U) - E(t)$ as x or t runs over a short interval. In [6] we proved asymptotic formulae for these mean squares, implying the upper estimate

$$(1.18)\quad \int_X^{X+H} (\Delta(x+U) - \Delta(x))^2 dx \ll (HU + X)X^{\varepsilon}$$

for $1 \leqq H, U \leqq X$, and analogously for $E(t)$. Here HUX^{ε} represents the expected bound for the integral on the left. The following theorem improves on (1.18) in some cases.

THEOREM 2. _For_ $1 \leqq H, U \leqq X$ _we have_

$$(1.19)\quad \int_X^{X+H} (\Delta(x+U) - \Delta(x))^2 dx \ll (HU + X^{2/3}U^{4/3})X^{\varepsilon},$$

and analogously

$$(1.20)\quad \int_T^{T+H} (E(t+U) - E(t))^2 dt \ll (HU + T^{2/3}U^{4/3})T^{\varepsilon}$$

for $1 \leqq H, U \leqq T$.

The starting point in the proof of (1.19) is the familiar approximate formula for $\Delta(x)$, and Atkinson's formula for $E(t)$ is applied likewise in the proof of (1.20).

Remark. The estimate (1.20) is a sharpening of Iwaniec's fourth moment estimate over the interval $[T, T+T^{2/3}]$ (the case $R = 1$, $T_0 = T^{2/3}$ of (1.16)) in the sense that (1.20) (for $U = T^{\varepsilon}$) implies the latter, but not vice versa. Indeed, the frequency of large values of $\zeta(1/2+it)$ in this interval can be estimated in terms of large values of $E(t+U) - E(t)$, and these cannot occur too often by (1.20).

1.6. _Estimates for exponential sums_. The mean value result (1.14) with $R = 1$, $T = F$, and $T_0 = T^{2/3}$ implies the pointwise estimate

$$(1.21)\quad \ll (G + MG')M^{1/2}F^{1/3+\varepsilon}$$

for the sum (1.1) under the assumptions in Remark 1, sec. 1.3. Such a deduction of a pointwise estimate from a mean value is familiar from the large sieve method (see [9] , Lemma 1.4). But the same estimate (1.21) holds even for sums

$$S(M,M') = \sum_M^{M'} d(m)g(m)e(f(m))$$

involving no parameter at all if f and g satisfy (i) - (iv) in the sense that only those conditions pertaining to x are taken into account and appropriately rephrased in Remark 1. Indeed, one may interpret $S(M,M')$ as the sum (1.1) with $f(m,t) = (t/F)f(m)$ for $t = F$. This estimate was proved in [8] (as Theorem 4.6), but only under the restriction that f' be approximately a power function.

Remark. The estimate (1.21) is trivial for $M \ll F^{2/3+\varepsilon}$. The case $F \ll M \ll F^2$ can in general be reduced to $M \ll F$ on "reflecting" the sum by Voronoi's summation formula to a sum of length $\asymp F^2M^{-1}$. In the case of Dirichlet polynomials, this phenomenon is a consequence of the approximate functional equation for $\zeta^2(s)$.

2. AN ANALOGUE OF GALLAGHER'S LEMMA

The proof of the mean value theorem for the Dirichlet polynomial (1.2) in [7] consisted of five steps, the first of which was an application of a lemma of P. X. Gallagher (Lemma 1.10 in [9]). In our present context, we need the following more general result of the same flavour.

LEMMA 1. *Let* $f(x,v)$ *be a function of the real variable* $x \in [M,M']$ *and of the complex variable* $v \in D = D(0,V;cV)$, *where* $1 \leq M < M' \leq 2M$, $V>0$, *and* $c > 0$ *is a constant. Put* $V_1 = -cV$, $V_2 = (1+c)V$. *Suppose that* f *is a holomorphic function of* v *in* D *for given* x, f_{vx} *exists in* $[M,M']\times D$, *and*

(2.1) $\quad f_{vx} \ll \lambda M^{-1}V^{-1}$ in $[M,M']\times D$,

(2.2) $\quad |f_{vx}| \asymp \lambda M^{-1}V^{-1}$ in $[M,M']\times[V_1,V_2]$,

where $\lambda \gg 1$. *Let* a_m *for* $m \in [M,M']$ *be any complex numbers with* $a_m \ll A$, *and set* $a_m = 0$ *for* $m \notin [M,M']$. *Write* $N = \lambda^{-1}M$. *Then*

$$\int_0^V \Big|\sum_M^{M'} a_m e(f(m,v))\Big|^2 dv \ll M^{\varepsilon}N^{-1}\int_{V_1}^{V_2} dv \int \Big|\sum_{\xi}^{\xi+N} a_m e(f(m,v))\Big|^2 d\xi + A^2V. \tag{2.3}$$

Proof. Let $w(v)$ be a smooth weight function such that $w(v) = 1$ for $v \in [0,V]$ and $w(v) = 0$ for $v \notin [V_1,V_2]$. Denote the sum on the left of (2.3) by $S(v)$. Then

$$S(v) = N^{-1}\int \sum_{\xi}^{\xi+N} a_m e(f(m,v))d\xi,$$

whence

$$\int_0^V |S(v)|^2 dv \leq \int w(v)|S(v)|^2 dv \tag{2.4}$$

$$= N^{-2}\iint\Big(\int w(v) \sum_{\xi\leq m\leq \xi+N}\ \sum_{\eta\leq n\leq \eta+N} a_m\bar{a}_n e(f(m,v)-f(n,v))dv\Big)d\xi d\eta.$$

We consider separately those pairs (ξ,η) with $|\xi-\eta| \leq N_0 = 2M^{\varepsilon}N$, and those with $|\xi-\eta| > N_0$. The contribution of the first mentioned pairs is estimated trivially by

$$\ll N^{-2} \iint_{|\xi-\eta|\leq N_0} d\xi d\eta \int_{V_1}^{V_2} \Big|\sum_{\xi}^{\xi+N} a_m e(f(m,v))\Big|\Big|\sum_{\eta}^{\eta+N} a_n e(f(n,v))\Big| dv$$

$$\ll N^{-2} \iint_{|\xi-\eta|\leq N_0} d\xi d\eta \int_{V_1}^{V_2} \Big(\Big|\sum_{\xi}^{\xi+N}\Big|^2 + \Big|\sum_{\eta}^{\eta+N}\Big|^2\Big)dv \ll M^{\varepsilon}N^{-1}\int_{V_2}^{V_1} dv \int \Big|\sum_{\xi}^{\xi+N}\Big|^2 d\xi.$$

If $|\xi-\eta| > N_0$, then $|m-n| \gg N_0$ in (2.4). We show that the integral

(2.5) $\int w(v)e(f(m,v)-f(n,v))dv$

is then small, say $\ll M^{-2}V$. We apply Theorem 2.3 from [8], which relates the size of this integral with that of the derivative of f(m,v) - f(n,v) with respect to v. If this is $\ll q$ for $v \in D$, and $\asymp q$ for $v \in [V_1,V_2]$, with $q \gg M^{\varepsilon}V^{-1}$, then the integral (2.5) is $\ll M^{-B}V$ for any fixed $B > 0$ if the weight function w(v) is suitably chosen. Now in the present case the derivative in question is for real $v \in [V_1,V_2]$ of the order

$$\asymp |m-n|\lambda M^{-1}V^{-1} \gg N_0\lambda M^{-1}V^{-1} \gg M^{\varepsilon}V^{-1}$$

by (2.2), and the same upper bound is valid for all $v \in D$ by (2.1). We conclude that the contribution of the pairs (ξ,η) with $|\xi-\eta| > N_0$ to the right hand side of (2.4) is $\ll A^2V$.

3. A LEMMA ON PAIRS OF RATIONALS

In a recent important paper, E. Bombieri and H. Iwaniec [1] encountered the following arithmetic problem: to estimate the number of pairs of rationals $(h_1/k_1,h_2/k_2)$ such that $h_i \asymp H$, $k_i \asymp K$, $(h_i,k_i) = 1$, and

(3.1) $\|\bar{h}_1/k_1 - \bar{h}_2/k_2\| \leqq \Delta_1,$

(3.2) $|h_1k_1 - h_2k_2| \leqq \Delta_2HK.$

Here $\|x\|$ denotes the distance of x from the nearest integer, and $\bar{h}_ih_i \equiv 1 \pmod{k_i}$. The number of such pairs turned out to be

(3.3) $\ll HK + \Delta_2(H^2 + K^2) + \Delta_1(\Delta_1 + \Delta_2)H^2K^2.$

The condition (3.2) is essentially the same as

$$|k_1/k_2 - (h_2/k_2)^{1/2}/(h_1/k_1)^{1/2}| \leqq \Delta_2.$$

This is a special case of the inequality

(3.4) $|k_1/k_2 - \omega(h_2/k_2)/\omega(h_1/k_1)| \leqq \Delta_2.$

It has been shown by M. N. Huxley and N. Watt [3] that the number of simultaneous solutions of (3.1) and (3.4) is still bounded by (3.3) if the function ω is supposed to satisfy certain natural conditions.

We generalize the above problem in one more aspect, letting ω depend on a parameter as well. Accordingly, we replace (3.4) by

(3.5) $|k_1/k_2 - \omega(h_2/k_2,y_{r_2})/\omega(h_1/k_1,y_{r_1})| \leqq \Delta_2,$

where the y_r constitute a Y_0-well-spaced system of points, in other words these numbers lie at least Y_0 apart (as in Theorem 1).

LEMMA 2. _Let_ $H, K \geqq 2$, $0 < \Delta_1 \leqq 1/2$, $0 < \Delta_2 \leqq 1/4$, _let_ J _be_ _a_ _sub-interval_ _of_ $[H/2K,2H/K]$, _and_ _let_ $\omega(x,y)$ _be_ _a_ _positive_ _continuously_ *differentiable function defined for* $x \in J$, $(0 <)\ Y \leq y \leq 2Y$. *Let* $\{y_r\}_{r=1}^{R}$ *be a* Y_0*-well-spaced system of numbers in the interval* $[Y, 2Y]$. *Suppose that*

(3.6) $\quad |\omega_x/\omega| \asymp K/H,$

(3.7) $\quad |\omega_y/\omega| \asymp Y^{-1}.$

Then the number of quadruples $(h_1/k_1, h_2/k_2, y_{r_1}, y_{r_2})$ *with* $h_i/k_i \in J$, $h_i \asymp H$, $k_i \asymp K$, *and* $(h_i, k_i) = 1$ *satisfying the inequalities* (3.1) *and* (3.5) *is at most*

(3.8) $\quad \ll (H^2 + K^2)(1 + \Delta_2 Y_0^{-1} Y)R + \Delta_1(\Delta_1 + \Delta_2)H^2K^2R^2.$

Proof. We follow closely the argument of the proof of Lemma 2.4 in [3]. The pairs (h_1, k_1) and (h_2, k_2) can be related by a unimodular matrix as follows:

$$\begin{pmatrix} h_2 \\ k_2 \end{pmatrix} = \begin{pmatrix} a & b \\ c & d \end{pmatrix}\begin{pmatrix} h_1 \\ k_1 \end{pmatrix}.$$

Consider first the "exceptional" matrices, for which $b = 0$ or $c = 0$. If $c = 0$, then $a = d = 1$, $k_1 = k_2$, $h_2 = h_1 + bk_1$, and supposing $\omega \asymp 1$, as we may, we can write the condition (3.5) as

$$|\omega(h_1/k_1, y_{r_1}) - \omega(h_1/k_1 + b, y_{r_2})| \ll \Delta_2.$$

For given values of h_1/k_1, b, and y_{r_1}, the number of possible values of y_{r_2} is $\ll 1 + \Delta_2 Y_0^{-1} Y$, by (3.7). The number of triplets $(h_1/k_1, b, y_{r_1})$ is $\ll HK(1 + H/K)R$, so the total number of quadruples under consideration is

(3.9) $\quad \ll H(H+K)(1+\Delta_2 Y_0^{-1} Y)R \ll (H^2+K^2)(1+\Delta_2 Y_0^{-1} Y)R.$

If $b = 0$, then $a = d = 1$, $h_1 = h_2$, $k_2 = ch_1 + k_1$, and the condition (3.5) can be written as

$$|\omega(h_1/k_1, y_{r_1}) - (1 + ch_1/k_1)\omega(\frac{h_1}{ch_1+k_1}, y_{r_2})| \ll \Delta_2.$$

For given h_1/k_1, c, and y_{r_1}, the number of possible values of y_{r_2} is $\ll 1 + \Delta_2 Y_0^{-1} Y$. The number of triplets $(h_1/k_1, c, y_{r_1})$ is $\ll HK(1+K/H)R$, and hence the total number of quadruples in the present class can be estimated by the latter bound in (3.9).

The same argument applies to matrices which are "almost exceptional" in the sense that $c \ll K/H$ for $K \leq H$, or $b \ll H/K$ for $H \leq K$, because then $a \ll 1$ and $d \ll 1$ (see [3]).

Consider finally pairs of rationals related by a matrix which did not occur above. In [3] , the number of such pairs satisfying (3.1) and (3.4) was found to be

$$(3.10) \quad \ll \Delta_1(\Delta_1 + \Delta_2)H^2K^2.$$

But the same argument works even if (3.4) is replaced by (3.5) if the y_{r_i} are fixed. Then, multiplying (3.10) by R^2, we get the second term in (3.8).

4. PROOF OF THEOREM 1

4.1. A special case. If we suppose, as we may (see Remarks 2 and 3 in sec. 1.3), that $F = T$ and $g = M^{-1/2}$, then the assertion (1.13) to be proved simplifies to

$$(4.1) \quad \sum_{r=1}^{R} \int_0^V |S|^2 dv \ll \{RV + R^{1/2}TV^{-1/2} \min(R^{1/2},(T/M)^{1/2})\}T^{\varepsilon}.$$

We consider separately the ranges

$$(4.2) \quad T^{2/3+\delta} \ll M \ll T^{1-\delta},$$

$$(4.3) \quad T^{1-\delta} \ll M \ll T,$$

where δ is a small positive constant. Our strategy in the proof is as follows. In this section, we settle the case (4.3), and in the next section we are going to show how the case (4.2) can be reduced to (4.3).

Our argument in the case (4.3) will be in principle the same as in [7]. The situation is different in two respects:

1) The sum (1.3) stands in place of the Dirichlet polynomial (1.2).

2) The functions f and g depend on the parameter y.

These complications will be coped with by Lemmas 1 and 2.

The proof of the mean value estimate in [7] was divided into five steps. The first step, involving an application of Gallagher's lemma, can now be repeated by use of Lemma 1, where $\lambda = V$, by (1.7), (1.9), (2.1), and (2.2). Then $N = MV^{-1}$, and we have for the left hand side of (4.1) the estimate

$$\ll M^{-2+\varepsilon}V^2 \sum_r \int \Big| \sum_{\xi}^{\xi+N} \chi(m)d(m)e(f(m,v_0,y_r))\Big|^2 d\xi + R$$

$$(4.4) \quad \ll M^{-2+\varepsilon}V^2 \sum_r \sum_{M\le m\le M'} \Big| \sum_{0\le n\le N} d(m+n)e(f(m+n,v_0,y_r))\Big|^2 + RM^{1+\varepsilon}V^{-1},$$

where $v_0 \in [V_1,V_2]$ and χ denotes the characteristic function of the interval $[M,M']$. This corresponds to (2.4) in [7].

The second step consists in writing the sum over m in (4.4) in terms of integrals over the unit interval as in the Hardy-Littlewood method. Exponential sums

$$S_q(\alpha) = \sum v_q(m)d(m)e(m\alpha),$$

where v_q is a smoothed characteristic function of the interval $[M_{q-1}, M_q]$ with $M_q = M + q\mu$, $\mu \asymp V$, appear in the integrands as in [7]. The rôle of the numbers α_q in [7] is now played by the numbers

$$(4.5) \qquad \alpha_{q,r} = f_x(M_q, v_0, y_r).$$

We may suppose that $f_x \asymp T/M$, in particular $\alpha_{q,r} \asymp T/M$, since the function f can be replaced by f + nx for any integer $n \ll T/M$ without affecting the sum S or the validity of our assumptions on f.

The next, third, step is transforming $S_q(\alpha)$ by a summation formula of the Voronoi type. Steps two and three are virtually identical with those in [7] (except that V now stands in place of T_0).

In the fourth step, a large sieve lemma of Bombieri and Iwaniec [1] is applied to a certain sum indexed by r, q, and rationals a/k. The only new feature is the appearance of the index r. The sum is restricted to a set of triplets (a/k,q,r), associated to a nonnegative integer ρ and a number K_0. The bounds for these are $\rho \ll \log T$ and $1 \leqq K_0 < K$ with

$$(4.6) \qquad K = V^{1/2}T^{-\delta},$$

and the conditions for the triplets are $(a,k) = 1$, $K_0 \leqq k \leqq 2K_0$ ($k \sim K_0$ for short), and

$$\|a/k + \alpha_{q,r}\| \ll 2^{\rho}T^{-1}V + (KK_0)^{-1}.$$

Then there are integers h and ν such that $(h,k) = 1$ and

$$(4.7) \qquad \alpha_{q,r} = h/k + (\nu + O(1))2^{\rho}T^{-1}V.$$

Compared with [7], the introduction of the parameter ν is a novelty here; otherwise we had to allow the additional term $O((KK_0)^{-1})$ to emerge on the right of (4.7).

The argument in [7] shows that up to a factor T^{δ}, the left hand side of (4.1) can be estimated by RV plus the following sum:

$$(4.8) \qquad I = 2^{-2\rho}K_0^{-1}\sum_r \sum_\nu \sum_q \sum_{k\sim K_0} \sideset{}{^*}\sum_{a \bmod k} \iint v_q(x)v_q(y)(xy)^{-1/4} \times$$

$$\times \sum_{\substack{m,n\sim N_0 \\ |m-n|\leq P}} d(m)d(n)(mn)^{-1/4}e((n-m)\bar{a}/k)e(2(\sqrt{mx}-\sqrt{ny})/k)\lambda(x-y)dxdy.$$

This corresponds to (4.14) in [7], up to the new sums over r and ν. The sum $\sum^*$ is restricted by $(a,k) = 1$, and the quadruples $(r,\nu,q,a/k)$ are supposed to satisfy (4.7), where $h/k \equiv -a/k \pmod 1$. Further, we have

$$N_0 \ll T^{1+2\delta}V^{-1},$$
$$P \asymp K_0 M^{-1/2} N_0^{1/2} T^{1+\delta} V^{-1} \ll K_0 T^{1+5\delta/2} V^{-3/2},$$
$$\lambda(\xi) \ll (KK_0)^{-1} \text{ for } |\xi| \leqq KK_0 T^{\delta},$$
$$\lambda(\xi) = 0 \text{ for } |\xi| > KK_0 T^{\delta}.$$

To simplify subsequent calculations, we introduce the notation A <<< B to mean that $A \ll BT^{c\delta}$ for some constant c.

To begin with, we dispose of the terms with m = n in (4.8). For given a/k and r, there are $\ll 2^{\rho} + (KK_0V)^{-1}T$ values of q such that the condition preceding (4.7) holds. Therefore the contribution of the terms in question is

$$<<< 2^{-2\rho} K_0 R(2^{\rho} + (KK_0V)^{-1}T) N_0^{1/2} T^{-1/2} V <<< RV.$$

Next, as in [7], the right hand side of (4.8) (where now $0 < |m-n| \leqq P$) is estimated by appealing to a large sieve lemma due to E. Bombieri and H. Iwaniec (Lemma 2.4 in [1], or Lemma 1 in [7]). Consider vectors

$$\bar{x} = (n-m, \sqrt{m} - \sqrt{n}, \sqrt{n}),$$
$$\bar{y} = (\bar{h}/k, 2\sqrt{x}/k, 2(\sqrt{x} - \sqrt{y})/k),$$

where it is understood that x and y lie in the support of a certain function v_q, and the triplets (h/k,r,q) satisfy (4.7) for a certain fixed value of ν. We write $x = M_q + u$, $y = x + \xi$ with $u \ll V$ and $\xi \ll KK_0$, keeping u and ξ fixed in the sequel. Let X_i and Y_i be numbers such that $|x_i| \leqq X_i$ and $|y_i| \leqq Y_i$, where the x_i (resp. y_i) are the components of $\bar{x}$ (resp. $\bar{y}$). We may choose

$$(X_1,X_2,X_3) = (P, O(P/\sqrt{N_0}), O(\sqrt{N_0})),$$
$$(Y_1,Y_2,Y_3) = (1, O(\sqrt{M}/K_0), O(KT^{\delta}/\sqrt{M})).$$

Let

$$A = \sum_{\substack{\bar{x},\bar{x}' \\ |x_i - x_i'| \leqq (2Y_i)^{-1}}} |a(\bar{x})a(\bar{x}')|,$$

$$B = \sum_{\substack{\bar{y},\bar{y}' \\ |y_i - y_i'| \leqq (2X_i)^{-1}}} |b(\bar{y})b(\bar{y}')|,$$

where

$$a(\bar{x}) = d(m)d(n)(mn)^{-1/4} <<< N_0^{-1/2},$$
$$b(\bar{y}) = v_q(x)v_q(y)(xy)^{-1/4}\lambda(x-y) \ll (KK_0\sqrt{M})^{-1}.$$

Then, by (4.8) and the large sieve lemma, we have

(4.9) $$I \lll (1+2^{-\rho}K_0^{-1}TV^{-3/2})2^{-2\rho}KV(AB\prod_{i=1}^{3}(1+X_iY_i))^{1/2} + RV$$

$$\lll (1+2^{-\rho}K_0^{-1}TV^{-3/2})2^{-2\rho}K_0^{-1/2}K^{3/2}PV(AB)^{1/2} + RV.$$

if u and ξ are chosen suitably. The first factor on the right represent the number of possible values of ν.

The quantity A is easy to estimate. Arguing as in [7] and taking moreover into account that the case m = n is excluded, we obtain

$$A \lll N_0^{-1}(N_0P + K_0M^{-1/2}N_0^{5/2}\log N),$$

thus

(4.10) $$A \lll P.$$

The conditions in the definition of B are

(4.11) $$\|\bar{h}_1/k_1 - \bar{h}_2/k_2\| \ll P^{-1},$$

(4.12) $$|\sqrt{x_1}/k_1 - \sqrt{x_2}/k_2| \ll \sqrt{N_0}/P,$$

$$|(\sqrt{x_1} - \sqrt{x_1+\xi})/k_1 - (\sqrt{x_2} - \sqrt{x_2+\xi})/k_2| \ll 1/\sqrt{N_0}.$$

The third condition is almost trivial and can be ignored. Note that the $\bar{h}_i$ are defined (mod k_i), so that the condition (4.11) can be taken (mod 1).

The condition (4.12) involves implicitely the indices q_i since $x_i = M_{q_i} + u$. We eliminate next the q_i, and reformulate (4.12) in terms of h_i/k_i and r_i. To this end, we define the function $\Omega(\beta,y)$ implicitely by the equation

(4.13) $$f_x(\Omega(\beta,y),v_0,y) = \beta + 2^\rho\nu T^{-1}V.$$

Then $|\Omega_\beta| \asymp M^2T^{-1}$ by (1.8), so (4.5) and (4.7) give together

$$M_{q_i} = \Omega(h_i/k_i,y_{r_i}) + O(2^\rho M^2T^{-2}V).$$

Hence

(4.14) $$x_i = \Omega(h_i/k_i,y_{r_i}) + O(2^\rho V).$$

We see that (4.12) holds if

(4.15) $$|k_1/k_2 - \omega(h_2/k_2,y_{r_2})/\omega(h_1/k_1,y_{r_1})|$$

$$\ll K_0M^{-1/2}N_0^{1/2}P^{-1} + 2^\rho M^{-1}V \lll 2^\rho T^{-1}V,$$

where $\omega(x,y) = \Omega(x,y)^{-1/2}$. By (4.14), the number of possible values of q_i is $\ll 2^\rho$ for given h_i/k_i and r_i. Thus, if X is an upper bound for

for the number of quadruples $(h_1/k_1, h_2/k_2, r_1, r_2)$ satisfying (4.11) and (4.15), then the number of admissible pairs of triplets $(h_i/k_i, r_i, q_i)$ $(i = 1, 2)$ in the definition of B is $\ll 2^{2\rho}X$. Therefore

(4.16) $B \ll 2^{2\rho}(KK_0)^{-2}M^{-1}X.$

We now estimate X by Lemma 2, where $\omega(x,y)$ is defined as above, and $K = K_0$, $H \lll K_0$, $\Delta_1 \asymp P^{-1}$, $\Delta_2 \lll 2^{\rho}T^{-1}V$. The assumptions of the lemma can be easily verified by (4.13) and the implicit function theorem by using (ii) - (iii) in sec. 1.2. Noting that $Y/Y_0 \ll TV^{-1}$ by (1.11), and that $\Delta_2 \lll 2^{\rho}\Delta_1$, we obtain

$$X \lll 2^{\rho}(K_0^2R + K_0^4P^{-2}R^2).$$

Then, by (4.16),

$$2^{-2\rho}B^{1/2} \lll K^{-1}T^{-1/2}(R^{1/2} + K_0P^{-1}R).$$

Substituting this and (4.10) into (4.9), we obtain finally

$$\begin{aligned} I &\lll (1+K_0^{-1}TV^{-3/2})K_0^{-1/2}K^{1/2}P^{3/2}T^{-1/2}V(R^{1/2}+K_0P^{-1}R) + RV \\ &\lll (K_0+TV^{-3/2})(R^{1/2}TV^{-1}+RV^{1/2}) + RV \\ &\ll RV + R^{1/2}TV^{-1/2}. \end{aligned}$$

This completes the proof of Theorem 1 in the case (4.3).

4.2. Reduction of the case (4.2) to (4.3). This reduction is done on transforming the sum S suitably. To begin with, we split up the interval $[M,M']$ into segments of length $\asymp M^2T^{-1}$, and the sum S is accordingly decomposed into $\asymp M^{-1}T$ subsums. Actually we prefer working with smoothed subsums, which are constructed as in [8]. Let $[x_1,x_2]$ be the range of summation in one of the subsums. We replace the x_j by $x_j + u_1 + \dots + u_J$, where

(4.17) $|u_j| \leqq U = MT^{-1/2+\delta}$,

and average with respect to the u_j to end up with the smoothed sums. Here J is a sufficiently large fixed integer.

In the course of the smoothing, the endpoints M and M' are somewhat displaced, whence the sum of the smoothed subsums (let us call these S_ν) is not exactly equal to S. However, by (4.17), the smoothing error is $\lll 1$, which can be omitted.

By arguments similar to those in the proof of Lemma 1, we have for fixed r

$$\int_0^V \Big|\sum_\nu S_\nu(v)\Big|^2 dv \ll \sum_\nu \int_{V_1}^{V_2} |S_\nu(v)|^2 dv + 1.$$

Hence it suffices to estimate right hand side summed with respect to r.

The sums $S_\nu = S_\nu(v,y_r)$ are now transformed by Theorem 3.3 in [8]. Let $[M_1,M_2]$ be the range of summation in S_ν (these M_j are not to be confused with those in sec. 1.2), and write $[M_1+2JU,M_2-2JU] = [M_1',M_2']$. Suppose that $f_{xx} > 0$. As before, we may suppose that $f_x > 0$ and $f_x \asymp T/M$. For a sum $S_\nu(v,y_r)$, choose an integer $h \asymp T/M$ such that

$$(4.18)\quad f_x(M_j,v,y_r) = h + \theta_j,\ j = 1,\ 2,$$

where $\theta_j > 0$ and $\theta_j \asymp 1$.(Here h and θ_j may depend on v and y_r). This is possible, for the change (increment) of f_x in the interval $[M_1,M_2]$ is $\asymp 1$ by (1.8). We choose $r = h$ in the above mentioned theorem. The condition (3.1.8) in [8] presupposes that $M^2T^{-1} >> M^{-1}T^{1+\delta}$, which holds since $M >> T^{2/3+\delta}$. Let

$$(4.19)\quad n_j = (h - f_x(M_j,v,y_r))^2 M_j,$$

and denote by n_j' the similar expression in which M_j is replaced by M_j'. Then $n_j \asymp M$, $n_j' \asymp M$. Let further

$$(4.20)\quad p_n(x) = p_n(x,v,y) = f(x,v,y) - hx - 2(nx)^{1/2} + 1/8,$$

and define $x_n = x_n(v,y)$ by

$$(4.21)\quad \frac{\partial p_n(x_n)}{\partial x} = 0.$$

Then the theorem gives the transformation formula

$$(4.22)\quad S_\nu = -\ i(2M)^{-1/2}\sum_{n_1}^{n_2} w(n)d(n)n^{-1/4}x_n^{-1/4}p_n''(x_n)^{-1/2}e(p_n(x_n)+1/8) + O(T^{-1/2+\delta}),$$

where $w(n) = 1$ for $n_1' < n < n_2'$, and $w(y) << 1$ in the intervals $[n_1,n_1']$ and $[n_2',n_2]$. Moreover, $w'(y) << |n_j - n_j'|^{-1}$, except at J-1 points at most. Actually (4.22) does not follow directly from Theorem 3.3 as it stands, for the sum S_ν must first be written as the difference of two sums of the type considered in the theorem, and the formula (4.22) comes out when the difference is written in the transformed shape. Theorem 3.3 deals with the case when the θ_j in (4.18) are of different sign. However, then the range in the transformed sum would be of the somewhat inconvenient form [1,N].

By summation by parts, we may confine ourselves to sums in which $w(n) \fallingdotseq 1$ throughout the interval $[n_1,n_2]$ if the meaning of the n_j is appropriately modified.

The error term in (4.22) is clearly negligible, and the sum on the right, for fixed h, is of the type (1.3) if the variability of the range of summation is ignored for a moment. But one may consider h as parameter as well, for fixed y_r. In this case, the numbers Y, Y_0, and

and R are $\asymp$ T/M, 1, and $\asymp$ T/M, respectively.

To analyze the transformed sum, note that

$$\left|\frac{\partial}{\partial n}p_n(x_n)\right| \asymp n^{-1/2}x_n^{1/2} \asymp 1$$

by (4.19) - (4.21) (we consider here n temporarily as a continuous variable). This means that $F \asymp M$ for the new sum. Hence we may apply the result just proved; it is easy to check that the new functions f and g satisfy the conditions of Theorem 1 for each choice (y_r or h) of the parameters. The order of G is now $T^{-1/2}$, and the rôle of T is the same as before.

The above mentioned complication that the n_j are not fixed can be eliminated by the usual Fourier transform device (see e. g. [1], Lemma 2.2). The characteristic function of the interval $[n_1,n_2]$ is first approximated by a function whose graph is a trapezoid with linear slopes in the intervals $[n_1 - M^{1/2}, n_1]$ and $[n_2, n_2 + M^{1/2}]$. This function is expressed as a Fourier integral, and the sum over $[n_1,n_2]$ is written approximately as a sum over a sufficiently wide interval (comprising all $[n_1,n_2]$) by using the trapezoidal function as a weight. The Fourier integral converges rapidly and can be truncated at ± 1 with an admissible error. This means that the terms in the sum will be provided with exponential factors $e(n\alpha)$ with $|\alpha| \leqq 1$. Then the respective function f will be replaced by $e(x\alpha)f$, which does not matter, for those properties of f which are of relevance in Theorem 1 remain unaffected. Hence, after all, we may argue as if the range of summation were the same in all transformed sums.

First we apply Theorem 1 keeping h fixed and summing then trivially over h, to obtain

$$\lll (T/M)T^{-1}M(RV + R^{1/2}M^{-1/2}T^{3/2}V^{-1/2})$$
$$= RV + R^{1/2}M^{-1/2}T^{3/2}V^{-1/2}.$$

Secondly, letting h play the rôle of the parameter, we arrive at the alternative estimate

$$\lll RT^{-1}M((T/M)V + (T/M)^{1/2}M^{-1/2}T^{3/2}V^{-1/2})$$
$$= RV + RTV^{-1/2}.$$

The minimum of these yields the right hand side of (4.1), and the proof of Theorem 1 is complete.

5. PROOF OF THEOREM 2

The proofs of (1.19) and (1.20) are analogous. Let us consider that of (1.20) which is somewhat more complicated. We may assume that

$H >> T^{2/3}$, for the assertion in the remaining case then follows trivially. In addition, since

$$\int_T^{T+T^{2/3}} E^2(t)dt << T^{7/6+\varepsilon}$$

(see [4], Corollary 15.2), we may assume that $U << T^{1/2}$.

The function $E(t)$ for $t \asymp T$ can be given by a formula due to F. V. Atkinson (Theorem 15.1 in [4]) as follows:

$$E(t) = \sum_1(t) + \sum_2(t) + O(\log^2 T),$$

where

$$\sum_1(t) = (2t/\pi)^{1/4} \sum_{n\leq N} (-1)^n n^{-3/4} e(t,n)\cos(f(t,n)),$$

$$e(t,n) = (1 + \frac{\pi n}{2t})^{-1/4}\{(\frac{2t}{\pi n})^{1/2}\operatorname{arsinh}((\frac{\pi n}{2t})^{1/2})\}^{-1},$$

$$f(t,n) = 2t \operatorname{arsinh}((\frac{\pi n}{2t})^{1/2}) + (2\pi nt + \pi^2 n^2)^{1/2} - \pi/4,$$

$$\sum_2(t) = -2 \sum_{n\leq N'} d(n)n^{-1/2}(\log(\frac{t}{2\pi n}))^{-1}\cos(t \log(\frac{t}{2\pi n}) - t + \pi/4),$$

$N \asymp T$, and $N' = t/2\pi + N/2 - (N^2/4 + Nt/2\pi)^{1/2}$. We fix $N = T$.

The sum $\sum_2(t)$ can be written, by partial summation, in terms of Dirichlet polynomials. These, in turn, can be expressed in terms of $\zeta^2(s)$ by Perron's formula. Then (1.16) implies that

$$\int_T^{T+T^{2/3}} |\sum_2(t)|^2 dt << T^{2/3+\varepsilon}.$$

Hence it remains to prove that

$$\int_T^{T+H} |\sum_1(t+U) - \sum_1(t)|^2 dt << (HU + T^{2/3}U^{4/3})T^{\varepsilon}. \tag{5.1}$$

We write $\sum_1(t)$ as the sum of two sums $\sum_{11}(t)$ and $\sum_{12}(t)$, in which the range of summation is $[1,TU^{-2}]$ and $(TU^{-2},T]$, respectively. Then (5.1) follows if we prove that

$$\int_T^{T+H} |\sum_{12}(t)|^2 dt << (HU + T^{2/3}U^{4/3})T^{\varepsilon}, \tag{5.2}$$

$$\int_T^{T+H} |\frac{d}{dt} \sum_{11}(t)|^2 dt << (HU^{-1} + T^{2/3}U^{-2/3})T^{\varepsilon}. \tag{5.3}$$

The sums $\sum_{12}(t)$ and $\frac{d}{dt}\sum_{11}(t)$ are of the type (1.1). Consider the proof of (5.2); that of (5.3) is similar. It suffices to prove the same for a segment of $\sum_{12}$, in which the sum is taken over integers $n \sim M \in [TU^{-2},T]$. We apply Theorem 1 in the variant (1.15) with $R = 1$, $T_0 = H$, $F = (MT)^{1/2}$, $G = M^{-3/4}T^{1/4}$, and $G' = G/M$. It should be noted that the restrictions as to the size of M and T_0 in Remark 1 after

Theorem 1 can be relaxed to $M \ll F$ and $T_0 \ll T$ in the case $R = 1$. Indeed, the validity of (1.15) for $T_0 \ll T$ follows immediately if it is known for $T_0 \asymp F^{-1/3}T$. This is the case for $F^{2/3+\varepsilon} \ll M \ll F$ by Theorem 1, and also for $M \ll F^{2/3+\varepsilon}$, by Lemma 1. Now (5.2) follows easily from (1.15).

REFERENCES

[1] E. Bombieri and H. Iwaniec, On the order of $\zeta(1/2+it)$, Ann. Scuola Norm. Sup. Pisa Cl. Sci. (4) 13 (1986), 449-472

[2] A. Good, The square mean of Dirichlet series associated with cusp forms, Mathematika 29 (1982), 278-295.

[3] M. N. Huxley and N. Watt, Exponential sums and the Riemann zeta function, Proc. London Math. Soc. (3) 57 (1988), 1-24.

[4] A. Ivić, "The Riemann zeta-function", John Wiley & Sons, New York, 1985.

[5] H. Iwaniec, Fourier coefficients of cusp forms and the Riemann zeta-function, Séminaire de Théorie des Nombres, Univ. Bordeaux 1979/80, exposé no 18, 36 pp.

[6] M. Jutila, On the divisor problem for short intervals, Ann. Univ. Turkuensis Ser. A I 186 (1984), 23-30.

[7] M. Jutila, The fourth power moment of the Riemann zeta-function over a short interval, Proc. Coll. Soc. János Bolyai, Coll. on Number Theory (Budapest 1987), North Holland, Amsterdam (to appear).

[8] M. Jutila, "Lectures on a method in the theory of exponential sums", Tata Institute of Fundamental Research, Lectures on Mathematics and Physics vol. 80, Bombay, 1987.

[9] H. L. Montgomery, "Topics in Multiplicative Number Theory", Lecture Notes in Mathematics vol. 227, Springer-Verlag, Berlin-Heidelberg-New York, 1971.

Department of Mathematics,
University of Turku,
SF-20500 Turku, Finland.

SOME RESULTS ON DIOPHANTINE APPROXIMATION RELATED TO DIRICHLET'S THEOREM

Hans G. Kopetzky, Leoben

Dirichlet's Theorem in its simplest form states that for any real number α and any positive integer N there exist integers p,q satisfying

$$|\alpha q - p| < 1/N, \quad 1 \le q \le N .$$

H. Davenport and W. M. Schmidt [1] have shown that an improvement of this theorem is possible in the sense that $1/N$ can be replaced for all sufficiently large N by c/N with $c < 1$ exactly for those numbers α which have bounded partial quotients in their regular continued fraction expansion $\alpha = [a_0, a_1, \ldots]$; these numbers are called "badly approximable". In particular, Davenport and Schmidt have proved the following

THEOREM. *For* $\alpha = [a_0, a_1, \ldots]$ *let*

$$\gamma(\alpha) = \liminf\, [0, a_{n+1}, a_{n+2}, \ldots] \cdot [0, a_n, \ldots, a_1] .$$

If $c > 1/(1 + \gamma(\alpha))$ *then for every sufficiently large* N *the inequalities*

$$|\alpha q - p| < c/N, \quad 1 \le q \le N$$

are soluble in integers p *and* q, *but this is not true if* $c < 1/(1 + \gamma(\alpha))$. *The value of* $\gamma(\alpha)$ *always lies between* 0 *and* $1/2(3 - \sqrt{5}) = 0.3819\ldots$.

We shall call the set of values of $\gamma(\alpha)$ the *Dirichlet spectrum* in analogy to the well known spectra of Markoff and Lagrange.

In this note we shall derive some results which are similar to those obtained by Markoff for his spectrum. In particular, the Dirichlet spectrum turns out to be discrete in its upper part and we get its largest accumulation point. Furthermore we shall characterize all α for which $\gamma(\alpha)$ has a value larger than this accumulation point. At last we shall determine one gap in the spectrum below this accumulation point which corresponds to Perron's gap in the Markoff spectrum.

Since equivalent numbers have the same partial quotients with at most finitely many exceptions the following holds

LEMMA. *If* α *and* α' *are equivalent, that is there exist integers* a, b, c, d *such that* $\alpha' = (a\alpha+b)/(c\alpha+d)$ *with* $|ad-bc| = 1$ *then* $\gamma(\alpha') = \gamma(\alpha)$.

Our main result is

THEOREM 1. *The largest accumulation point of the Dirichlet spectrum is* $\chi_1 = \sqrt{5} - 2 = 0.2360\ldots$. *All numbers* α *with* $\gamma(\alpha) > \chi_1$ *are equivalent to one of the following numbers*

$$\text{(i)} \qquad \alpha^* = \tfrac{1}{2}(1 + \sqrt{5}\,), \qquad\qquad \text{(ii)} \qquad \alpha_k = 1 + \sqrt{d_{k+2}/d_k}$$

with $k \equiv 1 (\mathrm{mod}\ 2)$, $k \geq 1$, *where* d_k *is the* k-th *element of the Fibonacci sequence, i.e.* $d_0 = d_1 = 1$, $d_{n+2} = d_{n+1} + d_n$, $n \geq 0$.

The corresponding values of γ *are given by*

$$\text{(i)} \qquad \gamma(\alpha^*) = \tfrac{1}{2}(3 - \sqrt{5}), \qquad\qquad \text{(ii)} \qquad \gamma(\alpha_k) = \frac{\sqrt{d_{k+2}} - \sqrt{d_k}}{\sqrt{d_{k+2}} + \sqrt{d_k}}$$

with $\gamma(\alpha^*) > \gamma(\alpha_1) > \gamma(\alpha_3) > \ldots > \chi_1$.

We introduce an abbreviated notation for continued fractions which allows a short formulation of the proof of this theorem. For $k \geq 2$ let $G_1 = (a_{01}, \ldots, a_{m_1,1}), \ldots, G_{k-1} = (a_{0,k-1}, \ldots, a_{m_{k-1},k-1})$ be finite sequences of natural numbers and let $G_k = (s_0, s_1, \ldots)$ be a finite or infinite sequence of natural numbers. Then we set

$$[G_1, \ldots, G_{k-1}, G_k] = [a_{01}, \ldots, a_{m,1}, \ldots, a_{0,k-1}, \ldots, a_{m_{k-1},k-1}, s_0, s_1, \ldots].$$

Analogously, $(G_1, \ldots, G_{k-1}, G_k)$ denotes the sequence obtained from $G_1, \ldots, G_{k-1}, G_k$ by juxtaposition. If $G = (a_0, a_1, \ldots, a_m)$ we put $G^* = (a_m, \ldots, a_1, a_0)$. The proof of Theorem 1 relies on

LEMMA 1. *Given a real number* α. *If there exists an infinite sequence* $m_1 < m_2 < m_3 < \ldots$ *of natural numbers with the following properties:*

(i) *For each* i (i = 1,2,...) *there exists a sequence* R_i *consisting of* m_i *natural numbers, finite sequences* G_i, H_i *(where* H_i *is allowed to be empty) and an infinite sequence* S_i *such that*

$$\alpha = [0, R_i, G_i, H_i, S_i] \tag{1}$$

(ii) *There exists a real number* $c > 0$ *such that for each* i

$$[0, G_i^*, R_i^*] \cdot [0, H_i, S_i] \le c \ . \tag{2}$$

Then we have

$$\gamma(\alpha) \le c \ . \tag{3}$$

Lemma 1 follows at once from the definition of $\gamma(\alpha)$ in the theorem of Davenport and Schmidt.

During the proof the following facts will be used frequently (cf. Perron [2], vol. 1, p. 34 f., Satz 2.7 and Satz 2.8):

Suppose G is a finite sequence with $L(G)$ elements and R, R_1, R_2 are arbitrary sequences. Now

$$\text{if } L(G) \equiv 1 \bmod 2 \text{ then } [G] < [G,R] \text{ if } R \neq \emptyset \ , \tag{4}$$
$$\text{if } L(G) \equiv 0 \bmod 2 \text{ then } [G] > [G,R] \text{ if } R \neq \emptyset \ , \tag{5}$$
$$\text{if } L(G) \equiv 0 \bmod 2 \text{ and } b_1 > b_2 \text{ then } [G, b_1, R_1] \ge [G, b_2, R_2] \ , \tag{6}$$
$$\text{if } L(G) \equiv 1 \bmod 2 \text{ and } b_1 > b_2 \text{ then } [G, b_1, R_1] \le [G, b_2, R_2] \ . \tag{7}$$

If a sequence G contains n consecutive elements b, we write b_n instead of $b, \dots, b$.

It is easily seen that

$$\lim_{n \to \infty} \gamma(\overline{2, 1_n}) = [0, 2, \overline{1}] \cdot [0, \overline{1}] = \sqrt{5} - 2 = 0{,}2360\dots \ .$$

Therefore $H_1 = \sqrt{5} - 2$ is an accumulation point of the Dirichlet spectrum.

Next we have

$$\gamma(\overline{3,1}) = [0, \overline{3,1}]\ [0, \overline{1,3}] = \tfrac{1}{6}\,(5 - \sqrt{21}) = 0{,}2087\dots =: c_{31} \ .$$

The proof of Theorem 1 requires the following two lemmas.

LEMMA 2. *For a number* $\alpha = [0, a_1, a_2, \dots]$ *with* $a_k \ge 3$ *for infinitely many* k *or with* $a_k = 2$, $a_{k+1} = 2$ *for infinitely many* k

$$\gamma(\alpha) \le c_{31} \quad . \tag{8}$$

Proof. We distinguish five cases

(i) $a_k \ge 5$ for infinitely many k.

We apply Lemma 1 with $G_i = (b_i)$ where $b_i \ge 5$ for each i and $H_i = \emptyset$. Thus for each i α has a representation of the shape

$$\alpha = [0, R_i, b_i, S_i] \ .$$

Now we have by (5)

$$[0, b_i, R_i^*] < [0, b_i] \le \frac{1}{5}$$

since $b_i \ge 5$ for each i.

Moreover (7) and (5) imply with the notation $(S_i) = (s_{0i}, S_i')$

$$[C, S_i] = [0, s_{0i}, S_i'] \le [0, 1, S_i'] < [0, 1] = 1$$

Therefore we obtain

$$[0, b_i, R_i] \cdot [0, S_i^*] \le \frac{1}{5} < c_{31}$$

and in view of Lemma 1 (8) follows. For the rest of the proof we have only to study numbers α with partial quotients $a_k \le 4$ for all k.

(ii) $a_k = 4$ for infinitely many k.

We apply Lemma 1 with $G_i = (4)$, $H_i = \emptyset$. Now if $R_i = (R_i', b_i, a_i)$, then

$$[0, 4, R_i^*] = [0, 4, a_i, b_i, R_i'] \le [0, 4, a_i, 1, R_i']$$

by (7) and since $1 \le b_i$. Moreover

$$[0, 4, a_i, 1, R_i'] \le [0, 4, 4, 1, R_i']$$

by (6) and since $a_i \le 4$. However by (5)

$$[0, 4, 4, 1, R_i'] < [0, 4, 4, 1] = \frac{5}{21}$$

Similary it is seen that

$$[0,S_i] \le [0,1,4,1] = \frac{5}{6}$$

Thus we obtain with $\alpha = [0,R_i,4,S_i]$

$$[0,4,R_i]\cdot[0,S_i] \le \frac{5}{21} \cdot \frac{5}{6} < c_{31}$$

and again by Lemma 1

$$\gamma(\alpha) < c_{31}.$$

For the rest of the proof we have only to study numbers α with $a_k \le 3$ for all k.

(iii) $a_k = 3,\ a_{k+1} = 2$ for infinitely many k
or $a_k = 2,\ a_{k+1} = 3$ for infinitely many k
or $a_k = 3,\ a_{k+1} = 3$ for infinitely many k.

Then in Lemma 1 we have

$$\alpha = [0,R_i,3,2,S_i]$$
$$\text{or}\quad \alpha = [0,R_i,2,3,S_i]$$
$$\text{or}\quad \alpha = [0,R_i,3,3,S_i].$$

Now $[0,3,R_i^*] \le [0,3] = \frac{1}{3}$ by (5), and $[0,2,R_i^*] \le [0,2] = \frac{1}{2}$. Similary $[0,3,S_i] \le \frac{1}{3}$ and $[0,2,S_i] \le \frac{1}{2}$. In view of Lemma 1 we obtain in all three cases

$$\gamma(\alpha) \le \frac{1}{6} < c_{31}.$$

In particular it follows that if α is such that $a_k = 3$ for infinitely many k and $a_k \le 3$ for all k then $\gamma(\alpha) > c_{31}$ would now be only possible if the continued fraction expansion of α contains infinitely many blocks of the shape $(1,3,1)$.

(iv) $a_k \le 3$ for all k and there exists infinitely many k with

$$a_k = 1,\quad a_{k+1} = 3,\quad a_{k+2} = 1.$$

This splits into three subcases:

(A) $a_k = 1,\ a_{k+1} = 3,\ a_{k+2} = 1,\ a_{k+3} = 1,$
(B) $a_k = 1,\ a_{k+1} = 3,\ a_{k+2} = 1,\ a_{k+3} = 2,$
(C) $a_k = 1,\ a_{k+1} = 3,\ a_{k+2} = 1,\ a_{k+3} = 3.$

We first treat (A). By (7) we get

$$[0,3,1,R_i^*]\cdot[0,1,S_i] \le [0,3,1,1]\cdot[0,1,1,1] = \frac{2}{7}\cdot\frac{2}{3} < c_{31}$$

Thus in this case $\gamma(\alpha) < c_{31}$. Using the same argument we see that numbers α with infinitely many blocks of the shape $(1,1,3,1)$ have

$$\gamma(\alpha) < c_{31}$$

Next we consider (B). Here we have to estimate the product $[0,3,1,R_i^*]\cdot[0,1,2,S_i]$ where $\alpha = [0,R_i,1,3,1,2,S_i]$. If we have $(R_i) = (R_i',1)$ for infinitely many i then α has infinitely many blocks $(1,2,3,1)$ and by the above remark we are done. Thus we may suppose that $(R_i) = (R_i',r)$ with $r = 2$ or $r = 3$. Now for such R_i we obtain by (7)

$$[0,3,1,R_i] \le [0,3,1,2].$$

Since moreover $[0,1,2,S_i] \le [0,1,2,1]$ by (7) we obtain

$$\gamma(\alpha) \le \frac{3}{11}\cdot\frac{3}{4} < c_{31}.$$

Similarly α with infinitely many blocks $(2,1,3,1)$ have $\gamma(\alpha) < c_{31}$.

There remains case (C). Here either α is equivalent to $[\overline{13}]$ and hence $\gamma(\alpha) = c_{31}$ or there are infinitely many k satisfying (C) and $a_{k+4} = 3$ or with $a_{k+4} = 2$ or with $a_{k-1} = 1$ or with $a_{k-2} = 2$. But then we are respectively in the situation of the cases (iii), (iv) (B) or (iv) (A).

Thus we finally have only to study those α with $a_k \le 2$ for all k.

(v) There exist infinitely many k with $a_k = a_{k+1} = 2$.

Then by (6) and (7)

$$[0,2,R_i^*]\ .\ [0,2,S_i] \le [0,2,2,1]\ .\ [0,2,2,1] = \frac{9}{49} = < c_{31}. \qquad \square$$

For the remainder of the paper in view of Lemma 2 we are only concerned with numbers consisting solely of blocks of the shape $(2,1_m)$, $m \ge 1$.

LEMMA 3. *Given* $\alpha = [0,F_1,F_2, \ldots]$ *with* $F_k = (2,1_{m_k})$, $m_k, m_{k+1} \geq 1$ and $m_k \neq m_{k+1}$ *for infinitely many* k. *Then we have*

$$\gamma(\alpha) \leq \chi_1 \quad . \tag{9}$$

Proof. We distinguish three cases.

(i) $m_k \not\equiv m_{k+1} \pmod 2$ for infinitely many k. Suppose $m_k \equiv 1 \pmod 2$. We apply Lemma 1 with $G_i = (2,1_{m_k},2)$ and $H_i = (1_{m_{k+1}},2)$. Now we have

$$[0,2,1_{m_k},2,R_i^*] \leq [0,2,\bar{1}]$$

by (7) and

$$[0,1_{m_{k+1}},2,S_i] \leq [0,1]$$

again by (7). Since $\chi_1 = [0,2,\bar{1}]\cdot[0,\bar{1}]$ we get (9). If $m_k \equiv 0 \pmod 2$ we apply Lemma 1 in the same way with $G_i = (1_{m_k},2)$ and $H_i = (2,1_{m_{k+1}},2)$ to obtain the result.

(ii) $m_k \equiv m_{k+1} \equiv 1 \pmod 2$, $m_k \neq m_{k+1}$, $m_k, m_{k+1} \geq 1$ for infinitely many k. Suppose first $m_k \leq m_{k+1}$. We apply Lemma 1 with $G_i = (2,1_{m_k},2)$ and $H_i = (1_{m_{k+1}},2)$ again. We have the estimates

$$[0,2,1_{m_k},2,R_i^*] \leq [0,2,1_{m_k},2,1,1]$$

by (7) since the case $a_j = a_{j+1} = 2$ is excluded now and we get

$$[0,1_{m_{k+1}},2,S_i] \leq [0,1_{m_{k+2}},2,S_i'] \leq [0,1_{m_{k+2}}]$$

first by (7) and then by (6). On the other hand, if we assume $m_{k+1} < m_k$ we put $G_i = (2,1_{m_k})$ and $H_i = (2,1_{m_{k+1}},2)$ in Lemma 1 and proceed in the same way to get the claim. Therefore we have to estimate the product

$$P_m = [0,2,1_m,2,1,1]\cdot[0,1_{m+2},2,1] \tag{10}$$

for $m \equiv 1 \pmod 2$ in both cases (with $m = m_k$ or $m = m_{k+1}$ respectively). Now more involved considerations are necessary.

If d_k is the k-th element of the Fibonacci sequence ($d_0 = d_1 = 1$, $d_k = d_{k-1} + d_{k-2}$, $k \geq 2$) then one obtains immediately by induction

$$[0,2,1_k] = \frac{d_k}{d_{k+2}}, \qquad [0,1_{k+2}] = \frac{d_{k+1}}{d_{k+2}}, \qquad k \geq 0 . \tag{11}$$

Using $[2,1,1] = 5/2$ and $[2,1] = 3/1$ we have

$$[0,2,1_k,2,1_2] = \frac{5d_k + 2d_{k-1}}{5d_{k+2} + 2d_{k+1}} =: \frac{A_k}{B_k},$$

$$[0,1_{k+2},2,1] = \frac{3d_{k+1} + d_k}{3d_{k+2} + d_{k+1}} =: \frac{A_k'}{B_k'}$$

for $k \geq 1$ and further

$$\begin{aligned} A_k &= A_{k-1} + A_{k-2}, \quad k \geq 2, \quad A_0 = 5, \quad A_1 = 7, \\ B_k &= B_{k-1} + B_{k-2}, \quad k \geq 2, \quad B_0 = 12, \quad B_1 = 19, \\ A_k' &= A_{k-1}' + A_{k-2}', \quad k \geq 2, \quad A_0' = 4, \quad A_1' = 7, \\ B_k' &= B_{k-1}' + B_{k-2}', \quad k \geq 2, \quad B_0' = 7, \quad B_1' = 11, \end{aligned} \tag{12}$$

Next, we define two seqences

$$C_k = A_{2k-1} \cdot A_{2k-1}', \quad C_0 = 6 \quad \text{and} \quad D_k = B_{2k-1} \cdot B_{2k-1}', \quad D_0 = 28$$

and consider first the sequence of the C_k. For abbreviation we write

$$Q_k = A_{2k-1}, \qquad Q_k' = A_{2k-1}' .$$

Using the relation (12) for A_k and A_k' twice we get

$$\begin{aligned} Q_k &= 3Q_{k-1} - Q_{k-2}, \quad Q_0 = 2, \; Q_1 = 7, \\ Q_k' &= 3Q_{k-1}' - Q_{k-2}', \quad Q_0' = 3, \; Q_1' = 7. \end{aligned} \tag{13}$$

By repeated use of these relations we obtain now a third order recurrence relation for the product $C_k = Q_k \cdot Q_k'$. Multiplikation of the terms in (13) gives

$$Q_k Q_k' = 9Q_{k-1}Q_{k-1}' + Q_{k-2}Q_{k-2}' - 3(Q_{k-1}Q_{k-2}' + Q_{k-2}Q_{k-1}'). \tag{14}$$

Using (13) on the left side of (14) we obtain further

$$Q_k Q_{k-1}' + Q_{k-1}Q_k' = 6Q_{k-1}Q_{k-1}' - (Q_{k-1}Q_{k-2}' + Q_{k-2}Q_{k-1}').$$

Elimination of the terms in parenthesis gives

$$3(Q_kQ'_{k-1} + Q_{k-1}Q'_k) = -Q_kQ'_k + 9Q_{k-1}Q'_k - Q_{k-2}Q'_{k-2} .$$

Inserting this expression with $k - 1$ instead of k in (14), we get the desired relation

$$C_k = 8(C_{k-1} - C_{k-2}) + C_{k-3}, \quad k \geq 3, \quad C_0 = 6,\ C_1 = 49,\ C_2 = 342. \qquad (15)$$

In the same way we obtain

$$D_k = 8(D_{k-1} - D_{k-2}) + D_{k-3}, \quad k \geq 3, \quad D_0 = 28,\ D_1 = 209,\ D_2 = 1450. \qquad (16)$$

Now we have to solve these recurrence relations, i.e. to find explicit expressions for the C_k and D_k. This is done in the usual way by solving first the charakteristic equation of each recurrence relation. This equation is in both cases

$$x^3 - 8x^2 + 8x - 1 = 0 .$$

The roots are

$$\omega_1 = 1, \quad \omega_2 = \tfrac{1}{2}(7 + 3\sqrt{5}), \quad \omega_3 = \tfrac{1}{2}(7 - 3\sqrt{5}) .$$

Now the C_k must be of the following form

$$C_k = c_1\omega_1^k + c_2\omega_2^k + c_3\omega_3^k$$

where the c_i are determind by solving the system of linear equations which are obtained putting $k = 0,1,2$. We get the explicit expressions

$$C_k = -1 + \tfrac{1}{10}(35+17\sqrt{5})\cdot\left(\tfrac{1}{2}(7+3\sqrt{5})\right)^k + \tfrac{1}{10}(35-17\sqrt{5})\cdot\left(\tfrac{1}{2}(7-3\sqrt{5})\right)^k . \qquad (17)$$

In the same way we obtain

$$D_k = -3 + \tfrac{1}{10}(155+69\sqrt{5})\cdot\left(\tfrac{1}{2}(7+3\sqrt{5})\right)^k + \tfrac{1}{10}(155-69\sqrt{5})\cdot\left(\tfrac{1}{2}(7-3\sqrt{5})\right)^k . \qquad (18)$$

After these preliminaries we can now simply exclude the numbers α of case (ii). With our notation the product (10) becomes $P_m = C_k/D_k$ with $m = 2k - 1$. Thus we only have to show

$$C_k/D_k < \chi_1 = \sqrt{5} - 2 \quad \text{or} \quad C_k < (\sqrt{5} - 2)\cdot D_k \,.$$

Substituting (17) and (18) in the last inequality gives

$$(-69 + 31\sqrt{5})\cdot\left[\tfrac{1}{2}(7 - 3\sqrt{5})\right]^k > -7 + 3\sqrt{5} \tag{19}$$

and this inequality is trivially true since the left side is positive and the right side is negative.

(iii) $m_k \equiv m_{k+1} \equiv 0 (\mathrm{mod}\ 2)$, $m_k \neq m_{k+1}$, $m_k, m_{k+1} \geq 1$ for infinitely many k. Suppose first $m_k < m_{k+1}$. We apply Lemma 1 with $G_i = (2, 1_{m_k})$ and $H_i = (2, 1_{m_{k+1}}, 2)$. We get

$$[0, 1_{m_k}, 2, R_i^*] \leq [0, 1_{m_k}, 2, 1, 1]$$

and

$$[0, 2, 1, m_{k+1}, 2, S_i) \leq [0, 2, 1_{m_{k+2}} 2, S_i'] < [0, 2, 1_{m_{k+2}}]$$

by (6) and (7). If $m_{k+1} < m_k$ we consider $G_i = (2, 1_{m_k}, 2)$ and $H_i = (1_{m_{k+1}}, 2)$. In both cases we ontain now the product

$$P_m = [0, 1_m, 2, 1, 1]\cdot[0, 2, 1_{m+2}].$$

which must be estimated for $m \equiv 0 (\mathrm{mod}\ 2)$ (again $m = m_k$ or $m = m_{k+1}$ respectively).

The remainder of the proof of this case follows exactly the same lines as in the previous case; the recurrence relations are the same, only the initial values and hence some derived numbers have to be altered. We use the same notations and obtain (corresponding expressions are indicated with a prime):

$$[0, 1_{2k}, 2, 1, 1] =: \frac{A_k}{B_k}, \quad [0, 2, 1_{2k+2}] =: \frac{A_k'}{B_k'}, \tag{11'}$$

$$A_0 = 2, \quad A_1 = 5, \quad B_0 = 5, \quad B_1 = 7\,, \tag{12'}$$
$$A_0' = 2, \quad A_1' = 3, \quad B_0' = 5, \quad B_1' = 8\,.$$

With $C_k = A_{2k}\cdot A_{2k}'$, $D_k = B_{2k}\cdot B_{2k}'$ we get

$$C_0 = 4, \quad C_1 = 35. \quad C_2 = 247 , \tag{15'}$$

$$D_0 = 25, \quad D_1 = 156, \quad D_2 = 1054 . \tag{16'}$$

The characteristic equation of the third order recurrence is clearly the same; only the coefficients in the explicit expressions have to be altered. We find

$$C_k = -\frac{6}{5} + \frac{1}{5}(35+6\sqrt{5})\cdot\left(\frac{1}{2}(7+3\sqrt{5})\right)^k + \frac{1}{5}(35-6\sqrt{5})\cdot\left(\frac{1}{2}(7-3\sqrt{5})\right)^k . \tag{17'}$$

and

$$D_k = \frac{13}{5} + \frac{1}{5}(56+25\sqrt{5})\cdot\left(\frac{1}{2}(7+3\sqrt{5})\right)^k + \frac{1}{5}(56-25\sqrt{5})\cdot\left(\frac{1}{2}(7-3\sqrt{5})\right)^k . \tag{18'}$$

For the last inequality we find in this case

$$(500+224\sqrt{5})\cdot\left(\frac{1}{2}(7-3\sqrt{5})\right)^k > 40 - 26\sqrt{5} . \tag{19'}$$

Again the left side is positive and the right side is negative and hence the claim follows. □

Now all numbers α with $\gamma(\alpha) < \chi_1$ are derived and the proof of our theorem can be finished.

Proof of Theorem 1. In view of Lemma 2 and Lemma 3 we are only concerned with numbers α which are equivalent to one of the numbers $\alpha_k := [\overline{2,1_k}]$ or $\alpha^* = [\overline{1}]$. We have

$$[2,1_k] = \frac{d_{k+2}}{d_k}, \quad \alpha_k = [\overline{2,1_k}] = \frac{d_{k+2}\alpha_k + d_{k+1}}{d_k\alpha_k + d_{k-1}}, \quad k \geq 1 .$$

This gives a quadratic equation for α_n of which only the positive root is possible:

$$\alpha_k = 1 + \sqrt{d_{k+2}/d_k} .$$

Next one obtains

$$\gamma(\alpha_k) = [0,\overline{2,1_k}]\cdot[0,1_k,\overline{2,1_k}] = \frac{1}{\alpha_k}(\alpha_k-2) = \frac{\sqrt{d_{k+2}} - \sqrt{d_k}}{\sqrt{d_{k+2}} + \sqrt{d_k}}$$

Now it is easy to show that $\gamma(\alpha_k) > \chi_1$ if and only if $k \equiv 1 (\mathrm{mod}\ 2)$. Suppose $k \equiv 1 (\mathrm{mod}\ 2)$. By rearranging and squaring the last expression we see that $\gamma(\alpha_k) > \chi_1$ is equivalent to

$$(7-3\sqrt{5})d_{k+2} > (3-\sqrt{5})d_k .$$

From (11) we get by (5)

$$\frac{d_{k+2}}{d_k} = [2,1_k] > \frac{3-\sqrt{5}}{7-3\sqrt{5}} = \frac{3+\sqrt{5}}{2}$$

since $[2,1] = (3+\sqrt{5})/2$ and k is odd. If k is even then $d_{k+2}/d_k < (3+\sqrt{5})/2$. On the other hand the quotient d_{2m+3}/d_{2m+1} is strictly decreasing; further we have $\gamma(\alpha^*) > \gamma(\alpha_1)$. Therefore all claims in the Theorem 1 are proved. □

THEOREM 2. *The interval*

$$\left(\frac{5-\sqrt{21}}{2}, \frac{3-\sqrt{3}}{6}\right) = (0.2087\ldots,0.2113\ldots)$$

contains no element of the Dirichlet spectrum. This interval cannot be enlarged - except possibly that the right endpoint which is an accumulation point of the spectrum must be included.

Proof. The left end of the interval is $c_{31} = \gamma(\overline{31})$ and we have seen that this is the largest value of $\gamma(\alpha)$ if $a_k \geq 3$ for infinitely many k or $a_k = a_{k+1} = 2$ for infinitely many k. On the other hand the smallest possible value of a product of two continued fractions which contain no elements $a_k \geq 3$ and no combination $a_k = a_{k+1} = 2$ gives using (6) and (7) the estimate

$$\gamma(\alpha) \geq [0,\overline{2,1}]\cdot[0,1,1,\overline{2,1}] = \frac{3-\sqrt{3}}{6} =: c_{21}.$$

Therefore we have determined an open interval which contains no element of the spectrum. This interval cannot be enlarged on the left side since the left endpoint is an element of the spectrum. We show now that the right endpoint is an accumulation point of the spectrum.

We consider now the numbers

$$\beta_n = [\overline{(2,1)_n,2,1,1}] .$$

Since $\gamma(\beta_n) = \liminf\, [0,R_i^*]\cdot[0,S_i]$ where the lim inf has to be taken over all decompositions $\beta_n = (R_i,S_i)$, we have the estimate

$$\gamma(\beta_n) \le \lim_{r\to\infty} [0,2,((1,2)_n,1,1,2)_r]\cdot[0,1,1,\overline{(2,1)_n,2,1,1}]$$
$$= [0,2,\overline{((1,2)_n,1,1,2)}]\cdot[0,1,1,\overline{(2,1)_n,2,1,1}].$$
$$=: L_n.$$

On the other hand we have obviously

$$\lim_{n\to\infty} L_n = [0,\overline{2,1}]\cdot[0,1,1,\overline{2,1}] = c_{21} .$$

Finally we have

$$\lim_{n\to\infty} \gamma(\beta_n) = c_{21} .$$

Therefore c_{21} is an accumulation point of the spectrum but we don't know whether this point belongs to the spectrum. □

Aknowledgement. The author is most grateful to the referee for considerable help in improving an earlier version of this paper.

REFERENCES

[1] H. Davenport, W. M. Schmidt, Dirichlet's Theorem on Diophantine Approximation. *Symposia math., 4, Inst. Naz. di alta Mat.* (Academic Press, London, 1970) 113-132.132.
also in: B. J. Birch, H. Halberstam, C. A. Rogers, The Collected Works of Harold Davenport (Academic Press, London, New York, San Francisco, 1977) Vol. II.

[2] O. Perron, Die Lehre von den Kettenbrüchen (B. G. Teubner, Stuttgart, 1954)

Hans Günther Kopetzky
Institut für Mathematik
und Angewandte Geometrie
Montanuniversität Leoben
A-8700 L e o b e n

On cliques of exceptional units and Lenstra's construction of Euclidean fields

Armin Leutbecher and Gerhard Niklasch

Mathematisches Institut der
Technischen Universität München
Postfach 202420, D–8000 München 2,
Federal Republic of Germany

Zusammenfassung

Im Gefolge einer von H. W. LENSTRA JR. 1977 angegebenen Methode, euklidische Zahlkörper mit Hilfe von Ausnahme-Einheiten zu gewinnen, wird eine Gruppenoperation auf Cliquen von Ausnahme-Einheiten studiert, die zugehörige Gruppe vollständig bestimmt und ihre Wirkung in einigen konkreten Ringen ausgenutzt. Damit konnten auch 37 bisher noch nicht bekannte euklidische Zahlkörper mit Graden 5, 6, 7, 8, 9 und 10 gefunden werden.

Summary

In the wake of a method for detecting Euclidean number fields with the aid of exceptional units, described in 1977 by H. W. LENSTRA JR., we study a group action on cliques of exceptional units, determine the corresponding group and exploit the action in some concrete rings. This has also yielded 37 new Euclidean fields in degrees 5, 6, 7, 8, 9, and 10.

1. Introduction

It is now ten years since HENDRIK W. LENSTRA JR. published his celebrated *Inventiones* article "Euclidean number fields of large degree" [L1]. Combining an idea of A. HURWITZ with an argument from the geometry of packings, he had shown that if one can find in an algebraic number field a sufficiently long and 'dense' sequence (in the sense that many differences of its members are units), then the usual norm provides a Euclidean algorithm for that field. LENSTRA's new examples nearly doubled the number of known Euclidean fields, and by now it has been doubled again, all but a few cases having been handled by the same method (see mainly [LM], [Me], [Le], and part 4 of this article).

LENSTRA's main result may be stated as follows. Let K be an algebraic number field, with r_1 real and r_2 complex places; let $\mathbb{Z}_K$ be its ring of integers, D its discriminant and $n = r_1+2r_2$ its degree over the field of rationals $\mathbb{Q}$. For k a positive integer,

---

AMS subject classification: 13F07, 11R27, 05C25, 20B25

Key words and phrases: Euclidean number fields, unit equation, arithmetic graphs, symmetric groups.

define the kth *Lenstra constant* $M_k = M_k(\mathbb{Z}_K)$ to be the maximal length m of a sequence $(\omega_1, \ldots, \omega_m)$ in K with the following property: Among any $k+1$ members of the sequence (with distinct subscripts, but not necessarily distinct values) there are at least two whose difference is a (Dirichlet) unit in $\mathbb{Z}_K$. Such sequences are called *exceptional*. By translation, we may assume all the ω_i to lie in $\mathbb{Z}_K$. Note that the M_k are finite since M_k/k cannot exceed the norm of any nontrivial $\mathbb{Z}_K$-ideal.

Theorem (LENSTRA).

(i) There exist positive constants $\alpha^{(k)}$, depending on k, r_1 and r_2 only, such that the inequality

$$M_k(\mathbb{Z}_K) > \alpha^{(k)} \sqrt{|D|}$$

implies that $\mathbb{Z}_K$ is Euclidean with respect to the field norm $N_{K|\mathbb{Q}}$.

(ii) An upper bound for $\alpha^{(k)}$ is $k \cdot \alpha(r_1, r_2)$ with MINKOWSKI*'s expression*

$$\alpha(r_1, r_2) = \frac{n!}{n^n} \left(\frac{4}{\pi}\right)^{r_2} .$$

An upper bound $\alpha(n)$ for $\alpha^{(1)}$, depending on n only, may be obtained from the theory of packings of spheres in $\mathbb{R}^n$. (This bound is sharper than $\alpha(r_1, r_2)$ if r_1 is small or n is very large.) □

If $k=1$, i.e. all differences $\omega_i - \omega_j$ are required to be units, a translation and a multiplication by a unit convert the exceptional sequence $(\omega_1, \ldots, \omega_m)$ to normal form $(0, 1, \omega'_3, \ldots, \omega'_m)$. Each ω'_i is then an *exceptional unit* in T. NAGELL's terminology [N4], i.e. both ω'_i and $1-\omega'_i$ are units. LENSTRA's theorem therefore suggests studying the following concepts:

- The solutions of the equation $X+Y=1$ in units of a number ring, or more generally of a commutative ring with a unit element—it is well known that the set of solutions is finite (and, in theory, effectively computable) for number rings,
- the graph structure on such a ring, obtained by declaring two elements connected if their difference is a unit.

Both are intimately related: The solutions of the unit equation are precisely the ring elements connected to both 0 and 1. Arithmetic graphs of this kind (using the opposite definition of connectedness) were introduced by K. GYŐRY (see [G3]) and have been used by him to obtain refined finiteness theorems. Our primary concern, however, will be with structural properties, focusing on cliques (complete induced subgraphs) consisting of exceptional units.

The natural equivalence relation ([LM], definition 2.3.2) on cliques of order N, coming from permutations and componentwise inversion, and from translations and the action of units on exceptional sequences of length $N+2$, was shown in [Ni] to correspond to the orbits under an action of the symmetric group S_{N+3} $(N>1)$ or S_3 $(N=1)$. In part 2, we study this "mesh" group action on the *universal ring* for N-cliques of exceptional units R_N (a function ring in N variables over $\mathbb{Z}$). After determining all the exceptional units of this ring, we can exhibit the mesh group as the automorphism group of R_N. That this is a symmetric group will be retrieved as a corollary.

Lower bounds for the Lenstra constants of a ring may be obtained by writing down a suitable exceptional sequence. In a cyclotomic field $\mathbb{Q}(\zeta_p)$ of prime level p one always has $M_1 = M_k/k = p$ ([L1], (3.1)), using cyclotomic units. For the maximal real subfield $\mathbb{Q}(\zeta_p+\bar{\zeta}_p)$, we shall show in section 3.1 that $M_1 \geq p-1$, improving on a result by LENSTRA; our bound is best possible if p is a Fermat prime. The case p=7 provides an illuminating example. Our construction, based on p-torsion on the unit circle, may be viewed as the cyclotomic counterpart to a modular one due to J.-F. MESTRE [Me] which exploits p-torsion on elliptic curves.

In section 3.2 we construct function rings in one variable with large M_2, with a view to mapping them homomorphically into suitable number fields. Section 3.3 presents examples of number fields for which M_2 strictly exceeds $2M_1$.

Part 4 is devoted to 37 new Euclidean fields. Six cases where M_1 did not suffice for an application of LENSTRA's criterion could be settled with a large enough M_2. (Three of our fields appear already in [Ni], and one was mentioned in [D3].)

Bibliographic notes: Related topics. The long history of research on Euclidean fields has been recounted by LENSTRA [L2] and by F. J. VAN DER LINDEN (in [vL]).

As a side result of the search for Euclidean fields by means of exceptional sequences, many number fields with absolute discriminants close to the analytic lower bounds have appeared. Many of these have indeed turned out to constitute the first successive minima of the discriminant (per signature). The systematic enumeration of number fields has now proceeded to degree 8, although extensions of the tables in degrees 5 and 6 [P-Z] still seem possible and desirable. (Large gaps remain in the mixed signatures of sextic fields.) J. MARTINET's article [M2] gives an almost up-to-date overview. Recent contributions are due to M. POHST [P1–3], POHST with P. WEILER and H. ZASSENHAUS [P-Z], MARTINET [M1] and F. DIAZ Y DIAZ [D1–4]. G. NIKLASCH is now revising and updating a table of fields in the appendix of [Ni] which for degrees ≥ 5 should represent the state of knowledge by the end of 1987.

The general unit equation $\lambda X + \mu Y = 1$, and the fact that it has only finitely many solutions over a number ring, were known in substance already to C. L. SIEGEL ([S2], second part, §1), but three decades were to pass before S. LANG [La] stated them explicitly. More recently, GYŐRY [G1] derived explicit effective bounds for the solutions, and together with J. H. EVERTSE established bounds for the number of solutions ([E], [EG]). Thus $M_1(\mathbb{Z}_K)$ is effectively computable by listing all the exceptional units of $\mathbb{Z}_K$. NAGELL has done this for fields of unit rank ≤ 1 and for the first two real cubic fields [N1–4]; for higher degrees and ranks the bounds become hopelessly large, and even when such a list has been drawn up, the computation of M_1 as the clique number of a finite graph is 'hard' (NP-complete). It is not yet known whether the other M_k are effectively computable.

The finiteness theorem for the pure unit equation $X + Y = 1$, conjectured by JULIA ROBINSON, was proved independently of [La] and of each other by S. CHOWLA and by NAGELL ([N1], Théorème 8). NAGELL uses results from SIEGEL's dissertation [S1] but misses a one-line proof based on [S1], Satz 7, Zusatz 1.—[N1] also shows that exceptional units in quartic CM fields are scarce, and GYŐRY follows this up in [G2] for general CM extensions.

2. The graph of a ring and exceptional units

Let R be a commutative ring having a unit element 1_R different from zero. $R^\times$ denotes the group of units of R. To R we associate the simple graph $\Gamma(R)$ whose set of vertices is R itself and whose edges are the pairs $(a,b) \in R\times R$ such that the difference $a-b$ is a unit. This graph structure is compatible with ring morphisms $\varphi: R \to R'$ (always assumed to send 1_R to $1_{R'}$). For integers $k>0$ we define the kth *Lenstra constant* $M_k(R)$ to be the supremum (possibly infinite) of the lengths m of sequences $(\omega_1, \omega_2, \ldots, \omega_m)$ in R such that among any $k+1$ distinct subscripts there are at least two, say, i and j for which $\omega_i - \omega_j \in R^\times$. $M_1(R)$ is the clique number of $\Gamma(R)$, i.e. the maximal order of a complete induced subgraph. By juxtaposition of finite sequences one finds immediately

$$kM_l(R) \leq M_{kl}(R) \qquad (2.1).$$

For ring morphisms $\varphi: R \to R'$ one always has

$$M_k(R) \leq M_k(\varphi(R)) \leq M_k(R') \qquad (k \in \mathbb{N}) \qquad (2.2).$$

For finite rings R' there is the trivial upper bound

$$M_k(R') \leq k \cdot \#R' \qquad (2.3).$$

Thus all the $M_k(R)$ are finite provided that R has a finite (nontrivial) homomorphic image R'.

For the remainder of this section, we consider the case $k=1$ of cliques in $\Gamma(R)$. Following NAGELL [N4], we call *exceptional* those units u of R for which $1-u$ is also a unit. The set $\mathrm{E}(R)$ of all exceptional units in R consists of those elements which are at the same time neighbours of 0 and of 1. $\mathrm{E}(R)$ will be considered as (the vertex set of) an induced subgraph of $\Gamma(R)$. As mentioned in the introduction, any clique of $\Gamma(R)$ can be normalized to consist of 0, 1, and exceptional units. Therefore $\mathrm{E}(R)$ has clique number $N = M_1(R)-2$.

NAGELL [N1] remarked that for any exceptional unit ω also $1/\omega$ and $1-\omega$ are exceptional units. Thus the group

$$\mathcal{H} = \langle i, j \mid i^2 = j^2 = (ij)^3 = 1 \rangle ,$$

isomorphic to the symmetric group S_3, acts via $\omega^i := 1/\omega$, $\omega^j := 1-\omega$ on the induced subgraph $\mathrm{E}(R)$. This action commutes with ring morphisms.—For any non-isolated $x \in \mathrm{E}(R)$ the Kleinian group

$$\mathcal{V} = \langle q, r, s \mid q^2 = r^2 = s^2 = 1, qr = s \rangle$$

acts on the neighbourhood of x (also viewed as an induced subgraph) in $\mathrm{E}(R)$ by

$$\begin{aligned} y^q &:= q_x(y) := x/y, \\ y^r &:= r_x(y) := (y-x)/(y-1), \\ y^s &:= s_x(y) := x(y-1)/(y-x). \end{aligned}$$

If the clique number of $\mathrm{E}(R)$ is at least $N\geq 2$, we can combine these two group actions with the natural right action of the symmetric group $\mathcal{S}_N$ on the set $\mathrm{E}_N(R)$ of cliques of order N in $\mathrm{E}(R)$, given for $\pi\in\mathcal{S}_N$ by

$$(\omega_*)^\pi = (\omega_1,\ldots,\omega_N)^\pi := (\omega_{\pi 1},\ldots,\omega_{\pi N})$$

in the following way: Firstly, componentwise application of the $h\in\mathcal{H}$ defines a right action of the direct product $\mathcal{S}_N\times\mathcal{H}$ on $\mathrm{E}_N(R)$. Secondly, we obtain a right action of $\mathcal{V}$ (note that $\mathcal{V}$ is abelian) on $\mathrm{E}_N(R)$ by putting

$$(\omega_*)^v := (\omega_1,\, v_{\omega_1}(\omega_2),\ldots,\, v_{\omega_1}(\omega_N))$$

for all $v\in\mathcal{V}$. This gives us formally an action of the free product

$$\mathcal{F}_N = (\mathcal{S}_N \times \mathcal{H}) \amalg \mathcal{V}$$

on $\mathrm{E}_N(R)$ for any ring R whose Lenstra constant $M_1(R)$ is at least $N+2$. This action commutes with (componentwise) application of ring morphisms to cliques. The $\mathcal{F}_N$-orbits in $\mathrm{E}_N(R)$ are called *N-meshes*, for reasons that become obvious when one begins to draw the subgraph of $\mathrm{E}(R)$ underlying a 2-mesh. We put $\mathcal{F}_1=\mathcal{H}$.

In fact a finite quotient of $\mathcal{F}_N$ acts effectively. In order to determine this factor group we introduce the *universal ring for N-cliques of exceptional units* $\mathbb{Z}\{X_1,\ldots,X_N\}$ as the ring of quotients of the polynomial ring $\mathbb{Z}[X_1,\ldots,X_N] =: \mathbb{Z}[X_*]$ in N variables X_k with respect to the multiplicative semigroup generated by the X_k, the X_k-1 and by all differences X_l-X_m $(1\leq l<m\leq N)$. This ring, for which we shall also write $R_N = \mathbb{Z}\{X_*\}$, is uniquely determined up to an isomorphism by the property that for any ring R having first Lenstra constant $M_1(R)\geq N+2$ and for any N-clique $(\omega_1,\ldots,\omega_N)$ of exceptional units in R there exists a unique ring morphism $\psi\colon R_N \to R$ sending X_k to ω_k for each k. Moreover the construction of R_N proves this ring to be factorial and to have first Lenstra constant $M_1(R_N)\geq N+2$ on account of the exceptional sequence $(0,1,X_1,\ldots,X_N)$.

In order to gain more information on the structure of R_N, we start by determining all its exceptional units. Clearly, $\mathcal{S}_N$ acts from the left on $\mathbb{Z}[X_1,\ldots,X_N]$ as a group of automorphisms of finite order via substitutions

$$\pi\colon X_k \mapsto \pi X_k := X_{\pi k} \qquad (1\leq k\leq N),$$

and by definition of R_N this action extends to the universal ring, so that we may identify $\mathcal{S}_N$ with a subgroup of the group of automorphisms $\operatorname{Aut} R_N$. Now any unit $u\in R_N^\times$ for which also $1-u\in R_N^\times$ corresponds to a triple of pairwise coprime polynomials $T_\diamond\in\mathbb{Z}[X_*]$, $\diamond\in\{-,0,+\}$, all of whose prime factors are taken from the generator set for denominators $P=\{X_k,\, X_k-1,\, X_l-X_m \mid 1\leq k,l,m\leq N,\ l<m\}$ and whose sum vanishes $(u=-T_+/T_0)$:

$$T_- + T_0 + T_+ = 0 \qquad (*).$$

Permutations of the $T_\diamond$ correspond to the different members of the orbit of u under the action of $\mathcal{H}$, while a change of sign in all $T_\diamond$ yields the same $\mathcal{H}$-orbit.

Proposition 1. *Let $N \geq 1$. The S_N-orbits of normalized unit equations* (∗) *in* $\mathbb{Z}\{X_*\}$ *are represented, up to changes of sign, by the following eight equations:*

$$\begin{aligned}
-X_1 + \quad & 1 + \quad & (X_1 - 1) &= 0 \\
-X_1 + \quad & (X_1 - X_2) + \quad & X_2 &= 0 \\
-(X_1 - 1) + \quad & (X_1 - X_2) + \quad & (X_2 - 1) &= 0 \\
-(X_1 - 1)X_2 + \quad & (X_1 - X_2) + \quad & (X_2 - 1)X_1 &= 0 \\
(X_1 - X_2) + \quad & (X_2 - X_3) + \quad & (X_3 - X_1) &= 0 \\
(X_1 - X_2)X_3 + \quad & (X_2 - X_3)X_1 + \quad & (X_3 - X_1)X_2 &= 0 \\
(X_1 - X_2)(X_3 - 1) + \quad & (X_2 - X_3)(X_1 - 1) + \quad & (X_3 - X_1)(X_2 - 1) &= 0 \\
(X_1 - X_2)(X_3 - X_4) + & (X_2 - X_3)(X_1 - X_4) + & (X_3 - X_1)(X_2 - X_4) &= 0
\end{aligned} \qquad (2.4).$$

Remark 1. It will be tacitly assumed that in case $N<4$ all equations which contain undefined variables X_k are to be dropped.

Remark 2. The first seven equations may be regarded as special cases of the last by formally substituting 0, 1, and/or ∞ for some of the indeterminates, using each of these values at most once. It is striking that SIEGEL's cross-ratio identity [S2]

$$\frac{(w-x)(y-z)}{(w-z)(y-x)} + \frac{(w-y)(x-z)}{(w-z)(x-y)} = 1$$

describes already the most general kind of exceptional unit occuring in R_N.

Remark 3. The total number of exceptional units in R_N turns out to be

$$\#\mathcal{H} \cdot \left(\binom{N}{1} + 3\binom{N}{2} + 3\binom{N}{3} + \binom{N}{4} \right) = 6 \cdot \binom{N+3}{4}.$$

This is obvious for $N \geq 4$, and remains true for $N = 1, 2, 3$.

Proof. (1) Let n be the number of variables X_k which actually enter into the unit equation (∗) $T_- + T_0 + T_+ = 0$; we may assume these variables to be $X_1, X_2, \ldots, X_n$. Each of them occurs either in two $T_\diamond$ with equal degrees, or in all three with two degrees equal and exceeding the third. If $n=1$ one of the $T_\diamond$ must be constant, say $T_0=1$, and up to permutations of summands we have

$$-X_1 + 1 + (X_1 - 1) = 0$$

since higher powers of the two prime polynomials are clearly impossible.

(2) If $n>1$, at least one $T_\diamond$ must be divisible by a mixed difference $(X_l - X_m)$. Otherwise the substitution $X_k \mapsto X_1$ $(1 \leq k \leq N)$ would produce an equation (∗) with $n=1$ and the same total degree; this degree would be >1 since at least one $T_\diamond$ had contained X_1 and X_2.

Assume $n=2$. The substitution $X_1\mapsto 0$, $X_2\mapsto 1$ sends one $T_\diamond$ to 0 and the others to ± 1. Its kernel meets P in the two elements X_1 and X_2-1. Therefore one or both of them occur in one $T_\diamond$ while the other two summands avoid them. Similarly, two $T_\diamond$ avoid the primes X_1-1, X_2. Therefore one summand (say T_0) avoids both sets and is a power of (X_1-X_2). If it were constant, the substitution $X_1\mapsto X_2$ would yield a contradiction; otherwise it shows that T_+ and T_- differ only by a change of sign and a swap of subscripts $1\leftrightarrow 2$. Thus we are left with an equation

$$\varepsilon X_1^a(X_2-1)^b + (X_1-X_2)^c - \varepsilon X_2^a(X_1-1)^b = 0 \qquad (2.5)$$

where $\varepsilon=\pm 1$, $a+b>0$, $c>0$. According as $b=0$, $a=0$ or $ab>0$ one finds the second, third and fourth equations of (2.4); higher powers of the primes can again be excluded, e.g. by suitable substitutions into the derivative of (2.5).

(3) If $n>2$, we know from the above that one summand, say T_0, is divisible by a mixed difference X_l-X_m and that there is a subscript k distinct from l and m such that X_k occurs in one of the other summands, say in T_-. Let $p_-\in P$ be a prime dividing T_- and containing X_k. The substitution $\varphi: X_m\mapsto X_l$ shows that there is a prime p_+, dividing T_+ and distinct from p_- but with the same φ-image (up to sign) $\varphi(p_-)=\pm\varphi(p_+)$. Inspecting images of primes, we find only two candidates for p_- and p_+, viz. X_k-X_l and X_m-X_k. After rearranging we may assume that the former divides T_- and the latter, T_+.

Similarly, if q_- is another prime occuring in T_-, applying φ gives a corresponding prime q_+ that divides T_+ and satisfies $\varphi(q_-)=\pm\varphi(q_+)$. Using instead the substitution $\psi: X_m\mapsto X_k$, we get a prime q_0 dividing T_0, distinct from q_- but with $\psi(q_-)=\pm\psi(q_0)$. Now if q_- did not contain X_m, we would have $\pm\varphi(q_+)=\varphi(q_-)=q_-=\psi(q_-)=\pm\psi(q_0)$; since the three primes are pairwise distinct, this would only be possible if both q_0 and q_+ contained X_m. But then q_- would contain both X_k and X_l, contrary to our assumptions. We conclude that any prime q_- other than X_k-X_l which occurs in T_- must contain X_m. Similar considerations apply to the other $T_\diamond$.

(4) In the case $n=3$ we are now left with the equation

$$\varepsilon_-(X_1-X_2)^a X_3^b(X_3-1)^c + (X_2-X_3)^a X_1^b(X_1-1)^c + \varepsilon_+(X_3-X_1)^a X_2^b(X_2-1)^c = 0$$

in integers $a>0$, $b\geq 0$, $c\geq 0$ and $\varepsilon_\pm=\pm 1$. There are exactly three solutions, viz. the fifth, sixth and seventh equations of (2.4).

If finally $n\geq 4$, we have to solve the equation

$$(X_1-X_2)^a g_- + (X_2-X_3)^a g_0 + (X_3-X_1)^a g_+ = 0$$

in integers $a>0$ and coprime polynomials $g_\diamond$ with prime factors in P but avoiding the set $\{(X_1-X_2),(X_2-X_3),(X_1-X_3)\}$. The substitutions employed in (3) (with $k=1$, $l=2$, $m=3$) show that X_4 must occur in each $g_\diamond$, in fact (X_3-X_4) must divide g_-, (X_1-X_4) divides $g0$ and (X_2-X_4) divides g_+. Putting now $k=2$, $l=1$ and $m=4$ in (3), we find that any prime dividing g_- must contain X_4. Therefore $n=4$, and the only possible solution is the eighth equation of (2.4). This completes the proof of proposition 1. □

Consider the mesh action of $\mathcal{F}_N$ on the universal ring R_N. For any $g\in\mathcal{F}_N$, let $\rho(g)$ be the unique endomorphism of R_N sending $(X_*)=(X_1,\dots,X_N)$ to $(X_*)^g$. If h is another group element, the endomorphism $\rho(g)\circ\rho(h)$ maps (X_*) to $\rho(g)((X_*)^h)=(\rho(g)(X_*))^h=(X_*)^{gh}$, whence it must coincide with $\rho(gh)$: ρ is a homomorphism of $\mathcal{F}_N$ into $\operatorname{Aut} R_N$. Call its image Γ. The following is now an immediate consequence of the universal property of R_N:

Theorem 1. *For any ring R with $\mathrm{E}_N(R)\neq\emptyset$, the right action of $\mathcal{F}_N$ on $\mathrm{E}_N(R)$ can be described in terms of the left action of Γ on R_N as follows. Let $(\omega_*)=(\omega_1,\dots,\omega_N)$ be an N-clique in R, and $\psi\colon R_N\to R$ the unique homomorphism sending (X_*) to (ω_*). Then for any $g\in\mathcal{F}_N$ we have $(\omega_*)^g=\psi\circ\rho(g)(X_*)$, and $\psi^g=\psi\circ\rho(g)\colon R_N\to R$ is the unique homomorphism sending (X_*) to $(\omega_*)^g$.* □

The assertion could obviously serve as the definition of a right action of the whole group $\operatorname{Aut} R_N$ on $\mathrm{E}_N(R)$. Once we have determined the group Γ, however, we shall see that the mesh group action is already as comprehensive as possible.

Remember that we have identified $\mathcal{S}_N$ with a subgroup of $\operatorname{Aut} R_N$, so that we may write $\rho(\pi)=\pi$ for $\pi\in\mathcal{S}_N$ if $N\geq 2$. Generators of the other factors of $\mathcal{F}_N$ correspond to the automorphisms

$$\begin{aligned}
&\mathcal{I}=\rho(i)\colon X_k\mapsto X_k^{-1},\\
&\mathcal{J}=\rho(j)\colon X_k\mapsto 1-X_k \qquad (1\leq k\leq N);\\
&\mathcal{Q}=\rho(q)\colon X_1\mapsto X_1,\ X_k\mapsto X_1/X_k, \qquad\qquad (2.6).\\
&\mathcal{R}=\rho(r)\colon X_1\mapsto X_1,\ X_k\mapsto (X_k-X_1)/(X_k-1),\\
&\mathcal{S}=\rho(s)\colon X_1\mapsto X_1,\ X_k\mapsto X_1(X_k-1)/(X_k-X_1) \qquad (2\leq k\leq N)
\end{aligned}$$

Since $\mathcal{R}=\mathcal{J}\mathcal{Q}\mathcal{J}$ and $\mathcal{S}=\mathcal{I}\mathcal{R}\mathcal{I}$, Γ is generated by $\mathcal{I}$, $\mathcal{J}$, $\mathcal{Q}$ and the permutations $\pi\in\mathcal{S}_N$ (or by $\mathcal{I}$ and $\mathcal{J}$ if $N=1$).

Proposition 2. *The group Γ acts transitively on $\mathrm{E}(R_N)$. If $N=1$, the graph $\mathrm{E}(R_N)$ consists of six pairwise unconnected vertices. If $N\geq 2$, the neighbourhood of X_1 in $\mathrm{E}(R_N)$ decomposes into four connected components, each a clique of order $N-1$, viz. $(X_2,\dots,X_N)$ and its images under $\mathcal{Q}$, $\mathcal{R}$ and $\mathcal{S}$.*

Corollary. *The ring R_N has first Lenstra constant $M_1(R_N)=N+2$.*

Proof. If $N=1$, there is just one $\mathcal{H}$-orbit in $\mathrm{E}(R_1)$ and each exceptional unit is obviously the image of X_1 under an element of $\langle\mathcal{I},\mathcal{J}\rangle=\Gamma$. If $N=2$, we find 30 exceptional units arranged into five $\mathcal{H}$-orbits, represented by X_1, X_2, $\mathcal{Q}(X_2)$, $\mathcal{R}(X_2)$ and $\mathcal{S}(X_2)$. (This is the essence of remark 3.4.1 in [LM].) Clearly $\mathcal{I}$, $\mathcal{J}$, $\mathcal{Q}$ and the permutation $X_1\leftrightarrow X_2$ suffice to reach each exceptional unit from X_1.

In the general case the transitivity of Γ follows from the observation that $\mathcal{I}$ transforms the 4th and 6th $\mathcal{S}_N$-orbits of unit equations from proposition 1 (2.4) into the 3rd and 5th respectively, $\mathcal{J}$ maps the 3rd and 7th onto the 2nd and 6th equations, while application of $\mathcal{Q}$ to the 2nd, 5th and (if $N\geq 4$) 8th equations yields equations of the 1st, 4th or 7th type.

The assertions about the graph $\Gamma(R_N)$ follow from a careful inspection of the exceptional units found in proposition 1. This is somewhat tedious but straightforward and will be left to the reader. Observe that two units of a ring are connected in its graph if and only if their quotient is an exceptional unit. □

Theorem 2. *Let N be an integer ≥ 2. The universal ring $R_N = \mathbb{Z}\{X_1,\dots,X_N\}$ for cliques of N exceptional units has exactly $N+3$ distinct subrings $\mathbb{Z}\{Y_1,\dots,Y_{N-1}\}$ generated by cliques $(Y_1,\dots,Y_{N-1})$ of order $N-1$ in $\mathrm{E}(R_N)$. These are:*

$$R^{(k)} = \mathbb{Z}\{\, X_l \mid 1 \leq l \leq N,\, l \neq k\} \quad (1 \leq k \leq N),$$
$$R^{(q)} = \mathbb{Z}\{\, X_1/X_l \mid 2 \leq l \leq N\},$$
$$R^{(r)} = \mathbb{Z}\{\, (X_l - X_1)/(X_l - 1) \mid 2 \leq l \leq N\},$$
$$R^{(s)} = \mathbb{Z}\{\, X_1(X_l - 1)/(X_l - X_1) \mid 2 \leq l \leq N\}.$$

The subgroup Γ of $\mathrm{Aut}\, R_N$ generated by $\{\mathcal{I}, \mathcal{J}, \mathcal{Q}\} \cup \mathcal{S}_N$ acts $(N+3)$-fold transitively on the set of these subrings. The identity id_{R_N} is the sole automorphism of R_N mapping each of these subrings into itself.

Corollary. $\Gamma = \mathrm{Aut}\, R_N \cong \mathcal{S}_{N+3}$.

Proof. (1) We begin by noting that Γ acts transitively on the sets $\mathrm{E}_\nu(R_N)$ of cliques of each level $1 \leq \nu \leq N$. This follows from proposition 2 since we can map any ν-clique of R_N onto one containing X_1 via a suitable element of Γ, and then to $(X_1,\dots,X_\nu)$ via $\mathcal{Q}$, $\mathcal{R}$ or $\mathcal{S}$ (if necessary) and a permutation that fixes X_1. (These automorphisms generate the stabilizer of X_1 in Γ, viz. the direct product of $\rho(\mathcal{V})$ with the stabilizer of X_1 in $\mathcal{S}_N$.) It follows that, firstly, Γ is the whole group $\mathrm{Aut}\, R_N$ (because any automorphism of R_N must map $(X_1,\dots,X_N)$ onto an N-clique and is determined by this clique) and, secondly, that Γ in particular acts transitively on $\mathrm{E}_{N-1}(R_N)$. Therefore Γ permutes the subrings of R_N which are generated by $(N-1)$-cliques, and even permutes them transitively.

(2) Let us investigate the effect of each generator of Γ in turn. For $\mathcal{Q}$ we find $\mathcal{Q} \circ \mathcal{Q} = id_{R_N}$, $\mathcal{Q}(R^{(1)}) = R^{(q)}$, $\mathcal{Q}(R^{(k)}) = R^{(k)}$ $(2 \leq k \leq N)$, and, because of $\mathcal{Q}((X_l - X_1)/(X_l - 1)) = X_1(X_l - 1)/(X_l - X_1)$, $\mathcal{Q}(R^{(r)}) = R^{(s)}$, proving that $\mathcal{Q}$ induces the permutation $1 \leftrightarrow q$, $r \leftrightarrow s$ on the subrings defined in the theorem.

In order to deal with the remaining generators, we introduce systems $U^{(t)}$ of (exceptional) units in $R^{(t)}$ $(t \in \{q, r, s\})$ which together with -1 generate the unit groups of the respective $R^{(t)}$ (and hence generate the appropriate $R^{(t)}$ themselves), and which are transformed in a comprehensible manner:

$$U^{(q)} = \{\, (X_l - X_m)/X_l = 1 - X_m/X_l \mid l \neq m\},$$
$$U^{(r)} = \{\, (X_m - X_l)/(X_m - 1) = 1 - (1 - X_l)/(1 - X_m) \mid l \neq m\},$$
$$U^{(s)} = \{\, (X_l(X_m - 1)/(X_m - X_l) = 1 - X_m(X_l - 1)/(X_l - X_m) \mid l \neq m\}$$

(subscripts l, m always ranging over $1,\dots,N$). For $\pi \in \mathcal{S}_N$ viewed as an automorphism of R_N we obviously have $\pi R^{(k)} = R^{(\pi k)}$ $(1 \leq k \leq N)$ and $\pi R^{(t)} = R^{(t)}$ $(t \in \{q, r, s\})$. In other words, π as a permutation of subring superscripts acts as usual on $\{1,\dots,N\}$

and fixes $\{q,r,s\}$. $\mathcal{I}$ and $\mathcal{J}$ are involutions and restrict to automorphisms of each $R^{(k)}$ whereas one easily finds $\mathcal{I}(U^{(q)}) = U^{(q)}$, $\mathcal{J}(U^{(s)}) = U^{(s)}$; $\mathcal{I}(U^{(r)}) = U^{(s)}$ and vice versa and $\mathcal{J}(U^{(q)}) = U^{(r)}$ and v.v. Thus $\mathcal{I}$ and $\mathcal{J}$ respectively induce the permutations $r \leftrightarrow s$ and $q \leftrightarrow r$ on the subrings. Clearly the permutations we have found generate the whole symmetric group on the set $\{1,\ldots,N,q,r,s\}$.

(3) To complete the proof we have to show that no automorphism $\sigma \in \Gamma$ other than the identity can leave all $N+3$ subrings invariant. Since each X_l is contained in the intersection of all $R^{(k)}$, $1 \le k \le N$, $k \ne l$, the same would hold for $\sigma(X_l)$. It follows that no indeterminate other than X_l can occur in $\sigma(X_l)$ if we write the latter as a quotient of coprime polynomials. Choose therefore the unique $h \in \mathcal{H}$ with $\sigma(X_1)^h = X_1$, then the image of $(X_2,\ldots,X_N)$ under $\rho(h) \circ \sigma$ must be one of the four $(N-1)$-cliques from proposition 2; in fact it must be the one contained in $R^{(1)} = \sigma(R^{(1)}) = \rho(h)(R^{(1)})$, viz. $(X_2,\ldots,X_N)$ itself. But then σ would have been $\rho(h^{-1})$, while the automorphisms in $\langle \mathcal{I}, \mathcal{J} \rangle \cong \mathcal{H}$ distinct from id_{R_N} do not induce the identity permutation on the last three subrings. □

Definition. *The factor group*

$$\mathcal{M}_N = \mathcal{F}_N / \operatorname{Ker} \rho \,,$$

isomorphic to the group of automorphisms of the universal ring R_N, is called the mesh group of level N. We also put $\mathcal{M}_1 = \mathcal{F}_1$. □

Remark 1. The kernel of ρ has trivial intersection with $\mathcal{S}_N \times \mathcal{H}$ as well as with $\mathcal{V}$.

Remark 2. Each of the subrings from theorem 2 is isomorphic to R_{N-1} and contains a unique $(N-1)$-mesh. Therefore we could have had the $\sigma \in \Gamma$ permute submeshes instead of subrings. This was the second author's original approach in [Ni]. There the action of $\mathcal{M}_N$ on $(N-1)$-submeshes was called *aspect action*, a metaphor referring to an $(N+2)$-dimensional simplex with coloured hyperfaces whose "aspect" when viewed from a fixed direction varies under its symmetry group. Considering submeshes rather than subrings permits finer distinctions when one wants to investigate homomorphic images of R_N.

Both the mesh action and the aspect action of $\mathcal{M}_N$ on the cliques and $(N-1)$-meshes of R_N are effective and have no fixed points. Neither need be true for their images on a mesh of another ring. The example treated in the following section shows something of what can happen in a number ring; further details can be found in section 7.4 of [Ni]. It is easy to list the cases of ineffective mesh action, which can occur only if $N \le 2$. As to fixed points, the stabilizer of a clique in $\mathcal{M}_N$ also fixes the corresponding $(N+3)$-tuple of submeshes. This is sometimes useful for listing whole meshes. On the other hand, a theorem of [Ni] shows that each universal ring R_N has among its homomorphic images cyclotomic fields and finite fields of arbitrarily prescribed characteristic which contain injective images of the set of exceptional units $\mathrm{E}(R_N)$, and thus yield faithful images of the "free" mesh group actions. □

For writing down a mesh $(\omega_*)^{\mathcal{M}_N}$ it is useful to possess a set of representatives for the cosets modulo $\mathcal{S}_N \times \mathcal{H}$ in $\mathcal{M}_N$. We shall present such a set in terms of Γ, using the conjugates $\mathcal{Q}_{(k)}$, $\mathcal{R}_{(k)}$, $\mathcal{S}_{(k)}$ of $\mathcal{Q} = \mathcal{Q}_{(1)}$, $\mathcal{R} = \mathcal{R}_{(1)}$, $\mathcal{S} = \mathcal{S}_{(1)}$ under the swaps $X_1 \leftrightarrow X_k$ $(2 \le k \le N)$.

Proposition 3. *A complete system of representatives for the cosets* $\sigma\cdot(S_N\times\langle\mathcal{I},\mathcal{J}\rangle)$ *in* Γ *is given by*

$$\begin{array}{ll} id_{R_N}, & \\ \mathcal{Q}_{(k)},\ \mathcal{R}_{(k)},\ \mathcal{S}_{(k)} & (1\le k\le N), \\ \mathcal{Q}_{(k)}\mathcal{R}_{(l)},\ \mathcal{R}_{(k)}\mathcal{S}_{(l)},\ \mathcal{S}_{(k)}\mathcal{Q}_{(l)} & (1\le k<l\le N), \\ \mathcal{Q}_{(k)}\mathcal{R}_{(l)}\mathcal{Q}_{(m)} & (1\le k<l<m\le N). \end{array}$$

Proof. The symmetric group Γ on the union of two disjoint finite sets $\mathcal{N}$ and $\mathcal{Z}$ contains the direct product of the symmetric groups $S_{\mathcal{N}}$ and $S_{\mathcal{Z}}$. The cosets modulo this product are represented by the permutations

$$\prod_{t\in\mathcal{T}}(t\leftrightarrow\pi_{\mathcal{U}}^{\mathcal{T}}(t)),$$

where $\mathcal{T}$ runs through all subsets of $\mathcal{N}$ with cardinality $0\le\#\mathcal{T}\le\min\{\#\mathcal{N},\#\mathcal{Z}\}$, for each $\mathcal{T}$, $\mathcal{U}$ runs through the subsets of $\mathcal{Z}$ of the same cardinality, and for each such pair $(\mathcal{T},\mathcal{U})$, $\pi_{\mathcal{U}}^{\mathcal{T}}$ is some fixed bijection from $\mathcal{T}$ onto $\mathcal{U}$. We apply this to our case with $\mathcal{N}=\{1,2,\dots,N\}$, $\mathcal{Z}=\{q,r,s\}$. Given the action of $\mathcal{Q}_{(k)}$, $\mathcal{R}_{(k)}$ and $\mathcal{S}_{(k)}$ on the subscripts in $\mathcal{N}\cup\mathcal{Z}$ as determined during the proof of theorem 2, the assertion can now be verified by a simple calculation. □

3. Lower bounds for the Lenstra constants of some special rings

3.1 Cyclotomic fields

For an odd prime p, let ζ_p be a primitive p^{th} root of unity in a fixed algebraic closure of $\mathbb{Q}$, which we shall assume embedded into $\mathbb{C}$. Using the exceptional sequence

$$(\zeta_p^n-1)/(\zeta_p-1)\qquad(0\le n\le p-1),$$

LENSTRA showed in [L1] that the ring $\mathbb{Z}_K=\mathbb{Z}[\zeta_p]$ of integers in the pth cyclotomic field $K=\mathbb{Q}(\zeta_p)$ has $M_1(\mathbb{Z}_K)=p$. For the maximal real subfield K^+ of K he found the lower bound

$$M_1(\mathbb{Z}_{K^+})\ge\frac{p+1}{2}.$$

This result can be sharpened as follows.

Theorem 3. *For every odd prime* p, *the ring of integers* $\mathbb{Z}_{K^+}$ *in the field* $K^+=\mathbb{Q}(\zeta_p+\zeta_p^{-1})$ *has first Lenstra constant*

$$M_1(\mathbb{Z}_{K^+})\ge p-1.$$

Proof. (1) The powers of any 2×2-matrix

$$A=\begin{pmatrix}a&b\\c&d\end{pmatrix}\in\mathbf{SL}_2(\mathbb{C})$$

are given by the formula

$$A^n=-g_{n-1}(a+d)\mathbf{1}_2+g_n(a+d)A,$$

where $\mathbf{1}_2$ is the unit matrix and where the polynomials $g_n \in \mathbb{Z}[X]$ are defined by

$$g_0 := 0, \quad g_1 := 1, \quad g_{n+1} := Xg_n - g_{n-1}.$$

They are closely related to the Chebyshov polynomials.

(2) We let $\mathbf{SL}_2(\mathbb{C})$ act on $\mathbb{C}\cup\{\infty\}$ by fractional linear transformations in the usual way. If $N>1$ is a rational integer and x is an algebraic integer such that for $1 \le n < N$ $g_n(x)$ is always an algebraic unit, then

$$\omega_n := \begin{pmatrix} x & -1 \\ 1 & 0 \end{pmatrix}^n (\infty) = g_{n+1}(x)/g_n(x) \qquad (1 \le n < N) \tag{3.1}$$

defines an exceptional sequence in $\mathbb{Q}(x)$ due to the formula

$$g_{m+1}g_n - g_m g_{n+1} = g_{n-m} \qquad (0 \le m \le n)$$

which is trivial if $m=0$ and otherwise follows inductively from

$$\begin{aligned} g_{m+1}g_n - g_m g_{n+1} &= (Xg_m - g_{m-1})g_n - g_m(Xg_n - g_{n-1}) \\ &= g_m g_{n-1} - g_{m-1}g_n . \end{aligned}$$

(3) For $n \ge 1$, g_n is monic of degree $n-1$. Using the special matrices

$$A_k = \begin{pmatrix} \zeta_{2n}^k & 0 \\ 0 & \zeta_{2n}^{-k} \end{pmatrix} \qquad (1 \le k \le n-1),$$

which have n^{th} powers equal to $(-1)^k \mathbf{1}_2$ but are not themselves multiples of $\mathbf{1}_2$, one finds the zeroes of g_n to be (assuming e.g. $\zeta_{2n} = \exp(\pi i/n)$)

$$\zeta_{2n}^k + \zeta_{2n}^{-k} = 2\cos(\pi k/n) \qquad (1 \le k \le n-1).$$

(4) Now we put $N=p$ and $x=\zeta_{2p}+\zeta_{2p}^{-1}$ in (3.1). According to (3) we find

$$g_n(x) = \prod_{k=1}^{n-1} \left(\zeta_{2p} + \zeta_{2p}^{-1} - \zeta_{2n}^k - \zeta_{2n}^{-k}\right).$$

Each factor of this product is a unit since it can be written as $\zeta_{2pn}^{-n}(\zeta_{2pn}^{n+kp}-1)(\zeta_{2pn}^{n-kp}-1)$ (with a suitably chosen primitive $2pn^{\text{th}}$ root of unity) and since the order of each root $\zeta_{2pn}^{n\pm kp}$ is a strict multiple of p dividing $2pn$. Thus all the required $g_n(x)$ are units, and (2) applies. □

Remark 1. The monic polynomials g_n are odd or even according as n is even or odd. From the explicit formula $g_n(2\cos\varphi) = \sin(n\varphi)/\sin\varphi$ one can read off

$$-\omega_n = (\zeta_p^{n+1} - \zeta_p^{-n-1})/(\zeta_p^n - \zeta_p^{-n});$$

one could also verify by direct computation that this yields an exceptional sequence.— Furthermore one has $g_{m+n} = g_{m+1}g_n - g_m g_{n-1}$ $(m \ge 0,\ n \ge 1)$. This gives, for odd primes p,

$$g_p = g_{(p+1)/2}^2 - g_{(p-1)/2}^2 .$$

On the other hand, g_p factors as $\Psi_p(X)\Psi_p(-X)$, where Ψ_p is the minimal polynomial of $-\zeta_p-\zeta_p^{-1}$ over $\mathbb{Q}$. As the second coefficient $\sum_{m=1}^{p-1}\zeta_p^m = -1$ of Ψ_p is negative, we conclude that

$$\Psi_p = g_{(p+1)/2} - g_{(p-1)/2}\,.$$

Remark 2. $\mathbb{Z}[\zeta_p+\zeta_p^{-1}]$ has a nontrivial ideal of norm p and none of smaller norm unless p is a Fermat prime, in which case $p-1$ is the least ideal norm. Therefore theorem 3 is sharp for Fermat primes. The following example shows, however, that for $p=7$ one has $M_1(\mathbb{Z}_{K^+})=p$, and the same is known for $p=11$ and $p=13$; cf. [L1], [LM], [Le]. □

Example ($p=7$). Let $K^+=\mathbb{Q}(\eta)$, where $\eta=\zeta_7+\zeta_7^{-1}$, $\eta^3+\eta^2-2\eta-1=0$. Fix the generators of $\mathcal{G}=\mathrm{Gal}(K^+|\mathbb{Q})$ by putting $\eta'=\eta^2-2=-1/(\eta+1)$, $\eta''=-\eta^2-\eta+1=(\eta\eta')^{-1}$. η and η' can be taken as a fundamental system of units. NAGELL [N4] has determined all 42 exceptional units of $\mathbb{Z}_{K^+}=\mathbb{Z}[\eta]$. Abbreviating ${}^NM$ for the Mth orbit of N-cliques under the N-mesh group action, and writing ${}^NM'$, ${}^NM''$ for its conjugates under Galois action, the orbits of exceptional units in $\mathbb{Z}[\eta]$ are represented by

$$\begin{array}{lll}
{}^11: & \eta\,, & \\
{}^11': & \eta'\,, & \\
{}^11'': & \eta''\,, & \\
{}^12={}^12'={}^12'': & -\eta & (-\eta'=1/(1+\eta))\,, \\
{}^13={}^13'={}^13'': & 2\eta^2-\eta-1 & =-\eta^3\eta'\,, \\
{}^14={}^14'={}^14'': & -3\eta^2+3\eta+2 & =-\eta^4\eta'\,, \\
{}^15={}^15'={}^15'': & 305\eta^2-244\eta-169 & =-\eta^{11}\eta'^3\,.
\end{array}$$

Starting from these, a moderate amount of computation produces the higher meshes. For each N-mesh we indicate below a representative clique, the isomorphism type of its stabilizer in the mesh group $\mathcal{M}_N$, and the $(N-1)$-submeshes (with multiplicities) belonging to the subrings $R^{(1)},\dots,R^{(s)}$ from theorem 2; their stabilizer under the aspect action will then be obvious.

^{21}: (η,η^{-1}) . . . . (conjugates ${}^21'$, ${}^21''$)
stabilizer $\cong S_2$ submeshes ^{11} (twice), ${}^11'$ (twice), ${}^11''$

^{22}: $(\eta,1-\eta')$ . . . . (self-conjugate)
trivial stabilizer submeshes ^{11}, ${}^11'$, ${}^11''$, ^{12}, ^{13}

^{23}: (η,η'^{-1}) . . . . (self-conjugate)
trivial stabilizer submeshes ^{11}, ${}^11'$, ${}^11''$, ^{13}, ^{14}

^{31}: $(\eta,\eta^{-1},1-\eta')$ . . . . (conjugates ${}^31'$, ${}^31''$)
stabilizer $\cong S_2$ submeshes ^{21} (twice), ${}^21'$ (twice), ^{22} (twice)

^{32}: $(\eta,\eta^{-1},-\eta'')$ . . . . (self-conjugate)
stabilizer $\cong S_2$ submeshes ^{21} (twice), ${}^21'$ (twice), ${}^21''$ (twice)

^{33}: $(\eta,\eta^{-1},2\eta^2-\eta-1)$ . . . . (conjugates ${}^33'$, ${}^33''$)
stabilizer $\cong S_2$ submeshes ^{21} (twice), ^{22} (twice), ^{23} (twice)

^{41}: $(\eta,\eta^{-1},1-\eta',-\eta'^{-1})$ . . . . (self-conjugate)
stabilizer $\cong S_2$ submeshes ^{31} (twice), $^31'$ (twice), $^31''$ (twice), ^{32}

^{42}: $(\eta,\eta^{-1},1-\eta',-\eta')$ . . . . (self-conjugate)
trivial stabilizer submeshes ^{31}, $^31'$, $^31''$, ^{32}, ^{33}, $^33'$, $^33''$

^{43}: $(\eta,\eta^{-1},-\eta'',-\eta''^{-1})$ . . . . (self-conjugate)
stabilizer $\cong \mathcal{D}_{14}$ submesh ^{32} (7 times)

^{51}: $(\eta,\eta^{-1},1-\eta',-\eta'^{-1},-\eta'')$ . . . . (self-conjugate)
stabilizer $\cong S_2\times S_2$ submeshes ^{41} (4 times), ^{42} (4 times).

Since there are cliques of order 5, we have in fact $M_1=7$, which is best possible as the field has a prime ideal of norm 7, generated e.g. by $\eta-2$. All the 10 080 cliques of maximal length belong to the same 5-mesh ^{51}.—The orbit ^{15} does not take part in any 2-mesh, its six exceptional units are isolated in the graph $\mathrm{E}(\mathbb{Z}_{K^+})$. Similarly, the mesh ^{43} (with trivial aspect action) does not take part in ^{51}; the corresponding exceptional sequences of length 6 are maximal without having maximal length. A quick calculation shows that the 'cyclotomic' sequence constructed in theorem 3 belongs precisely to this maximal mesh. It is no surprise that we cannot show $M_1=p$ with the cyclotomic theory alone, because there are (some) Fermat primes; but here we find that the sequence predicted by the theory is not even extensible to a maximal one. We do not know what happens if $p=11$ or 13; something similar may be expected.

3.2 Conditional exceptional sequences

Most of the Euclidean fields K obtained via LENSTRA's criterion are given by a special integral primitive element x whose minimal polynomial over $\mathbb{Q}$ already guarantees a reasonably (if not sufficiently) long clique of exceptional units in $\mathbb{Z}[x]\subseteq\mathbb{Z}_K$. This works as follows. Choose a finite set Q of irreducible polynomials $h_i\in\mathbb{Z}[X]$. Form the quotient ring of the polynomial ring $\mathbb{Z}[X]$ with respect to the multiplicative semigroup generated by Q. If Q is well chosen, the ring $\mathbb{Z}[X]_Q$ will have large Lenstra constants (e.g. in comparison with $\sum_i\deg(h_i)$). Then, if x is an algebraic integer such that all $h_i(x)$ are algebraic units, the substitution $X\mapsto x$ will extend to a homomorphism of $\mathbb{Z}[X]_Q$ into $\mathbb{Z}[x]$ and by (2.2) will provide lower bounds for the Lenstra constants of the ring of integers of $\mathbb{Q}(x)$. The conditions on x can be checked by computing resultants of the h_i with the minimal polynomial f of x. An exceptional sequence of the function ring $\mathbb{Z}[X]_Q$ intended to be mapped into a number ring in this fashion is called *conditional*, or *exceptional under the conditions 'h_i becomes a unit'*.

We shall discuss three choices for the set Q which have been used in [LM] and [Le] and which will be employed in part 4 for finding new Euclidean fields. To start with, consider again the field $\mathbb{Q}(\zeta_7+\zeta_7^{-1})$ of the preceding example, but this time using the generator $x=-\eta$ with minimal polynomial

$$\Psi_7=g_4-g_3=X^3-X^2-2X+1.$$

Here obviously x, $x-1$, $x+1$, x^2-x-1, $x-2$ are units, which shows that $(x,\ x+1,\ x^2,\ x/(x-1),\ 1/(2-x))$ is a 5-clique in $\mathrm{E}(\mathbb{Z}[x])$ and that $M_1(\mathbb{Z}[x])=7$. Generalizing this, we take $Q = \{X, X-1, X+1, X^2-X-1, X-2\}$ and conclude that the ring $\mathbb{Z}[X]_Q = \mathbb{Z}\{X, X+1, X^2, X/(X-1), 1/(2-X)\}$ (in obvious extension of a notation from part 2; the ring is universal for this type of 5-clique) has Lenstra constants $M_k = 7k$. (The known clique and (2.1) provide the lower bound, while the substitution $X \mapsto -\eta$ shows together with (2.2) and (2.3) that the M_k cannot be any larger.)

Proposition 4. *For the above choice of Q, the set of exceptional units $\mathrm{E}(\mathbb{Z}[X]_Q)$ consists of 16 orbits under the action of $\mathcal{H}$, each of length 6. They are given by the following unit equations in $\mathbb{Z}[X]$:*

$$\begin{array}{rlrl}
-1 + X - (X-1) &= 0 & & \\
1 + X - (X+1) &= 0 & 1 - (X-1) + (X-2) &= 0 \\
-1 + X^2 - (X-1)(X+1) &= 0 & -1 + (X-1)^2 - X(X-2) &= 0 \\
1 + \Theta - X(X-1) &= 0 & & \\
\Theta + X - (X-1)(X+1) &= 0 & \Theta - (X-1) - X(X-2) &= 0 \\
\Theta - X^2 + (X+1) &= 0 & \Theta - (X-1)^2 - (X-2) &= 0 \\
\Theta + X^3 - (X-1)(X+1)^2 &= 0 & \Theta - (X-1)^3 + X(X-2)^2 &= 0 \\
1 + X\Theta - (X-1)^2(X+1) &= 0 & 1 - (X-1)\Theta + X^2(X-2) &= 0 \\
1 - \Theta + (X+1)(X-2) &= 0 & & \\
1 - \Theta^2 + X(X-1)(X+1)(X-2) &= 0, & &
\end{array}$$

where we have used the abbreviation $\Theta = X^2-X-1$.

Proof. This is similar to the proof of proposition 1, and the details will be left to the reader. Observe that in the present case, the substitution $X \mapsto 1-X$ induces an automorphism of $\mathbb{Z}[X]_Q$ and exchanges $X+1$ and $-(X-2)$, so that the equations in the righthand column (which contain $X-2$ but not $X+1$) are immediately reduced to their lefthand counterparts. □

Remark. Using the automorphisms $\mathcal{R}_{(2)}$, $\mathcal{S}_{(2)}$, $\mathcal{Q}_{(3)}$, $\mathcal{R}_{(3)}$ and $\mathcal{Q}_{(1)}\mathcal{R}_{(4)}\mathcal{Q}_{(5)}$ of R_5, one finds that any exceptional unit of $\mathbb{Z}[X]_Q$ apart from those in the $\mathcal{H}$-orbit of Θ^2 occurs in some clique of the mesh

$$\left(X, X+1, X^2, X/(X-1), 1/(2-X)\right)^{\mathcal{M}_6}.$$

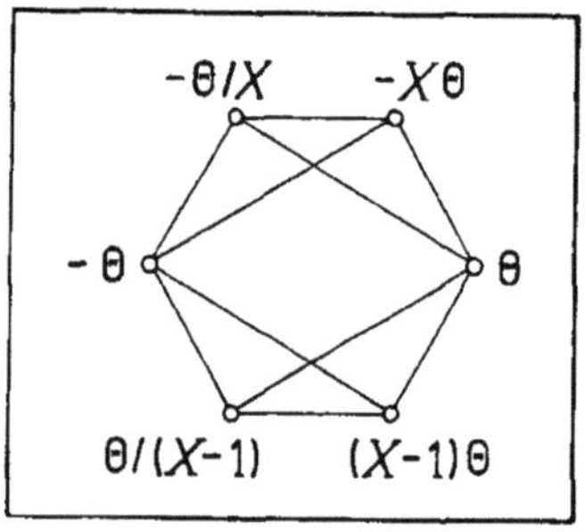

The neighbourhood of Θ^2 in $\mathrm{E}(\mathbb{Z}[X]_Q)$, on the other hand, looks like the hexagon on the right. Its clique number as an induced subgraph is three, so we see that Θ^2 takes part in 4-cliques but in no 5-cliques. □

We shall now discuss two conditional exceptional sequences associated to quotient rings that have $M_2 = 14$ while M_1 is only 6. Both are based on the totally complex quartic field of (minimal) discriminant 117, generated over the third cyclotomic field by a root β of $X^2 + \zeta_3 X - 1$. This is again one of the fields for which NAGELL has determined all exceptional units [N2]; $\mathbb{Z}_K=\mathbb{Z}[\beta]$ contains the $\mathcal{H}$-orbits of β, $\beta+1$ and β^2 of length 6 and the two-element orbit of $\zeta_6=\zeta_3+1$. One finds $M_1(\mathbb{Z}[\beta])=6$, a clique of maximal order being given e.g. by

$$(\beta,\ \beta+1,\ \beta^2,\ (\beta+1)/\beta)\,.$$

Another such clique is found by putting $B=\beta^2-\beta$ which is a root of $X^2-X-\zeta_3$, so B, $B-1$ and B^2-B+1 are units, and

$$(B,\ (B-1)/B,\ 1/(1-B),\ B-B^2)$$

is a 4-clique. As $\mathbb{Q}(\beta)$ has ideals of norm 7, we have $6\leq M_k/k\leq 7$ for all $k\in\mathbb{N}$. In fact, $M_2(\mathbb{Z}[\beta])=14$, and we can extract from each of the two cliques above a function ring with the same property:

Proposition 5. *In each of the following two cases*
(i) $Q=\{X,\ X-1,\ X+1,\ X^2-X-1,\ X^3-X-1\}$,
(ii) $Q=\{X,\ X-1,\ X^2-X+1,\ X^2+1,\ X^3-2X^2+X-1\}$,
the ring $\mathbb{Z}[X]_Q$ *has Lenstra constants* $M_1=6$, $M_2=14$.

Proof. In view of the above, we may write the two function rings involved as $\mathbb{Z}\{X,\ X+1,\ X^2,\ (X+1)/X\}$ and $\mathbb{Z}\{X,\ (X-1)/X,\ 1/(1-X),\ X-X^2\}$. The 4-cliques prove $M_1\geq 6$ in each case, while the substitutions $X\mapsto\beta$ and $X\mapsto B$ respectively show together with (2.2) that $M_1\leq 6$, $M_2\leq 14$. It remains to exhibit for each of the two rings a sequence of length 14 such that among any three of its members at least two are connected. Duplication of a member is thus allowed provided that the duplicate entry is connected with each of the remaining members. In fact, both of the sequences below contain twice 0 and twice 1, and the remaining members are exceptional units. In order to keep the pictures tidy, we have drawn them with arcs indicating *non-connections*, so we have to check that there are no triangles of arcs. A suitable sequence for case *(i)* appears on the left and one for case *(ii)* on the right. □

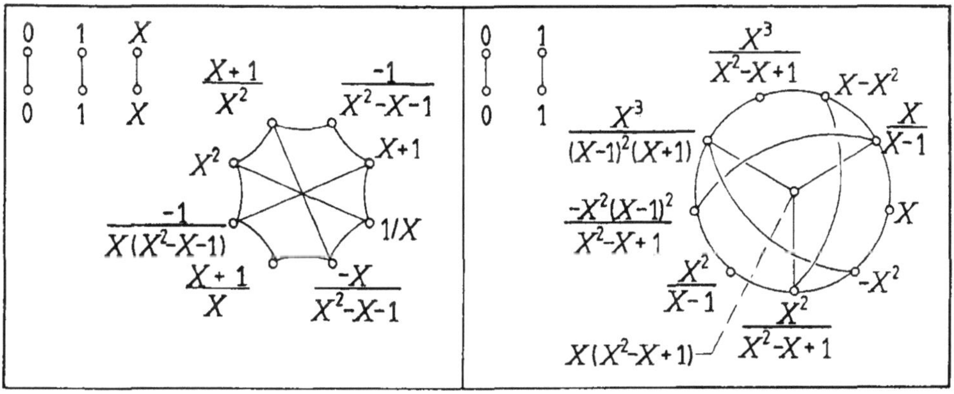

Remark. Arguments similar to those which established propositions 1 and 4 show that $E(\mathbb{Z}[X]_Q)$ consists of 18 or 20 $\mathcal{H}$-orbits respectively in the above two cases. □

3.3 Non-trivial second Lenstra constants in number rings

In proposition 5 we have encountered two function rings for which the second Lenstra constants M_2 are strictly bigger than $2M_1$. The simplest examples of this kind would be rings R with $M_1(R)=2$ and $M_2(R)>4$; in fact we must then have $M_2(R)=5$ since any simple graph of order 6 would contain either a complete or a discrete triangle among its induced subgraphs.

Proposition 6. *Among all quadratic and cubic number fields K of unit rank at most 1 and without exceptional units in their rings of integers $\mathbb{Z}_K$, there are (up to isomorphisms) precisely three fields having $M_2(\mathbb{Z}_K)=5$. These are the real quadratic field of discriminant 13, generated by a root of X^2-3X-1, the complex cubic field of discriminant -87, generated by X^3-2X^2-X-1, and the complex cubic field of discriminant -135, generated by X^3-3X^2-1.*

Proof. If $M_1(R)=2$, there is only one possibility for the induced subgraph of $\Gamma(R)$ underlying an M_2-sequence of length 5, viz. a pentagon (without diagonals). We are looking for number rings $R=\mathbb{Z}_K$ containing such a subgraph but no exceptional units. From NAGELL's work [N1] we know that the only fields of degree ≤ 3 and unit rank ≤ 1 containing exceptional units are the quadratic fields of discriminants -3 and 5 and the complex cubic fields of discriminants -23 and -31. By translation and unit group action, we may assume that our sequence contains 0 adjacent to 1, while the other members are of the forms u_1 (adjacent to 0), $1-u_2$ (adjacent to 1) and $1-u_2-u_3 = u_1+u_4$ (closing the cycle), with units u_i. Thus we have to solve the unit equation

$$u_1 + u_2 + u_3 + u_4 = 1 \qquad (3.2).$$

in $u_i \in \mathbb{Z}_K^\times$. If K is complex quadratic of discriminant <-3, there are no solutions since the lefthand side always lies in a prime ideal above 2. This leaves us with fields of unit rank 1. Furthermore, K must not possess a prime ideal of norm 2. Fix (one of) the real place(s) of K and let x be a fundamental unit >1 at that place. Expressing the u_i in terms of x, we may write the unit equation (3.2) as

$$x^{n_0} + \varepsilon_1 x^{n_1} + \varepsilon_2 x^{n_2} + \varepsilon_3 x^{n_3} + \varepsilon_4 = 0 \qquad (3.3)$$

with $\varepsilon_i \in \{\pm 1\}$ and, without loss of generality, decreasing exponents $n_0 \geq n_1 \geq n_2 \geq n_3 \geq 0$. No two summands cancel, or the remaining equation would give rise to an exceptional unit. From this we infer at our distinguished real place

$$x^{n_0} \leq 3x^{n_0-1} + 1 .$$

The lefthand side of (3.3), read as a polynomial in x, must be a multiple of the minimal polynomial of x over $\mathbb{Q}$. Therefore

$$n_0 \geq n = [K:\mathbb{Q}] .$$

Now if x were to exceed the real root of X^2-3X-1 $(n=2)$ or of X^3-3X^2-1 $(n=3)$, the last two inequalities would contradict each other. Thus for real quadratic fields, the trace of x must be 1, 2, or 3, with $N_{\mathbb{Q}(x)|\mathbb{Q}}(x)=-1$ in the former two cases; of the four minimal polynomials thus obtained, two have discriminant 5 and one (belonging to $x=1+\sqrt{2}$)

has 2 among its values.—If $n=3$, the minimal polynomial $f = X^3-sX^2+tX-1$ of x has one real zero >1, $f(\pm 1)$ can be neither -1 (or x would be exceptional) nor -2, so they must be ≤ -3, and the trace of x cannot exceed 4; this gives us

$$1-s \leq t \leq s-3 \leq 1 .$$

Nine polynomials are obtained. Three of them take values divisible by 2 but not by 8, while four of the remaining ones violate the upper bound for their real roots. We are left with the minimal polynomials mentioned in the proposition, from which the solutions of the unit equation (3.2) can easily be read off. □

Remark 1. The missing case of (totally complex) quartic fields with unit rank 1 can be handled in the same manner. Of course the fields containing $\mathbb{Q}(\sqrt{13})$ as a subfield yield trivial solutions (they have $M_1=2$ except for $\mathbb{Q}(\sqrt{13},\zeta_3)$), but apart from these, there are again only finitely many candidates. We omit the discussion of all possible cases and confine ourselves to listing four examples:

$$X^4-X^3+X^2+X+1 = N_{\mathbb{Q}(\sqrt{-11})|\mathbb{Q}}(X^2-\tfrac{1}{2}(1+\sqrt{-11})X-1), \qquad D_K=5\cdot 11^2=605$$

$$X^2-(1+2\zeta_4)X+1, \qquad D_K = 2^4\cdot 5\cdot 13 = 1040$$

$$X^2-(1-2\zeta_4)X+\zeta_4, \qquad D_K = 2^4\cdot 73 = 1168$$

$$X^2-3X+\zeta_4, \qquad D_K = 2^4\cdot 97 = 1552.$$

Remark 2. There are infinitely many abelian and non-abelian totally real cubic number fields K with $M_1(\mathbb{Z}_K)>2$, cf. [N3]. NAGELL ([N3], Théorème 4) showed that for a non-abelian one to contain exceptional units it is necessary that all primes dividing its discriminant D_K are $\equiv \pm 1 \pmod 8$. (An even sharper statement was proved in [LM], theorem 4.3.1.) This fact allows us to construct an infinite sequence of non-abelian real cubic fields with $M_1=2$, $M_2=5$, each generated by a zero x of

$$F = F_s(X) = X^3 - sX^2 + (3s-9)X + 1 \quad (s \in \mathbb{Z}) \qquad (3.4).$$

The discriminant of F

$$\Delta = \Delta(s) = 9s^4 - 158s^3 + 999s^3 - 2754s^2 + 2889$$

is positive unless $s\in\{5,6,7\}$. From $F(0)=F(3)=1$ and the sequence (0, 1, 2, 3, x) one gets a pentagon subgraph in $\mathbb{Z}_K$, which proves $M_2\geq 5$. Choosing $s=3t$, $t\not\equiv 1 \pmod 3$, one finds that

$$\Delta(s) \equiv 3^3(107+t^3) \not\equiv 0 \pmod{3^4}$$

is divisible by 3 to an odd power. Therefore D_K is a multiple of 3 and not a square, K is non-abelian and $\mathbb{Z}_K$ contains no exceptional units.

According to SIEGEL ([S1], Satz 7, Zusatz 1), the polynomial $X(X-3)$ takes unit values for only finitely many algebraic integers x in any given number field. Thus only finitely many choices of s in (3.4) can lead to roots x generating the same field. Letting s run through the integers $\equiv 0,-3 \pmod 9$, we obtain infinitely many fields $K=\mathbb{Q}(x)$ with non-trivial second Lenstra constants $M_2(\mathbb{Z}_K)$. □

4. New Euclidean fields

Here we are going to present 37 new Euclidean fields K discovered with the aid of LENSTRA's criterion. They are listed in table 1 at the end of this section. Each of them is defined by a primitive integral element x (in a fixed algebraic closure $\overline{\mathbb{Q}}$ of $\mathbb{Q}$) whose minimal polynomial f has been chosen so as to guarantee a nontrivial clique in $\mathrm{E}(\mathbb{Z}_K)$. If this does not yield a sufficiently large lower bound on M_1 or M_2, a longer sequence will be given explicitly. In six cases we have been unable to reach a large enough M_1, and had to make use of the second Lenstra constants.

Table 2, containing the numbers of known Euclidean fields per signature, is an update of table 11 of [L1], table 5 of [LM] and table 5 of [Le].

We conclude this article with a Hasse diagram of 138 (discriminants of) Euclidean fields, modelled on table 2 of [LM]. The degree n increases from left to right, and within each column the number r_1 of real embeddings grows from top to bottom. The lines indicate inclusion relations (from left to right). In some cases field generators are given; see the table in the bottom right corner of the diagram for the defining polynomials. The diagram includes all those fields of degree ≤ 4 and all those of their extensions for which the Lenstra constants are known to be large enough to guarantee a Euclidean algorithm. 42 of the fields listed were found after LENSTRA's paper [L1] had appeared.

4.1 Initial pieces of cliques

Let f and g be monic irreducible polynomials in $\mathbb{Z}[X]$ with respective zeroes x and y in $\overline{\mathbb{Q}}$. $g(x)$ is a unit in $\mathbb{Z}[x]$ iff $f(y)$ is a unit in $\mathbb{Z}[y]$ iff the resultant of f and g over $\mathbb{Z}$ is ± 1. For brevity we shall write $U(f)$ for the set of all integral $y \in \overline{\mathbb{Q}}$ such that $f(y)$ is a unit. For instance, if $0, 1, -1, \vartheta, \alpha \in U(f)$ (where $\theta^2-\theta-1=0$, $\alpha^3-\alpha-1=0$), a conditional sequence (see section 3.2 above) gives us the 4-clique of exceptional units

$$(x,\ x+1,\ x^2,\ (x+1)/x)$$

in $\mathbb{Z}[x]$. We can then try to extend the clique with further exceptional units to a longer one. Exceptional sequences starting with 0, 1 and this clique are said to be 'in the B-line' (a classification from [LM], also used in [Le]). Similarly,

$$(x,\ (x-1)/x,\ 1/(1-x),\ x-x^2)$$

is a 4-clique in $\mathrm{E}(\mathbb{Z}_K)$ provided that $0, 1, \zeta_6, \zeta_4, \alpha^2 \in U(f)$. Sequences with this initial piece belong to the 'D-line'. Remember that by proposition 5 we have $M_2(\mathbb{Z}_K) \geq 14$ in both cases.—Some of the sequences we have been using require the B-line or D-line conditions on $U(f)$ but depart from the above initial pieces.

4.2 Comments on table 1

For each signature appearing in the table, we give the relevant Lenstra bounds $\alpha(n)$ and/or $\alpha(r_1, r_2)$. The entries per field include the field discriminant and its prime factorization, the minimal polynomial f of a primitive element x over $\mathbb{Q}$ or over a proper subfield, and information on $U(f)$ and/or an exceptional sequence as required.

Five cases require further explanations. In all the diagrams below, arcs represent edges *missing* in $\Gamma(\mathbb{Z}_K)$, and the absence of triangles in the diagrams shows that M_2 is at least as large as the number of vertices.

4.2.1. The quintic field of discriminant $D=4897$ with one real embedding is generated by a zero x of

$$f = X^5 + X^4 - X^3 - X^2 - 1 .$$

In [L1] it remained open whether it is Euclidean or not; the bound $\alpha(5)\sqrt{D}$ is slightly larger than 4, while the conditions $0, 1, -1 \in U(f)$ allow only to conclude $M_1 \geq 4$. But obviously $x^4 - x^2 - 1$ is a unit, i.e. $\sqrt{\vartheta} \in U(f)$. This gives rise to the sequence shown to the left proving $M_2 \geq 9$, which is sufficient since $2\alpha(1,2)\sqrt{D} \leq 8.72$. Among the members of our sequence there is one connection that cannot be inferred from $0, \pm 1, \sqrt{\vartheta} \in U(f)$ alone, nor is it necessary for reaching the required value of M_2; it is attached to $-1/\gamma \in U(f)$ $(\gamma^3 + \gamma - 1 = 0)$, i.e. to the fact that $x^3 + x^2 + 1$ is a unit. We have drawn this edge as a dashed line. (The other diagonal would belong to the condition $1/\gamma \in U(f)$, which is not satisfied here.) □

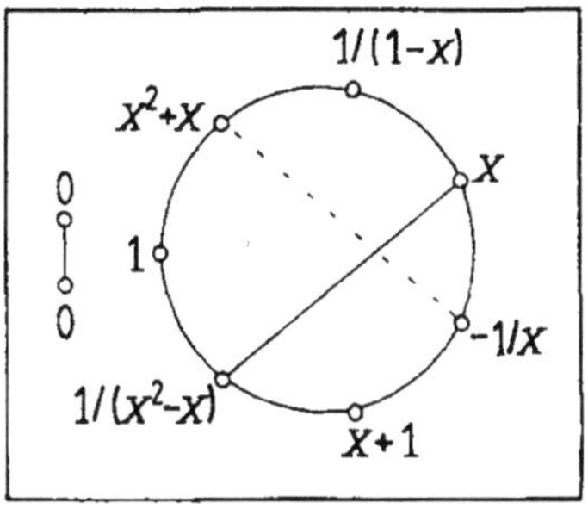

4.2.2. The sextic field of discriminant $D=50\,933$ with two real places is the ray class field over $\mathbb{Q}(\gamma)$ with conductor the prime $(3\gamma+2)$ of norm 53, where $\gamma^3 + \gamma - 1 = 0$. It can be defined by the polynomial

$$f = X^2 - \gamma^4 X - 1$$

of norm $F = X^6 - 2X^5 + 2X^4 + 3X^3 - 2X^2 - 2X - 1$. We need $M_1 \geq 6$. By substituting into f and taking norms from $\mathbb{Q}(\gamma)$ to $\mathbb{Q}$, one verifies $0, \pm 1, \gamma, -\gamma^2, \gamma^4, \gamma+1, \gamma^{-2} \in U(f)$ (not all of these are in $U(F)$!) and obtains the exceptional sequence

$$0,\ 1,\ \gamma(x-\gamma),\ 1/(\gamma(x-\gamma)),\ \gamma^2 x,\ x(\gamma-1)/(x-\gamma),$$

x being a zero of f. □

4.2.3. The sextic field K of discriminant $D=-174\,875$ with four real places is generated over $\mathbb{Q}(\vartheta)$ by a zero x of the polynomial $f = X^3 - \vartheta^2 X^2 - X + \vartheta$ of norm

$$F = X^6 - 3X^5 - X^4 + 4X^3 + 2X^2 - X - 1 .$$

The conditions $0, \pm 1, \vartheta, \alpha, \sqrt{2}, -1/\delta \in U(F)$ $(\delta^4 - \delta - 1 = 0)$ show by virtue of the following clique (cf. [LM], section 3.3)

$$(x,\ x+1,\ x^2,\ x^2/(x^2-1),\ -1/(x^2-x-1),\ (x+1)/x)$$

that $M_1(\mathbb{Z}_K) \geq 8$, which is not enough to apply LENSTRA's method. We obtain three more cliques by applying the mesh transformations $S_{(1)}$, $\mathcal{R}_{(1)}\mathcal{I}$ and $S_{(1)}\mathcal{Q}_{(6)}$ of Aut R_6

to the one above. Using the connections in $E(\mathbb{Z}_K)$ thus established and checking six further pairs of vertices, we obtain the sequence shown in the diagram and find $M_2 \geq 17$. This proves K to be Euclidean. □

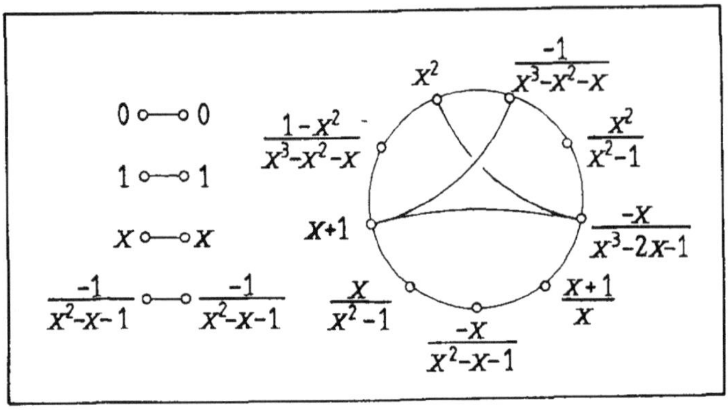

4.2.4. The Galois closure K of the totally complex quartic field $\mathbb{Q}(\xi_1)$ of discriminant 189 may be defined by the polynomial

$$F = X^2 + (\zeta_6\xi_1 - 1)X - \xi_1 ,$$

where $\xi_1^2 - \xi_1 - \zeta_6 = 0$. This can be verified in a completely elementary way. Firstly, the discriminant $\Delta_F = (3-\zeta_6)\xi_1$ differs from that of F', the image of F under the nontrivial automorphism $\xi_1 \mapsto \xi_1'$ of $\mathbb{Q}(\xi_1)|\mathbb{Q}(\zeta_6)$, by a factor $\xi_1'/\xi_1 = -\zeta_6/\xi_1^2$, which is a square in $\mathbb{Q}(\xi_1)$. Secondly, also the image $\bar{G}$ of the product

$$G = FF' = X^4 + (\zeta_6-2)X^3 + (1-\zeta_6)X^2 + (2\zeta_6-1)X - \zeta_6$$

under the automorphism $\zeta_6 \mapsto \bar{\zeta}_6 = \zeta_6^{-1}$ of $\mathbb{Q}(\zeta_6)$ splits over K because multiplication by ζ_6 sends the zeroes of G to those of $\bar{G}$.—The fact that this field is Euclidean was mentioned already by DIAZ Y DIAZ [D3]. □

4.2.5. The octic field K with four real places and discriminant 15 243 125 is the ray class field over $\mathbb{Q}(\vartheta)$ with conductor the product of a prime of norm 29 and one infinite prime. The intermediate field of the cyclic extension is the totally real quartic field of minimal discriminant 725, generated by σ, a root of $X^2 - \vartheta^{-1}X - 1$; a generator x of K has minimal polynomial $X^2 - (1+\vartheta^2\sigma)X + (1+\vartheta^2\sigma)$ over $\mathbb{Q}(\sigma)$. An automorphism of order 4 is given by $x \mapsto x' = 1+\vartheta-\sigma x$, $\sigma \mapsto \sigma' = -1/\sigma$.

The field $\mathbb{Q}(\sigma)$ probably has $M_1(\mathbb{Z}[\sigma]) = 10$, although the ideals of norm 11 would leave room for longer exceptional sequences. Here is an 8-clique made up from pairs of inverses:

$$\left(-\vartheta,\ -1/\vartheta,\ \sigma,\ 1/\sigma,\ 1+\vartheta^2\sigma,\ \sigma^2-\vartheta,\ \sigma/(1-\sigma),\ (1-\sigma)/\sigma\right).$$

The ray class field has $\alpha(4,2)\sqrt{D} \leq 15.22$, so we have to reach $M_1 \geq 16$ or $M_2 \geq 31$. We shall show that the latter inequality holds, exploiting some 13-cliques of exceptional units. First we need a few more elements of $E(\mathbb{Z}_K)$. We shall write them in terms of the integral base $(1, \vartheta, \sigma, \vartheta\sigma; x, \vartheta x, \sigma x, \vartheta\sigma x)$:

$$\begin{aligned}
1-x' &= (0,-1,\ 0,\ 0;\ 0,\ 0,\ 1,\ 0),\\
y &= (1,\ 0,\ 1,\ 1;-1,\ 0,\ 1,-1),\\
y'' &= (0,-1,\ 0,-1;\ 1,\ 0,-1,\ 1),\\
1/y &= (0,-1,\ 1,\ 1;\ 1,\ 1,-1,-1),\\
z &= (1,\ 0,\ 1,\ 0;\ 0,\ 0,-1,\ 0),\\
1/z &= (1,\ 1,\ 1,\ 1;\ 0,\ 0,\ 0,-1),\\
1/z'' &= (0,-1,\ 0,-1;\ 0,\ 0,\ 0,\ 1),\\
v &= (1,\ 0,\ 1,\ 1;-1,\ 0,-1,\ 0),\\
w &= (1,\ 1,\ 1,\ 2;\ 0,-1,-1,-1).
\end{aligned}$$

The elements $1-x'$, y, $1/y$, z, $1/z''$ combine with 0, 1 and the 8-clique in the intermediate field to an exceptional sequence of length 15. The same is true if we replace y, $1/z''$ with their conjugates y'', $1/z$. Another couple of sequences of the same length is obtained by taking 0, 1 and the first seven members of the above 8-clique, followed by $1-x'$, y, $1/y$, z, $1/z''$, v or by $1-x'$, y'', $1/y$, z, $1/z$, w. Observe, finally, that v and w are connected. We get the sequence of length 31 shown in the diagram; two more connections exist among its members (drawn again as dashed lines), but already without them we may conclude that K is Euclidean. □

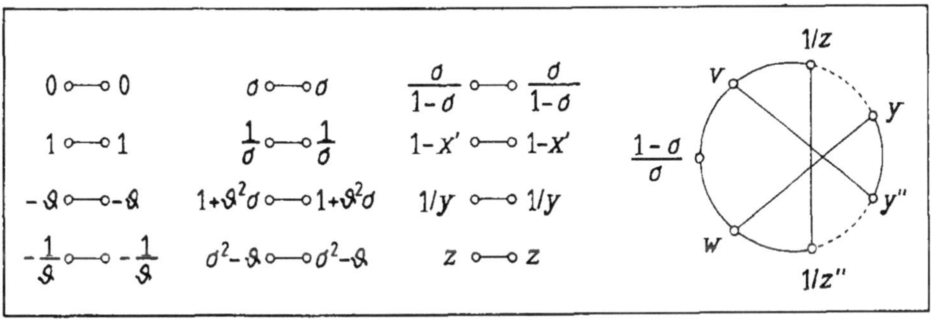

Table 1. New Euclidean number fields

$n = 5 \quad r_1 = 1$	$\alpha(n) \le 0.057\,170\,26 \qquad \alpha(r_1, r_2) \le 0.062\,251\,736$
$D = 4\,897 = 59 \cdot 83$	$f = X^5 + X^4 - X^3 - X^2 - 1$
$0, \pm 1, \sqrt{\vartheta} \in U(f) \Longrightarrow M_2 \ge 9$	(see 4.2.1 and [Ni] 10.14)

$n=6 \quad r_1=2$	$\alpha(n) \le 0.024\,037\,20 \qquad \alpha(r_1,r_2) \le 0.025\,017\,577$
$D = 50\,933 = 31^2 \cdot 53$	$f = X^2 - \gamma^4 X - 1 \qquad (\gamma^3+\gamma-1=0)$
$0, \pm 1, \gamma, -\gamma^2, \gamma^4, \gamma+1, \gamma^{-2} \in U(f) \Longrightarrow M_1 \ge 6$	(see 4.2.2)
$D = 63\,461 = 17 \cdot 3733$	$f = X^6 + X^3 - 2X - 1$
B-line $\Longrightarrow M_2 \ge 14$ (13 would already suffice)	
$D = 68\,389$ prime	$f = X^6 - 3X^5 + X^4 + 4X^3 - X^2 - 2X - 1$
B-line $\Longrightarrow M_2 \ge 14$	
$D = 75\,749 = 211 \cdot 359$	$f = X^6 + 2X^5 - 2X^4 - 3X^3 + 2X + 1$
B-line $\Longrightarrow M_2 \ge 14$	

$n=6 \quad r_1=4$	$\alpha(r_1,r_2) \le 0.019\,648\,758\,5$
$D = -174\,875 = -5^3 \cdot 1399$	$f = X^3 - \vartheta^2 X^2 - X + \vartheta$
$M_2 \ge 17$	(see 4.2.3)

$n=7 \quad r_1=3$	$\alpha(n) \le 0.009\,847\,38$
$D = 919\,969$ prime	$f = X^7 + X^6 - 3X^5 - X^4 + 3X^3 + X^2 - 2X - 1$
(B-line) sequ.: $0,\ 1,\ x,\ x+1,\ (x^3+x^2-x-1)/x,\ (x^3+x^2-x-1)/x^2,$ $(x^3+x^2-x-1)/x^3,\ (x^4+x^3-2x^2-x)/(x^3-x-1),\ (x^4-x^2-x)/(x^2-x-1),$ $(x^3-x-1)/(x^2-x-1) \Longrightarrow M_1 \ge 10$	

$n=8 \quad r_1=0$	$\alpha(n) \le 0.003\,954\,3$
$D = 1\,750\,329 = 3^6 \cdot 7^4$	$f = X^2 + (\zeta_6 \xi_1 - 1)X - \xi_1 \qquad (\xi_1^2 - \xi_1 - \zeta_6 = 0)$
sequ.: $0,\ 1,\ x,\ (x-1)/x,\ 1/(1-x),\ \zeta_6 \Longrightarrow M_1 \ge 6$	(see 4.2.4 and [D3])
$D = 2\,970\,513 = 3^4 \cdot 7 \cdot 13^2 \cdot 31$	$f = X^2 - X + B \qquad (B^2 - B - \zeta_3 = 0)$
sequ.: $0,\ 1,\ x,\ 1/(1-x),\ x^2/(x-1),\ 1/(x^2-2x+2),\ x/(x^4-2x^3+2x^2-x+1) \Longrightarrow M_1 \ge 7$	

$n=8 \quad r_1=2$	$\alpha(n) \le 0.003\,954\,3 \qquad f = \sum a_i X^i$
$D = -5\,293\,867 = -227 \cdot 23321$	$a_0, \ldots = 1, -3, 5, -7, 5, -4, 2, -1, 1$
D-line sequ.: $0,\ 1,\ x,\ (x-1)/x,\ 1/(1-x),\ x-x^2,\ x^2/(x-1),\ (x^3-x^2+2x-1)/x,$ $x^3/(x^3-x^2+x-1),\ (2x^3-2x^2+2x-1)/(x^3-x^2+x-1) \Longrightarrow M_1 \ge 10$	
$D = -6\,129\,779 = -43 \cdot 142553$	$a_0, \ldots = -1, -1, 2, -4, -1, 7, -1, -3, 1$
B-line sequ.: $0,\ 1,\ x,\ x+1,\ x^2,\ (x+1)/x,\ -1/(x^2-x-1),\ x^2/(x^2-1),\ x/(x^2-1),$ $(x^4-2x^2)/(x^4-x^3-1) \Longrightarrow M_1 \ge 10$	
$D = -6\,181\,019 = -13 \cdot 53 \cdot 8971$	$a_0, \ldots = -1, -4, 0, 3, 3, 1, -3, -1, 1$
(B-line) sequ.: $0,\ 1,\ x,\ x+1,\ (x+1)/x,\ -1/(x^2-x-1),\ x/(x^2-1),$ $(x^3+x^2-2x-1)/(x^3-x-1),\ (x^3-2x)/(x^3-x-1),\ -x^3+2x+1 \Longrightarrow M_1 \ge 10$	

$D = -6\,242\,419 = -1033 \cdot 6043$ $\quad a_0, \ldots = 1, 1, 0, 2, 0, -3, -2, 1, 1$

B-line sequ.: $0,\ 1,\ x,\ x+1,\ x^2,\ (x+1)/x,\ (x+1)/x^2,\ (x^3+x^2-x-1)/x,$
$x^3+x^2-x-1,\ x(x^3+x^2-x-1) \Longrightarrow M_1 \geq 10$

$D = -6\,284\,899 = -53 \cdot 118583$ $\quad a_0, \ldots = -1, 0, 2, -7, 11, -11, 8, -4, 1$

D-line sequ.: $0,\ 1,\ x,\ (x-1)/x,\ 1/(1-x),\ x-x^2,\ 1/(x^2-x+1),\ (x^3-x^2+x-1)/x^2,$
$(x^3-x^2)/(x^2-x+1),\ x/(x^4-2x^3+2x^2-x+1) \Longrightarrow M_1 \geq 10$

$D = -6\,397\,819 = -661 \cdot 9679$ $\quad a_0, \ldots = -1, 2, -4, 3, 0, -3, 4, -3, 1$

D-line sequ.: $0,\ 1,\ x,\ (x-1)/x,\ 1/(1-x),\ x-x^2,\ -1/x,\ -1/x^2,$
$(x^3-2x^2+x-1)/(x^3-x^2+x-1),\ (-x^4+2x^3-2x^2+2x-1)/x,\ (x^3-2x^2+x-1)/(x^3-x^2)$
$\Longrightarrow M_1 \geq 11$

$n = 8 \quad r_1 = 4$ | $\alpha(r_1, r_2) \leq 0.003\,896\,017\,2$

$D = 15\,908\,237 = 43 \cdot 369959$ $\quad f = X^8 - 4X^6 + 3X^4 + X^3 + X^2 - 1$

B-line sequ.: $0,\ 1,\ x,\ x+1,\ x^2,\ (x+1)/x,\ -1/(x^2-x-1),\ -1/(x^3-x^2-x),$
$x^2/(x^2-1),\ x/(x^2-1),\ (x^2+x)/(x^2+x-1),\ -x^2/(x^4-3x^2+1),\ -x/(x^4-3x^2+1),$
$x^4-x^2,\ (x^4-x^2-x-1)/(x^2-x-1),\ -(x+1)/(x^4+x^3-2x^2-2x-1) \Longrightarrow M_1 \geq 16$

$D = 15\,243\,125 = 5^4 \cdot 29^3$ $\quad f = X^2 - (1+\vartheta^2\sigma)X + (1+\vartheta^2\sigma) \quad (\sigma^2-\vartheta^{-1}\sigma-1=0)$

$M_2 \geq 31$ (see 4.2.5 and [Ni] 10.13)

$n = 9 \quad r_1 = 1$ | $\alpha(n) \leq 0.001\,563 \quad f = \sum a_i X^i$

$D = 37\,732\,753$ prime $\quad a_0, \ldots = 1, -1, 3, -3, 4, -4, 2, -1, -1, 1$

D-line sequ.: $0,\ 1,\ x,\ (x-1)/x,\ 1/(1-x),\ x-x^2,\ -x^3+x^2+1,\ (x^3-x^2)/(x^3-x^2-1),$
$(x^4-x^3-1)/(x^4-x^2+x-1),\ (x^4-x^3+x^2-x)/(x^4-x^3-1) \Longrightarrow M_1 \geq 10$

$D = 39\,388\,441 = 1093 \cdot 36037$ $\quad a_0, \ldots = -1, 0, -2, -4, 5, 5, -2, -2, -1, 1$

(*B*-line) sequ.: $0,\ 1,\ x,\ (x+1)/x,\ (x+1)/x^2,\ x^2-1,\ 1/(x^3-x^2),\ -x^3+2x+1,$
$(x^3+x^2-x-1)/x^2,\ (-x^3+x^2+x)/(x-1) \Longrightarrow M_1 \geq 10$

$D = 40\,643\,353 = 37 \cdot 1098469$ $\quad a_0, \ldots = -1, 4, -11, 20, -28, 29, -24, 15, -6, 1$

D-line sequ.: $0,\ 1,\ x,\ (x-1)/x,\ 1/(1-x),\ x-x^2,\ (x^2-x+1)/(x^2-x),$
$(-x^4+2x^3-x^2+x)/(x^2-x+1),\ (x^4-2x^3+2x^2-x+1)/(x^4-2x^3+2x^2-2x+1),$
$(x^4-3x^3+4x^2-2x+1)/(x^4-2x^3+x^2) \Longrightarrow M_1 \geq 10$

$D = 40\,756\,753 = 4523 \cdot 9011$ $\quad a_0, \ldots = -1, -2, -2, 1, 3, 4, -1, -4, 0, 1$

(*B*-line) sequ.: $0,\ 1,\ x,\ x+1,\ -x/(x^2-x-1),\ x^2/(x^2-1),\ 1/(x^2-x),$
$(x^3-x)/(x^3-x-1),\ (x^3-x-1)/(x-1),\ -x^3+2x+1 \Longrightarrow M_1 \geq 10$

$D = 41\,364\,413$ prime $\quad a_0, \ldots = 1, 0, -3, -1, 4, 4, -1, -4, 0, 1$

B-line sequ.: $0,\ 1,\ x,\ x+1,\ x^2,\ (x+1)/x,\ x^2+x,\ -1/(x^2-x-1),\ -1/(x^3-x^2-x),$
$(x^3-x-1)/(x^3-2x),\ (x^4-x^3-x^2)/(x^4-2x^2-x+1) \Longrightarrow M_1 \geq 11$

$D = 42\,818\,653 = 1873 \cdot 22861$ $\qquad a_0, \ldots = -1, 4, -9, 16, -20, 20, -16, 10, -4, 1$

(D-line) sequ.: $0,\ 1,\ x,\ (x-1)/x,\ 1/(1-x),\ (x^3-x^2)/(x^2-x+1),\ x^3-x^2+x,$
$(x^3-x^2+2x-1)/x,\ (x^3-x^2+x)/(x^2+1),\ (-x+1)/(x^2-x+1),$
$(x^4-x^3+x^2-x+1)/(x^2-x+1) \Longrightarrow M_1 \geq 11$

$D = 42\,934\,933 = 3067 \cdot 13999$ $\qquad a_0, \ldots = 1, -2, 4, -4, 4, -2, 0, 1, -2, 1$

D-line sequ.: $0,\ 1,\ x,\ (x-1)/x,\ 1/(1-x),\ x-x^2,\ (x^3-x^2+x-1)/x^2,$
$(-x^4+x^3+x)/(x^2+1),\ (-x^4+x^3+1)/(x^2-x+1),\ (x^4-x^3-1)/(x^3-x^2+x-1),$
$(x^4-x^3-1)/(x^4-x^2+x-1) \Longrightarrow M_1 \geq 11$

$D = 43\,187\,801 = 41 \cdot 1053361$ $\qquad a_0, \ldots = -1, 3, -4, 5, -6, 6, -7, 5, -3, 1$

D-line sequ.: $0,\ 1,\ x,\ (x-1)/x,\ 1/(1-x),\ x-x^2,\ x^2/(x-1),\ -x^3+x^2-x+1,$
$-x/(x^3-x^2+x-1),\ -x^2/(x^3-2x^2+x-1),\ (-x^3+x^2-2x+1)/(x^4-2x^3+2x^2-2x)$
$\Longrightarrow M_1 \geq 11$

$D = 43\,302\,353 = 23^3 \cdot 3559$ $\qquad a_0, \ldots = -1, 1, \alpha^2-\alpha-2, 1$

(D-line) sequ.: $0,\ 1,\ x,\ (x-1)/x,\ 1/(x^2-x+1),\ (x^2-x+1)/x^2,$
$(-x^2+x-1)/(x^3-2x^2+x-1),\ (x^3-x^2+x)/(x^2+1),$
$(-x^3+2x^2-x+1)/(x^4-2x^3+2x^2-x+1),\ (x^4-x^3-1)/(x^3-2x^2+x-1),$
$(-2x^3+3x^2-2x+2)/(x^4-3x^3+3x^2-2x+1) \Longrightarrow M_1 \geq 11$

$D = 43\,798\,753 = 31 \cdot 1412863$ $\qquad a_0, \ldots = -1, 0, 2, -2, -3, 2, 3, -1, -2, 1$

(B-line) sequ.: $0,\ 1,\ x,\ x^2,\ (x+1)/x,\ x/(x^2-1),\ x^2/(x^2-1),\ x^3/(x^2-1),$
$x^2/(x^3-x^2+1),\ x^3/(x^3-x^2+1),\ x^4/(x^3+x^2-x-1) \Longrightarrow M_1 \geq 11$

$D = 45\,007\,129$ prime $\qquad a_0, \ldots = 1, 3, -1, -7, 0, 9, 0, -5, 0, 1$

(B-line) sequ.: $0,\ 1,\ x,\ x+1,\ x^2,\ x^3-x,\ (x^3+x^2-x-1)/x,\ (x^3+x^2-x-1)/x^2,$
$(x^3+x^2-2x-1)/(x^2-1),\ (x^4+x^3-x^2-x)/(x^4-2x^2+x+1),$
$(x^4+x^3-x^2-2x-1)/(x^3+x^2-2x-1) \Longrightarrow M_1 \geq 11$

$D = 49\,358\,801 = 101 \cdot 488701$ $\qquad a_0, \ldots = 1, 1, -2, 0, 1, 1, 1, -2, -1, 1$

(B-line) sequ.: $0,\ 1,\ x,\ (x+1)/x^2,\ (x+1)/x^3,\ x^2-1,\ 1/(x^2-1),\ 1/(x^3-x^2),$
$(x^3+x^2-x-1)/x^2,\ -1/(x^4-x^3-x^2),\ -x^3+2x+1 \Longrightarrow M_1 \geq 11$

$n = 10 \quad r_1 = 0$ $\qquad \alpha(n) \leq 0.000\,609\,7 \qquad f = \sum a_i X^i$

$D = -240\,232\,739 = -467 \cdot 514417$ $\qquad a_0, \ldots = 1, -1, 2, -3, 7, -12, 15, -15, 11, -5, 1$

(D-line) sequ.: $0,\ 1,\ x,\ x^2/(x-1),\ x^2+1,\ x^3-x^2+x,\ -x^2/(x^3-2x^2+x-1),$
$(-x^2+x-1)/(x^3-2x^2+x-1),\ -x/(x^3-2x^2+x-1),\ (x^3-x^2)/(x^3-x^2-1)$
$\Longrightarrow M_1 \geq 10$

$D = -246\,944\,619 = -3^5 \cdot 73 \cdot 13921$ $\qquad a_0, \ldots = \zeta_6^2, -\zeta_6, \zeta_6, \zeta_6^2, -\zeta_6, 1$

(D-line) sequ.: $0,\ 1,\ x,\ (x-1)/x,\ x-x^2,\ (x^3-x^2)/(x^2-x+1),\ x+1,\ -1/x^2,$
$x/(x^2+1),\ (-x^4+x^3+1)/(x^2-x+1) \Longrightarrow M_1 \geq 10$

$D=-278\,645\,219=-317\cdot 879007$ $\quad a_0,\ldots = 1, 2, 1, -4, -6, 3, 9, -1, -5, 0, 1$

B-line sequ.: $0,\ 1,\ x,\ x+1,\ x^2,\ (x+1)/x,\ x^2/(x^2-1),\ (x^3-x)/(x^3-x-1),$
$-1/(x^3-x^2-x),\ (x^3-x-1)/(x^3-2x),\ x^4-x^3-x^2+x+1 \Longrightarrow M_1 \geq 11$

$D=-282\,748\,447$ prime $\quad a_0,\ldots = 1, 2, 0, -3, 0, 0, 0, 3, -1, -2, 1$

(B-line) sequ.: $0,\ 1,\ x,\ x^2,\ (x+1)/x,\ x/(x^2-1),\ x^2/(x^2-1),\ 1/(x^2-x),$
$(x^3-x^2)/(x^3-x-1),\ -1/(x^3-x^2-x),\ x^4-x^3-x^2+x+1 \Longrightarrow M_1 \geq 11$

$D=-291\,458\,939=-197\cdot 1479487$ $\quad a_0,\ldots = 1, 0, -2, 4, 5, -9, -4, 8, 0, -3, 1$

(B-line) sequ.: $0,\ 1,\ x,\ (x+1)/x,\ (x+1)/x^2,\ x/(x^2-1),\ x^2/(x^2-1),\ -1/(x^3-x^2-x),$
$(x^4-x^2)/(x^4-x^3-x^2+x+1),\ (x^5-x^3-x-1)/(2x^4-x^3-3x^2+1),$
$(x^5-x^4-2x^3+x^2+x)/(x^5-x^4-x^3+x^2-1) \Longrightarrow M_1 \geq 11$

$D=-310\,466\,763=-3^5\cdot 643\cdot 1987$ $\quad a_0,\ldots = -\zeta_6,\ \zeta_6,\ -2\zeta_6,\ 2\zeta_6,\ -1-\zeta_6,\ 1$

(D-line) sequ.: $0,\ 1,\ x,\ (x-1)/x,\ 1/(1-x),\ 1/(x^2-x+1),\ (x^3-x^2)/(x^2-x+1),$
$(x^4-x^3)/(x^3-2x^2+x-1),\ (x^5-x^4)/(x^4-2x^3+2x^2-x+1),$
$(x^4-x^3+x^2-x+1)/(x^2-x+1),\ (x^3-x^2-1)/(x^3-x^2+x-1) \Longrightarrow M_1 \geq 11$

$D=-316\,894\,187=-151\cdot 1181\cdot 1777$ $\quad a_0,\ldots = 1, -1, 2, -3, 6, -11, 15, -15, 11, -5, 1$

(D-line) sequ.: $0,\ 1,\ x,\ (x-1)/x,\ x-x^2,\ 1/(x^2-x+1),\ x/(x^2+1),$
$(x^4-x^3)/(x^3-2x^2+x-1),\ x/(x^4-2x^3+2x^2-x+1),\ -1/(x^5-3x^4+3x^3-2x^2+x),$
$(x^5-2x^4+x^3)/(x^5-2x^4+x^3+1) \Longrightarrow M_1 \geq 11$

$D=-386\,633\,299=-13^2\cdot 19\cdot 347^2$ $\quad f = X^2-\lambda X+1 \quad (\lambda^5-\lambda^4-3\lambda^3+2\lambda^2+\lambda-1=0)$

sequ.: $0,\ 1,\ \lambda,\ 1/\lambda,\ \lambda-1,\ 1/(\lambda-1),\ (\lambda^2-1)/\lambda,\ \lambda/(\lambda^2-1),\ (\lambda^2-\lambda-1)/(\lambda-1),$
$(\lambda-1)/(\lambda^2-\lambda-1),\ (\lambda^3-\lambda^2-2\lambda+1)/(\lambda^2-\lambda-1);\ x \Longrightarrow M_1 \geq 12$ (see [Ni] 10.12)

Table 2. Numbers of known Euclidean fields (November 1987)

r_1+r_2 \ n	1	2	3	4	5	6	7	8	9	10	total
1	1	5									6
2		16	52	35							103
3			57	11	13	28					109
4				9	10	37	39	45			140
5					1	12	26	65	92	50	246
6						3	3	3	0	0	9
total	1	21	109	55	24	80	68	113	92	50	613

Discriminants of some Euclidean Fields Having a Large Lenstra Constant

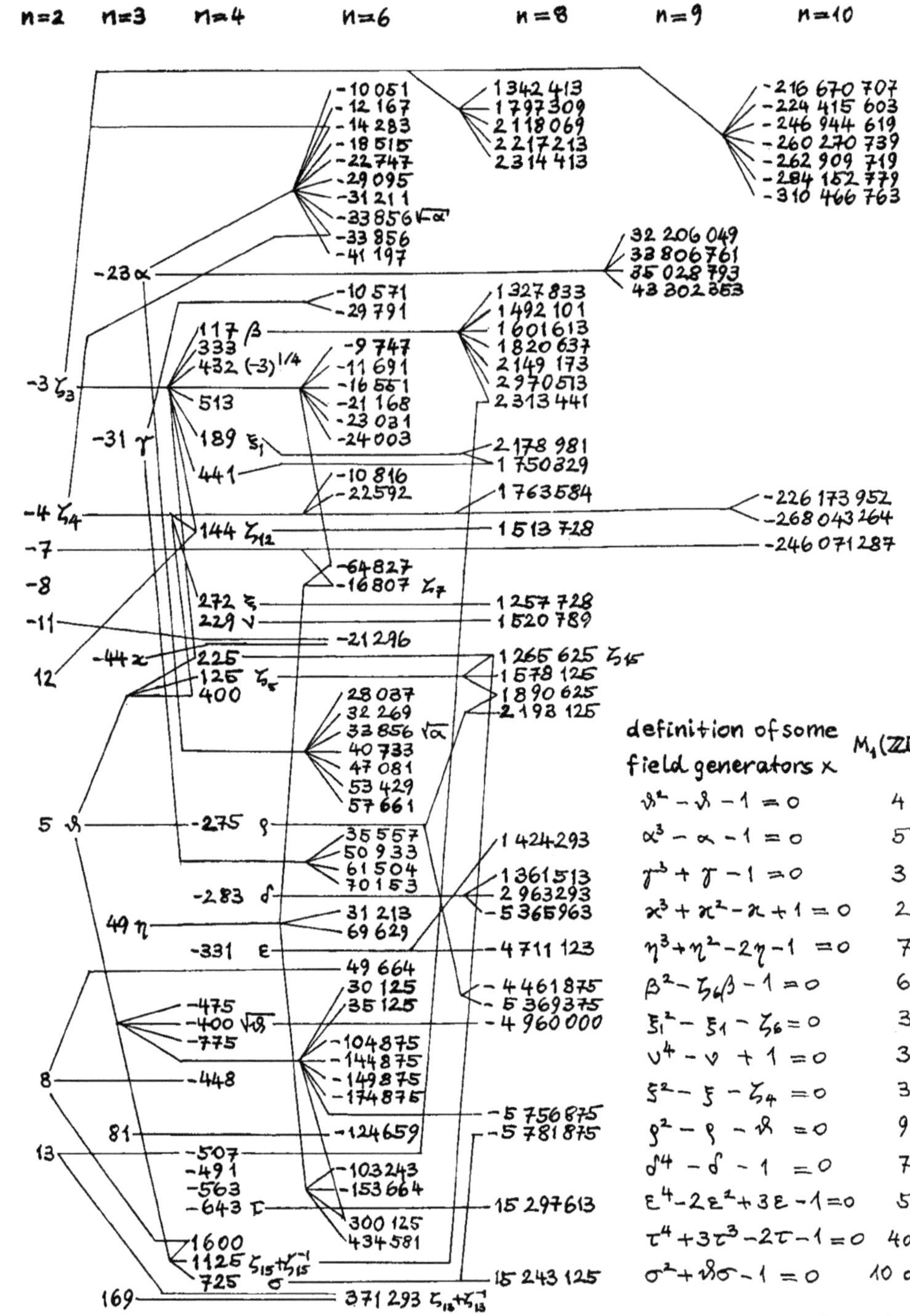

definition of some field generators x

	$M_1(\mathbb{Z}[x])$
$\vartheta^2-\vartheta-1=0$	4
$\alpha^3-\alpha-1=0$	5
$\gamma^3+\gamma-1=0$	3
$x^3+x^2-x+1=0$	2
$\eta^3+\eta^2-2\eta-1=0$	7
$\beta^2-\zeta_6\beta-1=0$	6
$\xi_1^2-\xi_1-\zeta_6=0$	3
$\nu^4-\nu+1=0$	3
$\xi^2-\xi-\zeta_4=0$	3
$\rho^2-\rho-\vartheta=0$	9
$\delta^4-\delta-1=0$	7
$\varepsilon^4-2\varepsilon^2+3\varepsilon-1=0$	5
$\tau^4+3\tau^3-2\tau-1=0$	4 or 5
$\sigma^2+\vartheta\sigma-1=0$	10 or 11

ζ_m = primitive m^{th} root of 1

References

[D1] F. Diaz y Diaz: Valeurs minima du discriminant des corps de degré 7 ayant une seule place réelle. *C. R. Acad. Sci. Paris* **296 I** (1983), 137–139.

[D2] —: Valeurs minima du discriminant pour certains types de corps de degré 7. *Ann. Inst. Fourier, Grenoble* **34**#3 (1984), 29–38.

[D3] —: Petits discriminants des corps de nombres totalement imaginaires de degré 8. *J. Number Theory* **25** (1987), 34–52.

[D4] —: Discriminant minimal et petits discriminants des corps de nombres de degré 7 avec 5 places réelles. *J. London Math. Soc.*, to appear.

[E] J. H. Evertse: Upper bounds for the number of solutions of diophantine equations. Mathematical Centre Tract 168, Mathematisch Centrum, Amsterdam 1983.

[EG] —, K. Győry: On unit equations and decomposable form equations. *J. reine angew. Math.* **358** (1985), 6–19.

[G1] K. Győry: Sur les polynômes à coefficients entiers et de discriminant donné II. *Publ. Math. Debrecen* **21** (1974), 125–144.

[G2] —: Sur une classe de corps de nombres algébriques et ses applications. *Publ. Math. Debrecen* **22** (1975), 151–175.

[G3] —: On certain graphs associated with an integral domain and their application to diophantine problems. *Publ. Math. Debrecen* **29** (1982), 79–94.

[La] S. Lang: Integral points on curves. *IHES Publ. Math.* **6** (1960), 27–43.

[L1] H. W. Lenstra jr.: Euclidean number fields of large degree. *Invent. Math.* **38** (1977), 237–254.

[L2] —: Euclidean number fields. *Math. Intelligencer* **2**#1 (1979), 6–15; **2**#2 (1980), 73–77; *ibid.* 99–103.

[Le] A. Leutbecher: Euclidean fields having a large Lenstra constant. *Ann. Inst. Fourier, Grenoble* **35**#2 (1985), 83–106.

[LM] —, J. Martinet: Lenstra's constant and Euclidean number fields. Journées Arithmétiques 1981, *Astérisque* **94** (1982), 87–131.

[M1] J. Martinet: Petits discriminants des corps de nombres. Journées Arithmétiques 1980, LMS Lecture Notes Series 56, Cambridge Univ. Press 1982, 151–193.

[M2] —: Méthodes géométriques dans la recherche des petits discriminants. Séminaire de Théorie des Nombres (Sém. Delange–Pisot–Poitou) Paris 1983–84, Progress in Mathematics 59, Birkhäuser Boston *et al.* 1985, 147–179.

[Me] J.-F. Mestre: Corps euclidiens, unités exceptionnelles et courbes elliptiques. *J. Number Theory* **13** (1981), 123–137.

[N1] T. Nagell: Sur une propriété des unités d'un corps algébrique. *Ark. Mat.* **5**#25 (1964), 343–356.

[N2] —: Sur les unités dans les corps biquadratiques primitifs du premier rang. *Ark. Mat.* **7**#27 (1968), 359–394.

[N3] —: Quelques problèmes relatifs aux unités algébriques. *Ark. Mat.* **8**#14 (1969), 115–127.

[N4] —: Sur un type particulier d'unités algébriques. *Ark. Mat.* 8#18 (1969), 163–184.

[Ni] G. Niklasch: Ausnahmeeinheiten und euklidische Zahlkörper. Diplomarbeit, Techn. Univ. München 1986.

[P1] M. Pohst: The minimum discriminant of seventh degree totally real algebraic number fields. In: H. Zassenhaus (ed.), Number Theory and Algebra, Acad. Press New York *et al.* 1977, 235–240.

[P2] —: On the computation of number fields of small discriminants including the minimum discriminants of sixth degree fields. *J. Number Theory* **14** (1982), 99–117.

[P3] —: On the determination of algebraic number fields of given discriminant. Computer Algebra, EUROCAM '82 Marseille, Lecture Notes in Computer Science 144, Springer Berlin *et al.* 1982, 71–76.

[P-Z] —, P. Weiler, H. Zassenhaus: On effective computation of fundamental units II. *Math. Comp.* **38** (1982), 293–329.

[S1] C. L. Siegel: Approximation algebraischer Zahlen. *Mathem. Zeitschr.* **10** (1921), 173–213. Also in: K. Chandrasekharan, H. Maaß (ed.s): C. L. Siegel. Gesammelte Abhandlungen I. Springer Berlin *et al.* 1966, 6–46.

[S2] —: Über einige Anwendungen diophantischer Approximationen. *Abh. Preuß. Akad. Wiss., phys.-math. Kl.* (1929). Also in: Gesammelte Abhandlungen I, *loc. cit.*, 209–266.

[vL] F. J. van der Linden: Euclidean rings with two infinite primes. CWI Tract 15, Centrum voor Wiskunde en Informatica, Mathematisch Centrum, Amsterdam 1985.

Note added in proof. Three further Euclidean fields have recently been announced to us. R. McKenzie from Michigan State University (personal communication) has proved $\mathbb{Q}(\zeta_{13})$ to be Euclidean; this is the first example in degree 12. H. Cohn and J. Deutsch (City University of New York, preprint) have treated the real quartic fields $\mathbb{Q}(\sqrt{2+\sqrt{2}})$ and $\mathbb{Q}(\sqrt{3+\sqrt{2}})$. In all three cases, computer scans based on geometric methods (subdivision of a fundamental cell for a suitable lattice) were employed. Our Table 2 should be updated accordingly.

We would like to thank K. Győry and H. W. Lenstra jr. for their valuable comments on a preliminary version of this paper.

SUMSETS CONTAINING k-FREE INTEGERS

Melvyn B. Nathanson

Provost and Vice President for Academic Affairs
Lehman College (CUNY)
Bronx, New York 10468

1. Introduction

Let $|A|$ denote the cardinality of the set A. Many problems in combinatorial number theory have the following character: For a given arithmetic property P, find a function f(n) such that if A is a subset of $\{1,2,\ldots,n\}$ and $|A| > f(n)$, then A has property P. If B is a finite set of integers, define the <u>subset sum</u> s(B) by

$$s(B) = \sum_{b \in B} b.$$

Erdös and Freud [2] recently asked the following two combinatorial questions concerning subset sums of finite sets of positive integers:

(1) If $A \subseteq \{1,2,\ldots,3m\}$ and $|A| > m$, does $s(B) = 2^t$ for some $t \geq 0$ and $B \subseteq A$? Note that this result would be best possible, since $A = \{3,6,\ldots,3m\}$ satisfies $|A| = m$, but 3 divides s(B) for every $B \subseteq A$.

(2) If $A \subseteq \{1,2,\ldots,4m\}$ and $|A| > m$, then is s(B) square-free for some $B \subseteq A$? This result would also be best possible, since $A = \{4,8,\ldots,4m\}$ satisfies $|A| = m$, but 4 divides s(B) for every $B \subseteq A$.

Problems (1) and (2) have been solved independently by several authors. Using analytic methods, Freiman [3] proved that the answer is yes to questions (1) and (2) if m is sufficiently large. If $|A| > m$, he showed that there is a subset B with $|B| > c \cdot \log m$ such that s(B) is a power of 2 in problem (1), and a subset B with $|B| > c \cdot \log m$ such that s(B) is square-free in problem (2). Alon [1] has obtained related results.

Freiman's theorems are not completely satisfactory, since the cardinality of the subset B goes to infinity as n tends to infinity. Nathanson and Sárközy [6] asked if there is an absolute constant h_1 with the property that, for m sufficiently large, if A satisfies the hypotheses of problem (1), then there is a subset $B \subseteq A$ with $|B| \leq h_1$

and $s(B) = 2^t$. Similarly, they asked if there is an absolute constant h_2 with the property that, for m sufficiently large, if A satisfies the hypotheses of problem (2), then there is a subset B of A with $|B| \leq h_2$ and s(B) square-free.

Nathanson and Sárközy [6] answered both these questions. Using elementary methods, they proved that there exist constants $h_1 \leq 30{,}961$ and $h_2 \leq 21$ that give the appropriate subset sums for large m.

The estimate for h_1 is almost certainly far too large. In an analogous infinite problem, Erdös, Nathanson, and Sárközy [4] proved that if A is an infinite sequence with lower asymptotic density $d_L(A) \geq 1/3$, and if some $a \in A$ is not divisible by 3, then there are infinitely many powers of 2, each of which is a sum of 5 or 6 distinct elements of A. In the opposite direction, Erdös (personal communication) observed that the set

$$A = \bigcup_{j=2}^{\infty} [2^j+1,\ 2^j + 2^j/3]$$

has the property that $d_L(A) = 1/3$, but $a + a' \neq 2^t$ for all $a, a' \in A$.

In the case of problem (2) for square-free numbers, Filaseta [5] showed that $h_2 = 2$ for m sufficiently large, and this result is clearly best possible.

In this paper I use Filaseta's method to solve the following generalization of problem (2):

(3) If $A \subseteq \{1,2,\ldots,2^k m\}$ and $|A| > m$, then is s(B) k-free for some $B \subseteq A$? The set $A = \{2^k, 2\cdot 2^k, 3\cdot 2^k,\ldots, m\cdot 2^k\}$ shows that this would be best possible.

Notation. Let $k \geq 2$. Let Q_k denote the set of all k-free positive integers, and let Q_k' denote the set of all odd k-free positive integers. Let U_k denote the set of all odd positive integers divisible by p^k for some prime p. Let V_k denote the set of all positive integers that are congruent to 2 modulo 4 and are divisible by p^k for some prime p. Let x be a real number. If S is any set of integers, let S(x) denote the number of positive elements of S not exceeding x. Let [x] denote the integer part of x.

2. k-free integers

Here is a simple example of a set A such that $s(B) \notin Q_k$ for all subsets $B \subseteq A$: Let $d \geq 2$ and let A be any set of multiples of d^k.

Then $d^k | s(B)$ for all $B \subseteq A$, and so $s(B) \notin Q_k$.

Let $h \geq 2$. If we wish to consider only subset sums $s(B)$ with $|B| = h$, then every set A, each of whose elements satisfies $a \equiv h^{k-1} \pmod{h^k}$, has the property that $s(B) \notin Q_k$ whenever $B \subseteq A$ and $|B| = h$. Another example, in the case $h = 2$, is any subset A of the set

$$\{n \geq 1 \mid n \equiv 2^{k-1} \text{ or } 2^{k-1}(m^k-1) \pmod{(2m)^k}\}$$

for any $m \geq 3$. Then $a+a' \notin Q_k$ for all $a, a' \in A$. I shall give an upper bound for the size of any set $A \subseteq \{1,2,\ldots,n\}$ with the property that $a+a' \notin Q_k$ for all $a, a' \in A$. It will follow from this result that if $|A| > m > 1$ and $A \subseteq \{1,2,\ldots,2^k m\}$, then there exist $a, a' \in A$ with $a \neq a'$ and $a+a' \in Q_k$. This solves problem (3).

LEMMA. Let $k \geq 2$. Then

$$V_k(2n) = U_k(n) < n/(k-1)2^k.$$

Proof.

$$U_k(n) = \sum_{\substack{1\leq m\leq n \\ m \text{ odd} \\ p^k | m}} 1 \leq \sum_{\substack{p^k\leq n \\ p \text{ odd}}} [n/2p^k + 1/2] \leq n \sum_{p\geq 3} p^{-k}$$

$$< n \sum_{\substack{m\geq 3 \\ m \text{ odd}}} m^{-k} < (n/2) \int_2^\infty y^{-k}dy = n/(k-1)2^k.$$

If $m \leq 2n$ and $m \equiv 2 \pmod 4$, then $m \in V_k$ if and only if $m/2 \in U_k$. Thus, $V_k(2n) = U_k(n)$. This completes the proof of the Lemma.

THEOREM 1. Let $k \geq 3$ and let $n = 2^k m$. Let A be a subset of $\{1,2,\ldots,n\}$ such that $a+a' \notin Q_k$ for all $a, a' \in A$ with $a \neq a'$. Then either $|A| \leq m$, or $n = 2^k$ and $A = \{a, n-a\}$ for some $a \in \{1,\ldots,n\}\backslash\{n/2\}$.

Proof. Let $A \subseteq \{1,2,\ldots,n\}$. If $A \subseteq \{a \equiv 0 \pmod{2^k}\}$ or $A \subseteq \{a \equiv 2^{k-1} \pmod{2^k}\}$, then $|A| \leq m$, and the Theorem is true. Therefore, we can assume that A satisfies both of the following conditions:

(i) $A \not\subseteq \{a \equiv 0 \pmod{2^k}\}$,

(ii) $A \not\subseteq \{a \equiv 2^{k-1} \pmod{2^k}\}$.

Let n be any positive integer, and let A be any subset of $\{1,2,\ldots,n\}$ such that A satisfies (i) and (ii) and such that

$a+a' \notin Q_k$ for all $a, a' \in A$ with $a \neq a'$. Let s be the largest integer such that 2^s divides a for all $a \in A$. Condition (i) implies that $s \le k-1$. Let $B = \{a/2^s \mid a \in A\}$. Then $B \subseteq \{1,2,\ldots,[n/2^s]\}$. It follows from the definition of s that B contains an odd integer.

There are two cases.

Case 1. Suppose that B also contains an even integer. Let B_0 consist of the even elements of B and let B_1 consist of the odd elements of B. Then B_0 and B_1 are disjoint, nonempty sets, and $|B_0| + |B_1| = |B| = |A|$. The cardinality of the sumset $B_0 + B_1$ satisfies

$$|B_0 + B_1| \ge |B_0| + |B_1| - 1 = |A| - 1.$$

Let $b_0 \in B_0$ and $b_1 \in B_1$. Then $b_0 + b_1$ is odd. If $b_0 + b_1$ is also k-free, then $s \le k-1$ implies that $2^s(b_0 + b_1)$ is k-free, which is impossible since $2^s b_0$ and $2^s b_1$ are distinct elements of the set A. Therefore, $b_0 + b_1 \in U_k$. Since $b_0 + b_1 \le 2[n/2^s]$, it follows that

$$|A| \le |B_0 + B_1| + 1 \le U_k(2[n/2^s]) + 1 \le U_k(2n) + 1.$$

Case 2. Suppose that b is odd for all $b \in B$. It follows from condition (ii) that $s \le k-2$. Let B_1 (resp. B_3) consist of all $b \in B$ such that $b \equiv 1 \pmod 4$ (resp. $b \equiv 3 \pmod 4$). Then $|B_1| + |B_3| = |B| = |A|$. Choose $i = 1$ or 3 such that $|B_i| \ge |A|/2$. If $b, b' \in B_i$, then $b + b' \equiv 2 \pmod 4$ and $b + b' \le 2[n/2^s]$. Suppose that $b + b' \in Q_k$, $b \neq b'$. Then $s \le k-2$ implies that $2^s(b + b') \not\equiv 0 \pmod{2^k}$, and so $2^s(b + b') \in Q_k$. Since $2^s b \in A$ and $2^s b' \in A$, however, it follows that $2^s(b + b') \notin Q_k$, hence $b + b' \notin Q_k$. Therefore, $b + b' \in V_k$ for all $b, b' \in B_i$, $b \neq b'$. There are at least $2|B_i| - 3$ distinct integers of the form $b + b'$, where $b, b' \in B_i$ and $b \neq b'$, hence

$$|A| \le (2|B_i| - 3) + 3 \le V_k(2[n/2^s]) + 3$$

$$\le V_k(2n) + 3 \le U_k(n) + 3.$$

Combining the conclusions of cases 1 and 2, we obtain

$$|A| \le \max\{U_k(2n) + 1, U_k(n) + 3\} \qquad (*)$$

for all $n \ge 1$.

Let $n = 2^k m$. Since $k \ge 3$, it follows from the Lemma that

$$U_k(2n) < 2n/(k-1)2^k = 2m/(k-1) \le m$$

and so

$$U_k(2n) + 1 \le m$$

for all $m \ge 1$. Similarly, since $U_k(n) < m/(k-1)$, it follows that $U_k(n) + 3 \le m$ if $m/(k-1) + 3 \le m + 1$, and the latter inequality holds whenever $m \ge 2 + 2/(k-2)$. We conclude from (*) that $|A| \le m$ if $k \ge 4$ and $m \ge 3$, or if $k = 3$ and $m \ge 4$.

This gives four special cases to consider. First, let $m = 1$ and $k \ge 3$. Then $n = 2^k$, and the sets A of the form $A = \{a, n-a\}$, where $a \in \{1,...,n\}\backslash\{n/2\}$, are the only sets with $|A| > 1$ that satisfy the conditions of the Theorem.

Next, let $m = 2$ and $k \ge 4$. Then $n = 2^{k+1}$. Let

$$A = \{a_1, a_2, a_3\} \subseteq \{1,...,2^{k+1}\},$$

where $a_1 < a_2 < a_3$. Then

$$1 < a_1 + a_2 < a_1 + a_3 < a_2 + a_3 < 2^{k+2} < 3^k.$$

Since the only integers in $(1, 2^{k+2})$ divisible by a k-th power are 2^k, $2\cdot 2^k$, and $3\cdot 2^k$, it follows that if each of the numbers $a_1 + a_2$, $a_1 + a_3$, and $a_2 + a_3$ is divisible by a k-th power, then $a_1 + a_2 = 2^k$, $a_1 + a_3 = 2\cdot 2^k$, and $a_2 + a_3 = 3\cdot 2^k$, which is impossible since

$$3\cdot 2^k = a_2 + a_3 < (a_1 + a_2) + (a_1 + a_3) = 3\cdot 2^k.$$

Similarly, if $m = 2$ and $k = 3$, it is easy to check that there is no set $A \subseteq \{1,..., 16\}$ with $|A| = 3$ that satisfies the conditions of the Theorem, and if $m = 3$ and $k = 3$, there is no set $A \subseteq \{1,...,24\}$ with $|A| = 4$ that satisfies the conditions of the Theorem. This completes the proof.

Definition. Let $\Phi_k(n)$ denote the cardinality of the largest subset $A \subseteq \{1,2,...,n\}$ such that $a + a' \notin Q_k$ for all $a, a' \in A$ with $a \ne a'$, and such that A satisfies both $A \not\subseteq \{a \equiv 0 \pmod{2^k}\}$, and $A \not\subseteq \{a \equiv 2^{k-1} \pmod{2^k}\}$.

In the case $k = 2$, Filaseta [5] proved that

$$\limsup \Phi_2(n)/n \le 1 - 8/\pi^2.$$

This can be generalized to k-free subset sums for any $k \ge 2$.

THEOREM 2. Let $k \ge 2$, and let $\zeta(k)$ denote the Riemann zeta function. Then

$$\limsup \Phi_k(n)/n \le 1 - 2^k/((2^k-1)\zeta(k)).$$

Proof. The asymptotic density of the set Q_k' of odd, square-free integers is $2^{k-1}/((2^k-1)\zeta(k))$. Also, $U_k(2n) = n - Q_k'(2n)$. Let $n \ge 2\cdot 3^k$. Then $U_k(2n) \ge U_k(n) + 2$. It follows from (*) that

$$\begin{aligned}\Phi_k(n) &\le \max\{U_k(2n) + 1,\ U_k(n) + 3\} \\ &= U_k(2n) + 1 \\ &= n - Q_k'(2n) + 1 \\ &= n\{1 - 2^k/(2^k-1)\zeta(k) + o(1)\}.\end{aligned}$$

This completes the proof.

It follows from the elementary estimate $\zeta(k) < (2^k/(2^k-1))^2$ that $\limsup \Phi_k(n)/n < 1/2^k$ for all $k \ge 2$.

It is an open problem to determine if $\lim \Phi_k(n)/n$ exists.

References

1. N. Alon, Subset sums, J. Number Theory 27 (1987), 196-205.
2. P. Erdös, Some problems and results on combinatorial number theory, in: Proc. First China-U.S.A. Conference on Graph Theory and its Applications (Jinan, 1986), Annals New York Acad. Sci., to appear.
3. P. Erdös and G. Freiman, On two additive problems, J. Number Theory, to appear.
4. P. Erdös, M. B. Nathanson, and A. Sárközy, Sumsets containing infinite arithmetic progressions, J. Number Theory 28 (1988), 159-166.
5. M. Filaseta, Sets with elements summing to square-free numbers, C. R. Math. Rep. Acad. Sci. Canada 9 (1987), 243-246.
6. M. B. Nathanson and A. Sárközy, Sumsets containing long arithmetic progressions and powers of 2, Acta Arith., to appear.

ON THE REPRESENTATION OF 1 BY BINARY CUBIC FORMS WITH POSITIVE DISCRIMINANT

Attila Pethő [*]
Mathematical Institute
Kossuth Lajos University
4010 Debrecen P.O. Box 12
Hungary

1. Introduction

Let $f(x,y)=ax^3+bx^2y+cxy^2+dy^3 \in Z[x,y]$, where Z denotes the ring of integers. We shall denote by D_f the discriminant of f, and by N_f the number of solutions in integers x and y of the diophantine equation

(1) $$f(x,y)=1.$$

A form with a=d=1 will be called <u>reversible</u>.

Two cubic forms $f_1(x,y), f_2(x,y) \in Z[x,y]$ are called equivalent, in the sequel $f_1 \approx f_2$, if there exist integers a_1, a_2, a_3, a_4 with $a_1a_4-a_2a_3 = \pm 1$ such that

$$f_2(x,y) = f_1(a_1x+a_2y, a_3x+a_4y).$$

It is easy to see that if $f_1 \approx f_2$ then $D_{f_1}=D_{f_2}$ and $N_{f_1}=N_{f_2}$.

Delone [6] and Nagell [12] proved independently (see also Delone and Faddeev [7] p. 402), that if $D_f<0$ then $N_f \le 5$. They were able to prove more precisely that

	$N_f=$	
if f is not equivalent to a reversible form	0, 1, 2	
	2, 3	, if f is equivalent to a reversible form
	4	, if $D_f=-44$ and -31
	5	, if $D_f=-23$

[*] Research supported by Hungarian National Fundation for Scientific Research Grant No. 273/86.

and there exist infinitely many inequvivalent forms with $N_f=3$.

If $f(x,y)$ has positive discriminant then much less is known. Siegel [16] proved that if D_f is large enough then $N_f \leq 18$. This was improved to 10 by A.E. Gel'man in a student's paper. For a proof we refer to Delone and Faddeev [7]. Evertse [9] proved unconditionally that $N_f \leq 12$. Evertse and Győry [10] found that for a fixed algebraic number field K there exist only finitely many inequivalent forms $g(x,y)$ of fixed degree $n \geq 3$ with splitting field K such that $N_g > 2$.

Up to now no cubic forms are known for which (1) has more than nine solutions or for which $N_f=7$ or 8. In Table 1 we collected representatives of all equivalence classes of cubic forms with $N_f \geq 6$.

$f(x,y)$	D_f	N_f	Reference
$x^3+x^2y-2xy^2-y^3$	49	9	[2], [15], [18]
$x^3-3xy^2+y^3$	81	6	[13], [18], [19]
$x^3-4xy^2+y^3$	148	6	[4], [15]
$x^3-5xy^2+3y^3$	257	6	[11]
$x^3-2x^2y-5xy^2-y^3$	361	6	[11]

Table 1.

Based on an idea of Baker and Davenport [1], Ellison at al [8], Pethő and Schulenberg [15] and Steiner [18] worked out a method to find, for a given f, all solutions of (1) with a computer. Unfortunately this method is too slow to solve a large number of equations.

In [14] we described a different, quick method for the resolution of the system of inequalities

$$|f(x,y)| \leq m \qquad\qquad |y| \leq y_0,$$

where m and y_0 are given integers.

In this note we report on results of a computer search which solved (1) with $|y| \leq 10^{41}$ for approximately 3000 cubic forms with positive discriminant.

2. Parametrization

To carry out the search we needed a suitable parametrization. Gaál [11] solved (1) for all cubic forms f with $0 < D_f \leq 1000$. We used in our search another listing. The first parametrization is described in

Theorem 1. Let $f(x,y)\in Z[x,y]$ *be a cubic form with* $D_f>0$ *If* (1) *is solvable then there exists* $f_1(x,y)=x^3+bx^2y+cxy^2+dy^3\in Z[x,y]$ *such that* $f_1\approx f$; $b=0$ *or* 1; $c<0$ *and*

$$0 < d < \left[\frac{2}{3} |c| \sqrt{\frac{|c|}{3}} \right] \quad ,\text{if } b=0 \tag{2}$$

$$\left[- \frac{2-9c+(2-6c)\sqrt{1-3c}}{27} \right] < d < \left[\frac{(2-6c)\sqrt{1-3c}-2+9c}{27} \right] \quad ,\text{if } b=1. \tag{3}$$

Proof. Assume that $x=\alpha$, $y=\beta$ is a solution of (1). Then $(\alpha,\beta)=1$ and there exist integers γ,δ such that $\alpha\gamma-\beta\delta=1$, whence

$$f_2(x,y) = f(\alpha x+\delta y, \beta x+\gamma y) = x^3+b_1x^2y+c_1xy^2+d_1y^3.$$

Let $b_1=3q+b_2$ with $|b_2|\le 1$, and $f_1(x,y)=f_2(x-\varepsilon qy, \varepsilon y)$, where

$$\varepsilon = \begin{cases} b_2 & , \text{ if } b_2\neq 0 \\ 1 & , \text{ if } b_2=0 \text{ and } f_2(-q,1)>0 \\ -1 & , \text{ otherwise.} \end{cases}$$

Then $f_1(x,y)=x^3+bx^2y+cxy^2+dy^3$, where $b=\varepsilon(b_1-3q)=\varepsilon b_2=1$, if $|b_2|=1$ and 0 otherwise. We have further $d=\varepsilon^3f_2(-q,1)>0$, if $b_2=0$ by the choice of ε.

It follows from the assumptions that $f_1(x)=f_1(x,1)$ has three real roots and so $f_1'(x)=3x^2+2bx+c$ has two real roots, but this is possible only if $c\le 0$.

Assume now that $b=0$. Then $c\neq 0$ and the roots of $f_1'(x)$ are $\pm\sqrt{-\frac{c}{3}}$. $f_1(0)>0$, hence $f_1\left[\sqrt{-\frac{c}{3}} \right] = \frac{2c}{3} \sqrt{-\frac{c}{3}} + d < 0$, which proves (2).

Let finally $b=1$. Then the roots of $f_1'(x)$ are $\alpha = \frac{-1+\sqrt{1-3c}}{3}$ and $\beta = \frac{-1-\sqrt{1-3c}}{3}$. We get by an easy computation that

$$f_1(\beta) = \frac{2-9c+(2-6c)\sqrt{1-3c}}{27} + d > 0 \quad , \text{and}$$

$$f_1(\alpha) = \frac{2-9c-(2-6c)\sqrt{1-3c}}{27} + d < 0,$$

which proves (3), and at the same time the second inequality $c\neq 0$ too.

It is natural to expect, and our computation strengthened this, that the forms with "large" N_f , i.e for which $N_f\ge 4$, are reversible. The second parametrization was based on this observation.

3. Reduction of the prescribed upper bound.

In order to establish all solution of (1) with $|y| \le y_0$, where y_0 is a given number we used the method desribed in [14]. Let $F(x,y) = a_0x^n + a_1x^{n-1}y + \ldots + a_ny^n \in Z[x,y]$ be of degree $n \ge 3$, such that $a_0 \ne 0$ and the roots $\alpha_1, \ldots, \alpha_n$ of $F(x,1)$ are all distinct and irrational. Let $m \in Z$; take $T_j = |\alpha_j - \alpha_1|$ $(j=2,\ldots,n)$, and for $k>0$

$$H_k(t) = \prod_{j=2}^{n} (T_j - t) - t^{n-2} \left|\frac{m}{f_0}\right|^{2/n} \frac{1}{k} .$$

$\beta = [b_0; b_1, \ldots]$ shall denote the simple continued fraction expansion of the irracional number β, while $\frac{p_n}{q_n}$ the n-th convergent to β. We have with this notation

Theorem P [14]. Let y_0 be a given real number, and $(x,y) \in Z^2$ a solution of the inequality

$$|F(x,y)| \le m$$

with $|x - \alpha_1 y| \le |x - \alpha_i y|$ $(i=2,\ldots,n)$, such that $y \ne 0$, $(x,y)=1$ and $|y| \le y_0$. Let $\alpha_1 = [b_0; b_1, \ldots, b_u, \ldots]$, where u is chosen so that $q_{u-1} > y_0$. Let $h \ge 1$, $B = \max_{h \le j \le u} b_j$, and let $T = T_{\alpha_1}$ be the smallest positive root of $H_{1/2}(t)$. Then either

$$|y| \le \min \left\{ y_0, \frac{1}{T} \left|\frac{m}{a_0}\right|^{1/n} \right\}$$

or $\frac{x}{y}$ is a convergent to α_1 with

$$|y| \le \max \left\{ q_{h-1}, \frac{1}{T} \left|\frac{m}{a_0}\right|^{1/n} \left[\frac{B+2}{2}\right]^{1/(n-2)} \right\} .$$

Using this theorem it is possible to reduce a prescribed large upper bound to a much smaller one computing only the partial quotients to the roots of $F(x,1)$. It proved to be useful to compute the small solutions by applying the following

Theorem 2. Assume that $f(x,y) = x^3 + bx^2y + cxy^2 + dy^3 \in Z[x,y]$ with $D_f > 0$. Let $(x,y) \in Z^2$ be a solution of (1) with $y \ne 0$, and with $|x - \alpha_1 y| \le |x - \alpha_i y|$ $(i=2,3)$. If $D_f > 229$, then $\frac{x}{y}$ is a convergent to α_1.

Proof. We use an idea of Bombieri and Schmidt [3]. Let $L_i=\alpha_i-\frac{x}{y}$ $(i=1,2,3)$, and suppose that $|L_1|\leq|L_2|\leq|L_3|$. We have

$$D_f = \prod_{1\leq i<j\leq 3}(\alpha_i-\alpha_j)^2 = L_3^4\, L_2^2\left[1-\frac{L_2}{L_3}\right]^2\left[1-\frac{L_1}{L_3}\right]^2\left[1-\frac{L_1}{L_2}\right]^2 .$$

The function $g(u,v)=(1-u)(1-v)\left[1-\frac{u}{v}\right]$ is on the domain $-1\leq u,v\leq 1$, $|v|\leq|u|$ non-negative and its maximum is 2. Hence

$$D_f \leq 4L_3^4L_2^2. \tag{4}$$

Multiplying both sides of (4) by $L_2^2L_1^4$ and using that $|L_2|\geq|L_1|$ and $|L_1L_2L_3|=\frac{1}{|y|^3}$, we get

$$|L_1| \leq \left[\frac{4}{D_f}\right]^{1/6}\frac{1}{y^2} . \tag{5}$$

If $D_f>256$, then the right hand side of (5) is less then $\frac{1}{2y^2}$, hence by Lagrange's theorem $\frac{x}{y}$ is a convergent to α_1.

The lower bound for D_f is verified by the observation, that there does not exist a cubic form f with $229<D_f<256$.

4. Computation of the continued fractions.

To compute the simple continued fraction expansion of the roots of $f(x,1)$, where $f(x,y)$ satisfies the assumptions of Theorem 2 we used - to the actual problem specialized form - the method of Cantor, Galyean and Zimmer [5] (see also Zimmer [20]). We shall suppose in addition that $[\alpha_i]\neq[\alpha_j]$ if $i\neq j$.

Let $\alpha_{i,0}=\alpha_i$ $(i=1,2,3)$, $b_0=[\alpha_{1,0}]$ and $f_0(x)=f(x,1)$. Assume that we have already defined for $n\geq 0$ $\alpha_{i,n}$ $(i=1,2,3)$, b_n and $f_n(x)$. If $\alpha_{i,n}=b_n$ then stop, otherwise put $\alpha_{i,n+1}=(\alpha_{i,n}-b_n)^{-1}$; $b_{n+1}=[\alpha_{1,n+1}]$, and $f_{n+1}(x)=x^3f_n\left[\frac{1}{x}+b_n\right]$, if $n=1$ and $f_n(b_n)>0$ while $f_{n+1}(x)=-x^3f_n\left[\frac{1}{x}+b_n\right]$ otherwise.

It is clear that $\alpha_1=[b_0;b_1,\ldots]$, and if the algorithm terminates then α_1 is rational.

Let $f_n(x)=a_{3,n}x^3+a_{2,n}x^2+a_{1,n}x+a_{0,n}$ for $n\geq 0$. Assume that the above procedure does not terminate in the n-th step. By induction it is easy to see that $\alpha_{i,n+1}$ $(i=1,2,3)$ are distinct, and are zeros of $f_{n+1}(x)$, hence $a_{3,n+1}\neq 0$. We have further $\alpha_{1,n+1}>1$, and so $b_{n+1}\geq 1$.

On the other hand, the assumptions $[\alpha_i]\neq[\alpha_j]$, $1\leq i<j\leq 3$, imply

$$\alpha_{i,1} < \begin{cases} 0, & \text{if } \alpha_{i,0} < \alpha_{1,0} \\ 1, & \text{if } \alpha_{i,0} < \alpha_{1,0} \end{cases}$$

hence

(6) $$-1 < \alpha_{i,n} < 0 \quad \text{for } i=2,3;\ n\geq 2.$$

The well-known identity $\alpha_{1,n}+\alpha_{2,n}+\alpha_{3,n}=-\frac{a_{2,n}}{a_{3,n}}$, $\alpha_{1,n+1}>1$ and (6) imply that

(7) $$u_n = \max\left\{1,\left[-\frac{b_{2,n}}{b_{3,n}}\right]\right\} \leq b_n \leq 2+\left[-\frac{b_{2,n}}{b_{3,n}}\right]$$

for $n\geq 2$. We have $\alpha_{1,n}>\alpha_{i,n}$ for $n>1$, hence $f_n(b_n)<0$, and so $b_{3,n+1}>0$.

This observation and (7) imply that

$$b_{n+1} = \begin{cases} u_{n+1}, & \text{if } f_{n+1}(u_{n+1}+1) < 0 \\ u_{n+1}+1, & \text{if } f_{n+1}(u_{n+1}+1) > 0 \text{ but } f_{n+1}(u_{n+1}+2) < 0 \\ u_{n+1}+2, & \text{otherwise.} \end{cases}$$

Hence, if $n\geq 2$ then one can establish b_n computing only two values of f_n.

5. An example.

We demonstrate the method described in sections 3 and 4 by solving the equation

(8) $$g(x,y) = x^3 - 17xy^2 + 13y^3 = 1$$

with $|y|<6*10^{99}$. We have $D_g = 15089$. The first 200 partial quotients of the roots - α,β and γ - are

α = [3;1,2,69,1,2,1,1,3,1,2,1,8,14,1,1,1,4,1,107,25,1,30,4,5,15,5,
4,4,12,1,4,13,7,5,2,19,1,1,1,5,1,3,2,2,22,1,11,1,12,10,2,43,4,1,22,
2,3,1,1,2,27,2,1,21,1,27,1,2,6,1,3,3,1,2,1,1,20,12,1,2,3,19,2,12,1,
3,2,1,12,3,1,6,1,57,10,3,1,2,6,6,1,4,1,104,1,14,4,1,2,2,1,2,1,3,2,3,
1,2,1,286,3,1,1,15,1,3,1,4,3,30,3,2,2,3,1,1,7,5,1,2,2,1,9,6,1,3,2,1,
1,1,3,5,4,3,1,3,3,2,1,1,6,4,1,1,1,155,2,2,12,1,11,1,1,2,10,57,1,1,
13,1,3,1,1,2,5,5,5,1,2,5,2,1,23,4,2,1,2,13,7,9]

β = [0;1,3,1,6,16,2,2,1,2,1,22,1,1,2,4,11,3,1,25,1,3,2,4,6,3,15,1,
1,1,1,4,1,1,1,1,8,1,1,2,2,1,6,1,1,2,1,2,2,4,2,1,2,26,1,2,17,11,2,3,
69,1,12,3,1,17,3,13,2,19,1,7,1,1,1,4,1,107,4,2,1,54,1,1,6,1,3,4,13,
2,2,16,1,1,2,1,1,1,1,136,1,12,1,5,1,4,2,2,190,9,1,2,4,1,2,30,1,1,6,
2,1,2,1,5,23,1,3,2,1,5,1,16,1,1,2,12,3,4,2,1,1,1,1,2,2,91,1,1,4,2,
4,1,4,28,4,190,10,1,11,1,10,1,2,1,10,3,2,1,1,3,1,1,5,1,2,1,4,2,387,
1,3,1,8,2,1,32,2,1,4,2,1,5,1,1,1,3,165,4,1,2,2]

γ = [-5;1,1,6,6,1,1,5,1,48337,1,2,3,1,2,1,1,2,1,2,1,2,2,5,1,30,4,2,2,2,1,1,1,2,14,27,1,6,1,6,8,5,2,5,35,1,3,3,1,1,1,4,1,5,1,1,1,5,1,4,88,4,6,2,2,1,1,2,1,4,4,1,1,19,2,4,10,1,3,2,6,1,3,1,9,3,1,2,1,3,3,29,6,1,2,2,10,1,3,1,2,5,4,6,1,1,3,2,1,1,5,1,2,10,11,1,2,3,3,3,1,2,20,1,6,1,1,1,2,1,4,3,1,3,1,4,12,1,4,1,1,4,18,1,1,4,8,1,5,2,1,4,1,3,1,2,7,5,1,1,1,1,4,10,1,1,1,3,1,2,4,2,31,28,2,1,1,1,2,4,3,9,5,1,3,1,1,6,1,15,3,1,1,1,19,4,7,1,1,11,1,1].

The numerators p_n and the denominators q_n of the starting convergents to the roots are

	n	0	1	2	3	4	5	6	7	8
α:	p_n	3	4	11	763	774	2311	3085	5396	19273
	q_n	1	1	3	208	211	630	841	1471	5254
	n	0	1	2	3	4	5	6	7	8
β:	p_n	0	1	3	4	27	436	899	2234	3133
	q_n	1	1	4	5	34	549	1132	2813	3945
	n	0	1	2	3	4	5	6	7	8
γ:	p_n	-5	-4	-9	-58	-357	-415	-772	-4275	-5047
	q_n	1	1	2	13	80	93	173	958	1131

Table 2.

$\frac{p_8}{q_8}$ gives a lower bound, while $\frac{p_7}{q_7}$ gives an upper bound for the corresponding roots, which are: $3.6682528<\alpha<3.6682529$, $0.7941698<\beta<0.7941699$ and $-4.462422<\gamma<-4.462421$. Using this values one can compute a lower bound for T_α, T_β and T_γ. A simple computation showes that $q_{199}>6*10^{99}$ holds for each root. You can see the effect of the reduction procedure in Table 3., where y_{new} stays for the improved upper bound.

		$T_\alpha>2.63$	$T_\beta>2.44$	$T_\gamma>4.6$
1. Step	u	200	200	200
	B	286	387	48337
	y_{new}	54	79	5254
2. Step	u	4	6	10
	B	69	16	48337
	y_{new}	13	3	5254

Table 3.

Using the last row of Table 3, and keeping in mind Theorem 2, we get from Table 2, that the candidates for solutions of (8) are the following pairs: $(\pm 3,\pm 1)$; $(\pm 4,\pm 1)$; $(\pm 11,\pm 3)$; $(0,\pm 1)$; $(\pm 1,\pm 1)$; $(\pm 5047,\mp 1131)$ and eventuelly some others which correspond to the convergents of γ with $q_n\leq 1130$. Applying again the reduction procedure to γ with $y_0=1130$ we get u=9, B=6 and $y_{new}=1$. This gives the candidates: $(\pm 5,\mp 1)$. Computing g(x,y) at this points we get that the solutions of (8) with $|y|<6*10^{99}$ are: (0,1); (-11,-3); (5047,-1131).

6. Technical remarks on the computer search.

The continued fraction algorithm - CF - from 4. and the reduction algorithm - R - from 3. were implemented on an IBM PC AT compatible computer. The only real numbers with periodic continued fraction expansions are, by Lagrange's theorem, the quadratic irrationalities. Hence, if $f(x,1)$ is irreducible over $Q[x]$, then $\max_{0\leq i\leq 3} |a_{i,n}| \to \infty$ if $n\to\infty$. For this reason, the implementation of the CF algorithm requires multiprecision arithmetic. We used the programming language muMATH.

The R algorithm was implemented by I. Gaal on PASCAL. The author thanks him for his selfless assistance.

In the actual search we does not prescribe a uniform upper bound for $|y|$, but we rather computed the first 200 partial quotients to the zeros of $f(x,1)$. One has a trivial lower bound for q_{199} taking $b_i=1$, $i=1,\dots,199$, which means $q_{199}>10^{41}$. We mention that the partial quotients were often much larger then 1. In some cases we computed q_{199}, and their size lay about 10^{100}.

7. Results of the computer search.

Using the CF and R algorithms we established all solutions of the equations

(9) $$x^3-cxy^2+dy^3=1$$

(10) $$x^3+x^2y-cxy^2+dy^3=1$$

(11) $$x^3-ax^2y-bxy^2+y^3=1$$

with $|y|<10^{41}$. We solved equations of type (9) for the ranges $0<c\leq 30$ and $46\leq c\leq 50$, while d satisfies (2), equations of type (10) for the ranges $0<c\leq 20$ and $c=50$, while d satisfies (3); and equations of type (11) for the range $1\leq a\leq 60$, $0\leq b\leq a$. These are approximatelly 3000 equations. In the sequel N_f' shall denote the number of solutions, found in the computer search.

a.) Our first observation was that *apart from the examples of* Table 1. *we does not find equations with more then five solutions.*

b.) The second observation was that *if the number of solutions of* (1) *is at least four, then* $f(x,y)$ *is equivalent to a reversible form.*

Although we have found a lot of examples with $N'_f=3$ where f inequivalent to a reversible form, we was unable to found a parametrized infinite class of such polynomials.

Let $f(x,y)= (x-Ay)(x-By)(x-Cy)+y^3$, where A, B, C denote distinct integers. For such polynomials $N_f\geq 4$. We found many polynomials with $N'_f=4$ not of this type, which we call exceptional. It seems possible that *there exist infinitely many equivalent classes of non-reversible forms with three solutions and infinitely many exceptional forms with four solutions.*

c.) It is easy to see that there exist infinitely many forms with $N_f\geq 5$. Already Ljunggren [13] pointed out that if $h\geq 1$ and $f_h(x,y)=x^3-(h+1)x^2y+hxy^2+y^3=1$ then at least the pairs (1,0); (1,1); (1,-(h+1)); (0,1); (h,1) are solutions of (1). Another infinite class with $N_f\geq 5$ is defined by $f_g(x,y)=x^3-g^2x^2y+y^3=1$, for which (1) has the trivial solutions (1,0); (1,g); (1,-g); (0,1); $(g^2,1)$. We prove that these two classes are essentially distinct.

Theorem 3. There exists only finitely many integers g,h such that $f_g\approx f_h$.

Proof. If $f_g\approx f_h$ then $D_{f_g}=D_{f_h}$. We have $D_{f_g}=4g^6-27$ and $D_{f_h}=h^4+2h^3-5h^2-6h-23$, hence if $f_g\approx f_h$ then

$$4g^6=h^4+2h^3-5h^2-6h+4.$$

This implies that h is even, say $h=2h_1$. Substituing $2h_1$ for h, and dividing by 4 we get

$$g^6=4h_1^4+4h_1^3-5h_1^2-3h_1+1. \tag{12}$$

The polynomial on the right hand side of (12) has four distinct real roots, hence by a theorem of Siegel [15] the number of solutions of (12) is finite.

We found five other examples of type (11) which do not belong to the above classes. Because all these forms are reversible, we omitted (1,0) and (0,1) from the list of solutions.

Coefficients of f		D_f	solutions
a	b		
1	3	148	(3,-2); (-14,-45); (-2,-1)
0	7	1345	(1,7); (-19,7); (18,7)
0	5	473	(7,-3); (1,5); (-2,-1)
8	20	322597	(-2,1); (-351,172); (10,1)
40	18	810661	(1,-2); (364,9); (1,20)

Table 4.

d.) Summarizing the observations *we conjecture the following connection between cubic forms* f(x,y) *with* $D_f>0$ *and the number of solutions* N_f *of* (1)

	$N_f=$	
if f is not equivalent to a reversible form	0	
	1	
	2	, if f is equivalent to a reversible form
	3	
	4	
	5	
	6	, if D_f=81, 148, 257, 361, ?
	7	, none
	8	, none
	9	, if D_f=49 .

(Values 0, 1, 2 are bracketed for "if f is not equivalent to a reversible form"; values 2, 3, 4, 5 are bracketed for "if f is equivalent to a reversible form".)

e.) So far we focused our attention to the number of solutions. We shall report now about their size. We found altogether 19 equations such that at least one of the coordinates of one of their solutions is larger then 100 in absolute value. The largest solution we found was (5047,-1131), which satisfies equation (8). In Table 5 we displayed all polynomials $x^3+ax^2y+bxy^2+cy^3$ for which (1) had a solution $(x,y)\in Z^2$ with max $\{|x|,|y|\} \geq 200$, apart from those occourring in section 5 and in the last two rows of Table 4.

a	b	c	solution	partial quotients
0	-48	1	(-791,114)	10992
0	-49	9	(787,-111)	11299
-1	-47	1	(779,-122)	10746
0	-4	1	(508,273)	1743
1	-17	-4	(393,104)	3473
-7	-3	1	(-362,-49)	2807
0	-28	9	(292,57)	2890
1	-15	-5	(-68,207)	3173

Table 5.

We found another equation, not of type (1), with a very large solution:

$$x^3+x^2y-20xy^2-10y^3=10 \qquad (x,y)=(103411,24309).$$

Comparing the height H(f) of f(x,y) (i.e the maximum of absolute values of the coefficients of f) , and the size of the solutions $(x,y)\in Z^2$ of (1) we get the following observation:

$$\max\{|x|,|y|\} \le H(f)^5.$$

f.) Computing the continued fraction expansion of the roots of polynomials of type x^3+x^2-cx+d we paid attention to the occourring large partial quotients too. We found the following examples larger than 100000.

c	d	n-th	part. quotient	c	d	n-th	part. quotient
-50	-79	176	111221	-20	-6	185	118747
-50	63	22	141550	-19	-12	36	578022
-50	94	64	230065	-18	-17	106	104445
-20	-10	12	104037				

Table 6.

References

[1] BAKER A., DAVENPORT H.: The equations $3x^2-2=y^2$ and $8x^2-7=z^2$, Quart. J. Math. Oxford, 20, 129-137 (1969).

[2] BAULIN V.I.: On an indeterminate equation of the third degree with least positive discriminant (Russian), Tul'sk Gos. Ped. Inst. Ucen. Zap. Fiz. Math. Nauk. Vip. 7, 138-170 (1960).

[3] BOMBIERI E., SCHMIDT W.M.: On Thue's equation, Invent. Math. 88, 69-82 (1987).

[4] BREMNER A.: Integral generators in a certain quartic field and related Diophantine equations, Michigan Math. J. 32, 295-319 (1985).

[5] CANTOR D.G., GALYEAN P.H., ZIMMER H.G.: A continued fraction algorithm for real algebraic numbers, Math. Computation 26, 785-791 (1972).

[6] DELONE (DELAUNAY) B.N.: Über die Darstellung der Zahlen durch die binäre kubischen Formen von negativer Diskriminante, Math. Zeitschr. 31, 1-26 (1930).

[7] DELONE B.N., FADDEEV D.K.: The theory of irrationalities of the third degree, Am. Math. Soc. Transl. of Math. Monographs 10. Providence, 1964.

[8] ELLISON W.J., ELLISON F., PESEK J, STAHL C.E., STALL D.S.: The diophantine equation $y^2+k=x^3$, J. Number Theory 4, 107-117 (1972).

[9] EVERTSE J.H.: On the representation of integers by binary cubic forms of positive discriminant, Invent. Math. 73, 117-138 (1983).

[10] EVERTSE J.H., GYŐRY K.: Thue-Mahler equations with a small number of solutions, to appear.

[11] GAÁL I.: Private communication.

[12] NAGELL T.: Darstellung ganzer Zahlen durch binäre kubische Formen mit negativer Diskriminante, Math. Zeitschr. 28, 10-29 (1928).

[13] LJUNGGREN W.: Einige Bemerkungen über die Darstellung ganzer Zahlen durch binäre kubische Formen mit positiver Diskriminante, Acta Math. 75, 1-21 (1943).

[14] PETHŐ A.: On the resolution of Thue inequalities, J. Symbolic Computation 4, 103-109 (1987).

[15] PETHŐ A., SCHULENBERG R.: Effektives Lösen von Thue Gleichungen, Publ. Math. Debrecen 34, 189-196 (1987).

[16] SIEGEL C.L.: Über einige Anwendungen diophantischer Approximationen, Abh. Preuss. Akad. Wiss. Phys. Math. Kl. 1, (1929).

[17] SIEGEL C.L. (under the pseudonym X): The integer solutions of the equation $y^2=ax^n+bx^{n-1}+...+k$, J. Lond. Math. Soc. 1, 66-68 (1926).

[18] STEINER R.P.: On Mordell's equation $y^2-k=x^3$: A problem of Stolarsky, Math. Computation 46, 703-714 (1986).

[19] TZANAKIS N.: The diophantine equation $x^3-3xy^2-y^3=1$ and related equations, J. Number Theory 18, 192-205 (1984).

[20] ZIMMER H.G.: Computational Problems, Methods and Results in Algebraic Number Theory, Lecture Notes in Math. 262, Springer Verlag, Berlin 1972.

A LINEAR RELATION BETWEEN THETA SERIES OF DEGREE AND WEIGHT 2

Rainer Schulze-Pillot
Institut für Mathematik II
Freie Universität Berlin
Arnimallee 3
D - 1000 Berlin 33

ANDRIANOV conjectured [A] that the theta series of degree n of the quadratic forms in a genus of positive definite integral quadratic forms of rank $m \leq 2n$ are linearly independent. This conjecture was taken up by YOSHIDA in [Y2]. Some support was given to the conjecture by work of HSIA and HUNG [HH 1,2], who showed that in certain genera of quaternary quadratic forms the theta series of degree two of those forms having an improper automorphism are indeed linearly independent. On the other hand, GROSS [G] gave an example of a nontrivial linear relation between ordinary theta series of ternary quadratic forms, thus coming fairly close to the conjected bound $m = 2n$.

It is the purpose of this talk to show that the linear relation from [G] (which has been explicitly computed by KRAMER [Kr]) leads to a nontrivial linear relation between theta series of degree and weight 2, thus giving a counterexample to the conjecture.

Let D be the definite quaternion algebra over $\mathbb{Q}$ which is ramified only at ∞ and at $p = 389$, let tr and n denote the trace and norm respectively. D is known to have 22 types of maximal orders, let $\mathcal{O} = \mathcal{O}_1, \mathcal{O}_2, \ldots, \mathcal{O}_{22}$ be representatives of these types. Let

$$L_i = \{x \in \mathcal{O}_i \mid tr(x) = 0\}$$

$$L_i' = L_i \cap (\mathbb{Z}\cdot 1 + 2\mathcal{O}_i)$$

It was shown by GROSS [G] that the theta series

$$\vartheta(L_i', \tau) = \sum_{x \in L_i'} e(n(x)\tau) \qquad (e(\tau) = \exp(2\pi i\tau), \tau \in H).$$

are linearly dependent. The only nontrivial linear relation between them was computed by KRAMER [Kr]. Moreover, Kramer showed, using the correspondence between modular forms of weight 3/2 and Jacobi forms of weight 2 and index 1, that the same linear relation holds for the Jacobi theta series

$$\vartheta(\mathcal{O}_i,1,\tau,z) = \sum_{x\in\mathcal{O}_i} e(n(x)\tau)e(\mathrm{tr}(x)z).$$

With

$$\vartheta^{(2)}(\mathcal{O}_i,Z) = \sum_{(x,y)\in\mathcal{O}_i\times\mathcal{O}_i} e\left(\mathrm{tr}\begin{pmatrix} n(x), \frac{1}{2}\,\mathrm{tr}(x\bar{y}) \\ \frac{1}{2}\,\mathrm{tr}(x\bar{y}), n(y)\end{pmatrix} Z\right)$$

($Z \in \mathcal{H}_2$, the Siegel upper half plane of genus 2) and denoting by $\Phi_1(\vartheta^{(2)}(\mathcal{O}_i,Z))$ the first Fourier-Jacobi coefficient of $\vartheta^{(2)}(\mathcal{O}_i,Z)$ one has therefore:

$$\sum_{i=1}^{22} \frac{\alpha_i}{e_i}\,\Phi_1(\vartheta^{(2)}(\mathcal{O}_i,Z)) = 0 \qquad (1),$$

where the α_i are the coefficients in Kramer's linear relation $\sum_{i=1}^{22}\alpha_i\,\vartheta(L_i',z) = 0$. The idea for obtaining the desired linear relation between theta series of degree 2 is now quite simple:
Given any Jacobi cusp form

$$\Phi(\tau,z) = \sum_{n,r} c(n,r)e(n\tau)e(rz)$$

of weight k and index 1 for $\Gamma_0(p)$ we can put

$$\Phi|V_\ell(\tau,z) = \sum_{n,r}\ \sum_{\substack{d|(n,r,\ell)\\(d,p)=1}} d^{k-1}\, c\left(\frac{n}{d^2},\frac{r}{d}\right)e(n\tau)e(rz)$$

$$=: \sum_{n,r} c_\ell(n,r)e(n\tau)e(rz).$$

This is a Jacobi cusp form of index k and the same weight and level as Φ. (Note that we have modified the operator V_ℓ from [EZ],p.41, formula (2) for our present situation of level p by summing only over those double cosets

$$\Gamma_0(p)\begin{pmatrix} a & b\\ c & d\end{pmatrix}\Gamma_0(p) \qquad (ad-bc=\ell) \quad \text{with } c \equiv 0 \bmod p,\ (a,p)=1).$$

$$M\Phi := \sum_{\ell,r,n} c_\ell(n,r)e\left(\mathrm{tr}\begin{pmatrix} \ell & r/2\\ r/2 & n\end{pmatrix} Z\right)$$

is then ([EZ], Th. 6.2.) a Siegel cusp form of weight k and degree 2 belonging to the generalized Maaß Spezialschar.

What we want to show is that for a nontrivial linear combination Φ of Jacobi theta series of index 1, $M\Phi$ is again a nontrivial linear combination of theta series of degree 2. Equation (1) will then imply the vanishing of this combination if one takes for Φ Kramer's linear combination of Jacobi theta series.

To this end, let y_i $(i=1,\ldots,h)$ be a set of representatives of the double cosets $\mathcal{O}_A^{\times} y_i D^{\times}$ in $D_A^{\times}$ with $n(y_i)=1$ and let the y_i be ordered such that

$$y_i^{-1}\,\mathcal{O}_A\,y_i\cap D=\mathcal{O}_i \qquad (i=1,\ldots,22),$$

put

$$K_{ij}=y_i^{-1}\,\mathcal{O}_A\,y_j\cap D$$

and define $\varphi : D_A^{\times}\longrightarrow \mathbb{R}$ by

$$\varphi(y_i)=e_i\alpha_i \qquad (i=1,\ldots,t)$$

$$\varphi(y_i)=0 \qquad (i>t)$$

Since (K_{ij},n) represents 1 if and only if $i=j$, the first Fourier-Jacobi coefficient of

$$F(1,\varphi):=\sum_{i,j=1}^{h}\frac{\varphi(y_j)}{e_ie_j}\,\vartheta^{(2)}(K_{ij},Z)$$

is just

$$\sum_{i=1}^{t}\frac{\alpha_i}{e_i}\,\Phi_1(\vartheta^{(2)}(\mathcal{O}_i,Z))=0,$$

and since the $K_{ii}=\mathcal{O}_i$ appear in $F(1,\varphi)$ with coefficient $\frac{\alpha_i}{e_i}$, this linear combination is nontrivial. Finally, Theorem 4.3 of [Y2] implies that the ℓ-th Fourier-Jacobi coefficient of $F(1,\varphi)$ is obtained from the first by applying V_ℓ, thus vanishes, and by [Y1] $F(1,\varphi)$ is cuspidal. We have therefore proved:

Theorem: With the previous notations,

$$\sum_{i=1}^{h}\ \sum_{j=1}^{t}\frac{\alpha_j}{e_i}\,\vartheta^{(2)}(K_{ij},Z)=0$$

is a nontrivial linear relation between the theta series of degree 2 of the quadratic forms in the genus of $(\mathcal{O},n)$.

Remark 1: A much simpler counterexample to the conjecture can be obtained by considering theta series of spinor genera. By a theorem of KNESER and WEIL [Kn], [W], the theta series of degree $\leq m-3$ of a spinor genus are the same for all spinor genera in a fixed genus of integral quadratic forms of rank m. Thus, taking the difference of the theta series of two distinct spinor genera in the same genus yields a nontrivial linear relation. In fact, with some care one can even construct examples where the theta series of degree $m-2$ of different spinor genera coincide, thus showing that KITAOKA's bound $n \geq m-1$ [Ki] is actually best possible in general.

Whereas these last examples can easily be avoided by restricting attention to forms inside one spinor genus, it is not so clear how one should modify Andrianov's conjecture in order to avoid the counterexample of the theorem. A common feature of both types of counterexample is certainly that they can be described as theta liftings of functions on groups that are smaller than the orthogonal group of the forms in question. The connection of our example with that of Gross, however, indicates that questions of vanishing of special values of certain L-series should be considered also - a fact common in the theory of the theta correspondence.

Remark 2: S. BÖCHERER points out that in the above example one should also consider the theta series of degree 3. In fact, it is easy to see that this is a cusp form of degree 3 whose HECKE eigenvalues are determined by those of an elliptic modular form. We will come back to this interesting example in another paper.

References

[A] A.N. ANDRIANOV, Degenerations of rings of Hecke operators on spaces of theta series, Proc. of the Steklov Institute 1983, no. 3, 1-18 (russ. Original: no. 157 (1981)).

[EZ] M. EICHLER, D. ZAGIER, The Theory of Jacobi Forms, Basel 1985.

[G] B. GROSS, Heights and the special values of L-series, Sem. Math. Sup., Presses Univ. Montreal.

[HH1] J.S. HSIA and D.C. HUNG, Theta series of ternary and quaternary quadratic forms, Inv. Math. 73 (1983), 151-156.

[HH2] J.S. HSIA and D.C. HUNG, Theta series of quaternary quadratic forms over Z and $Z[(1+\sqrt{p})/2]$, Acta arith. 45 (1985), 75-91.

[Ki] Y. KITAOKA, Representations of quadratic forms and their application to Selberg's zeta functions, Nagoya Math. J. 63 (1976), 153-162.

[Kn] M. KNESER, Darstellungsmaße indefiniter quadratischer Formen, Math. Z. 77 (1961), 188-194.

[Kr] J. KRAMER, On the linear independence of certain theta series Preprint Max-Planck-Institut Bonn MPI - SFB 85/40.

[W] A. WEIL, Sur la theorie des formes quadratiques, collected works vol. 2, 471-484, Springer Verlag, Berlin-Heidelberg-New York 1979.

[Y1] H. YOSHIDA, Siegel's modular forms and the arithmetic of quadratic forms, Inventiones math. 60 (1980), 193-248.

[Y2] H. YOSHIDA, On Siegel modular forms obtained from theta series, J.f.d. reine und angew. Mathematik 352 (1984), 184-219.

Integral Points on Curves and Surfaces

Joseph H. Silverman*

Introduction

Let $V \subseteq \mathbf{P}^n$ be a smooth quasi-projective variety defined over a number field. If $R \subset \overline{\mathbf{Q}}$ is a finitely generated ring with quotient field K, then one can talk about R-integral sets of points in $V(K)$. We will use the somewhat ambiguous notation $V(R)$ to denote such a set. ($V(R)$ really depends on choosing equations defining V. But note that if V is projective, then $V(R)$ coincides with the set of rational points $V(K)$.) One of the fundamental problems in Diophantine Geometry is to describe the set $V(R)$ in terms of the arithmetic of the ring R and the geometry of the variety V.

In order to somewhat simplify this difficult problem, we will take an approach which concentrates attention on the geometry of V, while relegating the arithmetic of R to a role of lesser importance. To do this, we will fix the variety V and study sets of R-integral points $V(R)$ under the assumption that R is *sufficiently large,* (always subject to the condition that R be finitely generated.) We will make this notion more precise later (cf. Sections 1 & 3,) but the idea is that after equations have been chosen for all varieties and maps being studied, we will allow finite extensions and inversion of finitely many elements of R.

Our goal is to characterize the sets $V(R)$ by their "size." Of course, we will want to distinguish the two cases when $V(R)$ is finite or infinite; but in the case that $V(R)$ is infinite, we will also measure how the points of $V(R)$ are distributed. To do this, let $H_V\colon V(\overline{\mathbf{Q}}) \to [1,\infty)$ be the (absolute, multiplicative) height on V obtained by restricting the usual height function $H_{\mathbf{P}^n}\colon \mathbf{P}^n(\overline{\mathbf{Q}}) \to [1,\infty)$ on $\mathbf{P}^n$. (See [6, Chapter 3] for the precise definition.) Then we define a counting function $\mathcal{N}_{V(R)}$ on $V(R)$ by

$$\mathcal{N}_{V(R)}(H) = \left|\{P \in V(R) : H_V(P) \leq H\}\right|.$$

Main Goal. For a given variety V and all sufficiently large rings R, describe the growth rate of $\mathcal{N}_{V(R)}(H)$ (as a function of H) in terms of simple geometric properties of V.

---

* This research supported by NSF grant DMS-8612393

In the case that V is a curve (i.e. has dimension 1,) we will be able to realize this goal by utilizing theorems of Dirichlet, Néron, Siegel, and Faltings. Thus let V be a smooth curve, and let $\chi(V)$ be the Euler characteristic of the (not necessarily complete) Riemann surface $V(\mathbf{C})$. Then

$$\begin{array}{lll} \chi(V) > 0 & \Longrightarrow & \mathcal{N}_{V(R)} \text{ grows like a power of } H; \\ \chi(V) = 0 & \Longrightarrow & \mathcal{N}_{V(R)} \text{ grows like a power of } \log H; \\ \chi(V) < 0 & \Longrightarrow & \mathcal{N}_{V(R)} \text{ is bounded.} \end{array}$$

(For details, see Section 2.) Notice that it is easy to produce a ring R and a curve V with $\chi(V) = 0$ such that $V(R)$ is finite. However, it is always true that given such a curve, one can find a ring so that $\mathcal{N}_{V(R)}(H)$ grows like a power of $\log H$. This illustrates how by allowing the ring R to grow, we obtain a unified geometric characterization of the counting function $\mathcal{N}_{V(R)}$.

We next turn our attention to surfaces. But now a new phenomenon occurs. It may happen that a surface V contains curves which have many more integral points than one would expect from the surface as a whole. For example, various conjectures predict that the Fermat hypersurface $X^5 + Y^5 + Z^5 + W^5 = 0$ in $\mathbf{P}^3$ should have "few" integral points. On the other hand, it has a "lot" of integral points lying on various lines, such as $[s, -s, t, -t]$ for $[s, t] \in \mathbf{P}^1(\mathbf{Q})$. Clearly the thing to do is to define a new counting function $\mathcal{N}^*_{V(R)}$ which ignores points lying on certain "bad" subvarieties of V. Of course, it may be a delicate matter to decide just which subvarieties are "bad;" and one must also specify to what extent the "bad" subvarieties are to be allowed to depend on the (sufficiently large) ring R. For the moment we will assume that we have managed to define $\mathcal{N}^*_{V(R)}(H)$, and will discuss its behavior as a function of H. (The reader will find a precise definition in Section 3, where $\mathcal{N}^*_{V(R)}$ is the function $\mathcal{N}^{V^*}_{V(R)}$ given by Proposition 3.6. Note it is not known that this function exists for all varieties; using the terminology of Section 3, $\mathcal{N}^*_{V(R)}$ is defined if V has an *arithmetic order.*)

In section 4 we make a case by case study of the counting function $\mathcal{N}^*_{V(R)}$ for surfaces. We start with (smooth) projective surfaces, so integral points coincide with rational points. A fairly coarse geometric invariant of a variety V is its Kodaira dimension $\kappa(V)$. (We refer the reader to [5, Chapter V, Section 6] or [2] for the definition of $\kappa(V)$; but for illustration we mention that if V is a curve, then $\kappa(V) = -1, 0, 1$ if $\chi(V) > 0, = 0, < 0$ respectively.) Unfortunately, $\kappa(V)$ does not in general determine the growth rate of $\mathcal{N}^*_{V(R)}$; but using various well-known results, we collect the following information for smooth projective surfaces:

$\kappa(V) = -1 \implies \mathcal{N}_{V(R)}$ grows like a power of H;

$\kappa(V) = 0 \implies \mathcal{N}_{V(R)}$ $\begin{cases} \text{for abelian and hyperelliptic surfaces,} \\ \quad \text{grows like a power of } \log H; \\ \text{for Enriques and K3 surfaces,} \\ \quad \text{not known in general;} \end{cases}$

$\kappa(V) = 1 \implies \mathcal{N}_{V(R)}$ $\begin{cases} \text{grows at least like a power of } \log H; \\ \text{in some cases, grows exactly} \\ \quad \text{like a power of } \log H; \\ \text{in some cases, grows like a power of } H; \end{cases}$

$\kappa(V) = 2 \implies \mathcal{N}_{V(R)}$ is conjecturally bounded.

Next we consider non-complete surfaces. Specifically, let $\overline{V}$ be a smooth projective surface, and let D be an ample effective divisor on $\overline{V}$. Then we can look at sets of R-integral points on the affine surface $V \overset{\text{def}}{=} \overline{V} \setminus D$. There is a conjecture of Vojta ([13]) which gives a general geometric condition under which $\mathcal{N}^*_{V(R)}$ should be finite. (Vojta's conjecture is actually more general.)

Conjecture. (Vojta) *With notation as above, let $\mathcal{K}_{\overline{V}}$ be the canonical bundle on $\overline{V}$, and let $\mathcal{L}(D)$ be the line bundle associated to the divisor D. Assume that D is a normal crossings divisor. (I.e. Any singularities of D consist of transversal intersections.) If $\mathcal{L}(D) \otimes \mathcal{K}_{\overline{V}}$ is ample, then $\mathcal{N}^*_{V(R)}(H)$ is bounded as $H \to \infty$.*

A striking feature of Vojta's conjecture is that for a given variety $\overline{V}$, the requirement on the divisor D which ensures that $V_D \overset{\text{def}}{=} \overline{V} \setminus D$ has few integral points depends only on the linear equivalence class of D. (I.e. It depends on the line bundle $\mathcal{L}(D)$.) A natural question to ask is whether the growth rate of $\mathcal{N}^*_{V_D(R)}(H)$ is determined in general by the linear equivalence class of D. (This is always subject to the condition that D have normal crossings.) Somewhat surprisingly, the answer is no. We take $\overline{V} = \mathbf{P}^2$, and consider various divisors of degree 3. If D consists of three lines, then one easily checks that $\mathcal{N}^*_{\mathbf{P}^2_D}(H)$ grows like a power of $\log H$; but if D is the sum of a conic and a line, then we show in Section 5 that $\mathcal{N}^*_{\mathbf{P}^2_D}(H)$ grows more rapidly than this. Precisely, we show that

$$D = \text{three lines} \implies \log\log \mathcal{N}^*_{\mathbf{P}^2_D}(H) \sim \log\log\log H \quad \text{as } H \to \infty.$$

$$D = \text{conic} + \text{line} \implies \log\log \mathcal{N}^*_{\mathbf{P}^2_D}(H) \sim \log\log\log H \quad \text{as } H \to \infty.$$

$$D = \text{nodal cubic} \implies \log\log \mathcal{N}^*_{\mathbf{P}^2_D}(H) \sim \log\log\log H \quad \text{as } H \to \infty.$$

$D =$ smooth cubic $\Longrightarrow$

growth rate of $\mathcal{N}^*_{\mathbf{P}^2_D}(H)$ not known

The proof is relatively elementary; the underlying reason that certain of the varieties $\mathbf{P}^2 \setminus D$ have "many" integral points is the fact (proven in Section 5) that a number of the form $\alpha^n - 1$ often has more divisors than one would expect from a number its size. (Cf. Corollary 5.3(ii).)

We conclude in Section 6 by listing a number of open questions concerning the counting functions $\mathcal{N}_{V(R)}(H)$. One of the most elementary to state is suggested by [12], in which the author proves that if V is an affine open subset of an abelian surface, then $\mathcal{N}_{V(R)}(H)$ grows no faster than $\log\log H$. (Earlier, Mumford [7] proved a similar estimate for rational points on curves of genus at least 2. Lang has conjectured that affine open subsets of abelian varieties have only finitely many integral points, so it is natural to ask whether there are any varieties whose counting functions grow this slowly (aside from the trivial ones whose counting functions are bounded.)

Question.. Does there exist a smooth quasi-projective variety $V/\overline{\mathbf{Q}}$ and a finitely generated ring $R \subset \overline{\mathbf{Q}}$ with the following two properties:

(*i*) $$\mathcal{N}_{V(R)}(H) \to \infty \qquad \text{as } H \to \infty$$

(*ii*) There is a constant $c > 0$ such that for all sufficiently large H,

$$\mathcal{N}_{V(R)}(H) \le c \log\log H.$$

(Such varieties do exist over function fields, cf. [7, remark on page 1007]; but even for function fields it is not known if there are varieties in which the bound in (*ii*) is replaced by $\log\log\log(H)$.)

If such varieties exist, then possibly someone with sufficient ingenuity can find one. But if there are no such varieties, then a proof of their non-existence would appear to lie quite deep. (For example, if one had an independent proof that there are no such curves, then Mumford's relatively elementary result [7] would imply the Mordell conjecture.)

Finally, we mention that the reader will find most of this paper concerned with studying what we call the ***arithmetic order*** of a variety. The precise definition of the arithmetic order is a bit convoluted due to the necessity of balancing the "bad set" to be discarded with the size of the "sufficiently large" ring. (Cf. Section 3.) But the basic idea is the following. A variety V has arithmetic order $\overline{\alpha}$ if for all sufficiently large rings R, the counting function $\mathcal{N}^*_{V(R)}(H)$ grows like an $\overline{\alpha}$-fold iterate of the logarithm of H; more precisely, if for some $k \ge 2$,

$$\overbrace{\log\circ\cdots\circ\log}^{k \text{ times}} \mathcal{N}^*_{V(R)}(H) \sim \overbrace{\log\circ\cdots\circ\log}^{k+\overline{\alpha} \text{ times}} H \qquad \text{as } H \to \infty.$$

If $\mathcal{N}^*_{V(R)}(H)$ is bounded, then we define the arithmetic order of V to be ∞. Thus a variant of our earlier question is to ask if there are any varieties whose arithmetic order lies strictly between 1 and ∞.

The arithmetic order of a variety is a fairly coarse measure of the growth rate of its counting function. Never-the-less, the arithmetic order provides a reasonable first estimate as to the distribution of the integral points on the variety; and further, calculating even such a coarse invariant is often quite a difficult (and in many cases still unsolved) problem.

The author would like to thank P. Vojta for his extensive comments on an early draft of this paper, B. Birch and J. Colliot-Thélène for their suggestions during the Ulm conference concerning some of the questions raised in this paper, and, most of all, G. Henniart, E. Wirsing, and the Journées Arithmétiques for providing such a congenial atmosphere in which to do mathematics.

§1. Integral Points on Varieties

Let K be a number field, and let V be a quasi-projective variety defined over K. We will be interested in studying the distribution of the integral points on V. For an abstractly given variety, there is in general no way to choose a particular subset of the rational points to be the set of integral points. Instead, following Vojta ([13]), one defines what it means for a *set* of points to be integral. Then one proves theorems which apply to sets of integral points. Notice in particular that given any one rational point of V, it does not make sense to ask whether that point is integral; integrality is a property of sets of points.

In order to define a particular set of R-integral points in $V(\overline{K})$ for some given ring R, one must fix a particular scheme over R whose generic fiber is the original variety V/K. One can do this for curves (take a minimal regular model) and for abelian varieties (take a Néron model); but in general there is no canonical way to choose such a scheme.

We will use the following concrete description of sets of integral points. (For fancier definitions, which are useful for studying more refined properties of sets of integral points, such as defects, see [13].) By definition, a quasi-projective variety V is a Zariski open subset of a projective variety, so it is given by a finite set of homogeneous polynomials

$$f_1, f_2, \ldots, f_r, g_1, g_2, \ldots, g_s \in K[x_0, \ldots, x_n];$$

V is then obtained by taking the variety defined by the f_i's and discarding the variety defined by the g_i's. Thus the set of points of V defined over an algebraic closure $\overline{K}$ of K is

$$V(\overline{K}) = \{P \in \mathbf{P}^n(\overline{K}) : f_1(P) = \cdots = f_r(P) = 0 \\ \text{and at least one } g_1(P), \cdots, g_s(P) \neq 0\}.$$

For example, the affine line can be described as the subset of $\mathbf{P}^1$ defined by taking no f_i's and $g_1(x_0, x_1) = x_0$. Similarly, the set

$$\{f_1 = x_2^2 x_0 - x_1^3 - x_0^3, \text{ no } g_i\text{'s}\}$$

defines a projective (elliptic) curve; while the set consisting of the same f_1 together with $g_1 = x_0$ defines an affine curve given (in non-homogeneous coordinates) by the equation $y^2 = x^3 + 1$.

Now let R be the ring of integers of K; or, more generally, let R be a finitely generated subring of $\overline{K}$. We are interested in studying sets of R-integral points of V. Since our eventual goal is to study such sets for "sufficiently large" rings R, it makes sense to take R to be "large" now, since this will simplify our definition of integrality. ([13] contains a good discussion of sets of integral points in general.) In what follows, we will say that *R is sufficiently large (relative to the given equations for V)* if it satisfies the following conditions:

R1. R is a finitely generated subring of $\overline{K}$.

R2. R is a principal ideal domain.

R3. All of the polynomials $f_1, f_2, \ldots, f_r, g_1, g_2, \ldots, g_s$ defining V have R-content 1. (The *R-content* of a polynomial is the fractional ideal of R generated by its coefficients; since R is a PID from (R2), we are requiring that the polynomials defining V have integral, relatively prime coefficients.)

Note that these are relatively innocuous assumptions. Given any finitely generated subring R of $\overline{K}$ and any equations for V/K, we can always find a non-zero $D \in R$ so that $R[1/D]$ is sufficiently large.

How should $V(R)$ be defined? In the second example above, we reduced to the non-homogeneous equation $y^2 = x^3+1$; so in this example it would be natural to define $V(R) = \{(x, y) \in R^2 : y^2 = x^3 + 1\}$. We obtained the non-homogeneous equation for V by using the fact that the definition of V includes the condition $g_1 = x_0 \neq 0$, allowing us to invert x_0. Another way to describe $V(R)$ would be as all points $P = [x_0, x_1, x_2] \in \mathbf{P}^2(K)$ satisfying

$$x_0, x_1, x_2 \in R, \qquad x_2^2 x_0 - x_1^3 - x_0^3 = 0, \qquad x_0 \in R^*.$$

(The condition $x_0 \neq 0$ becomes the requirement that x_0 be invertible in R.) More generally, if there are many g_i's, then we should require that the $g_i(P)$'s generate the unit ideal in R. This leads to the following definition.

Definition. Let V/K be a quasi-projective variety given by a set of equations as above, and let $R \subset \overline{K}$ be a sufficiently large ring. The *set of R-integral points of V*, denoted $V(R)$, is defined by

$$V(R) = \{P = [x_0, \ldots, x_n] \in \mathbf{P}^n(K) : x_0, \ldots, x_n \in R, \\ f_1(P) = \cdots = f_r(P) = 0, \quad g_1(P)R + \cdots + g_s(P)R = R\}.$$

If R is not a PID, then one defines $V(R)$ locally. For each prime $v \in \mathrm{Spec}(R)$, one multiplies each g_i by a constant so that its v-adic content is (1). One then requires that there exist homogeneous coordinates for P which are v-adically integral and such that some $g_i(P)$ is a v-adic unit. Note that the constants used to multiply the g_i's and the homogeneous coordinates chosen for P are allowed to change for different v's.

Remark 1.1. If V is projective, then $V(K)$ itself is a set of integral points. (I.e. Since there are no g_i's, it suffices to note that every point in $V(K)$ can be written with homogeneous coordinates in R.) Thus for projective varieties, the notions of integral point and rational point coincide.

Remark 1.2. Suppose that we take two different sets of equations for V/K, and denote the resulting sets of R-integral points by $V(R)$ and $V(R)'$. The two sets of equations give the same variety over K. If we take the polynomials giving the isomorphism between the two sets of equations, and enlarge R so that these polynomials have R-content 1, then the integral points (over the enlarged ring) will match up. In other words, we can find a $D \in R$ so that $V\left(R\left[\frac{1}{D}\right]\right) = V\left(R\left[\frac{1}{D}\right]\right)'$. Hence when talking about more than one set of equations for a given variety, we will add to the notion of ***sufficiently large ring*** R the assumption:

R4. There is an isomorphism between the given sets of equations given by polynomials of content 1.

We will also need a way of ordering the points in $V(R)$. To do this, we take a fixed embedding of V/K into some projective space $\mathbf{P}^n_K$, and define a height function

$$H_V\colon V(\overline{K}) \longrightarrow [1,\infty)$$

to be the restriction of the usual (absolute, multiplicative) height $H\colon \mathbf{P}^n(\overline{K}) \to [1,\infty)$. (For the precise definition, see [6]. We remind the reader that on $\mathbf{P}^n(\mathbf{Q})$, H is defined as follows: Write $P \in \mathbf{P}^n(\mathbf{Q})$ as $P = [x_0,\ldots,x_n]$ with $x_0,\ldots,x_n \in \mathbf{Z}$ and $\gcd(x_0,\ldots,x_n) = 1$; then $H(P) = \max\{|x_0|,\ldots,|x_n|\}$.) In particular, there are only finitely many points in $V(K)$ with height less than any given constant. This makes the height function a good tool for counting points, and leads to the following definition.

Definition. Let V/K be a quasi-projective variety given by equations as above, let R be a sufficiently large ring, and let $V(R)$ be the set of R-integral points of V. The ***counting function for*** V/R, denoted $\mathcal{N}_{V(R)}$, is defined by

$$\mathcal{N}_{V(R)}(H) = \bigl|\{P \in V(R) : H_V(P) \le H\}\bigr|.$$

Remark 1.3. The definition of $\mathcal{N}_{V(R)}$ clearly depends on the choice of the height function H_V, although our notation does not reflect this fact. However, if H'_V is the height for some other projective embedding of V, then elementary properties of height functions (cf. [6, Chapter 4, Proposition 5.4]) imply that there are constants $c_1, c_2, c_3, c_4 > 0$ so that

$$c_1 H'_V(P)^{c_2} \le H_V(P) \le c_3 H'_V(P)^{c_4} \qquad \text{for all } P \in V(\overline{K}).$$

Substituting this estimate into the definition of $\mathcal{N}_{V(R)}$, we obtain the following inequality.

Lemma 1.4. *Let $V(R)$, H_V, and H'_V be as above, and let $\mathcal{N}_{V(R)}$ and $\mathcal{N}'_{V(R)}$ be the counting functions for $V(R)$ corresponding to H_V and H'_V respectively. Then there are constants $c_1, c_2, c_3, c_4 > 0$ so that*

$$\mathcal{N}_{V(R)}(c_1 H^{c_2}) \subseteq \mathcal{N}'_{V(R)}(H) \subseteq \mathcal{N}_{V(R)}(c_3 H^{c_4}) \qquad \textit{for all } H \ge 1.$$

Example 1.5. Let $V = \mathbf{P}^n$, and let H_V be the usual height function on $\mathbf{P}^n$. Then Schanuel [8] has given a precise asymptotic formula for $\mathcal{N}_{V(R)}$:

$$\mathcal{N}_{V(R)}(H) \sim \frac{hR/w}{\zeta_K(n+1)} \left(\frac{2^{r_1}(2\pi)^{r_2}}{\sqrt{D_K}} \right)^{n+1} (n+1)^r H^{(n+1)d_K} \qquad \text{as } H \to \infty.$$

Here h,R,w,$\zeta_K(n+1)$,r,r_1,r_2,D_K and d_K are constants associated to K (cf. [8] or [6, Chapter 3, Theorem 5.3]); in any case, they are independent of H.

Example 1.6. Let V/K be an abelian variety. Then Néron proved that

$$\mathcal{N}_{V(R)}(H) \sim C_{V/K}(\log H)^{\frac{1}{2}\operatorname{rank} V(K)} \qquad \text{as } H \to \infty$$

(cf. [6, Chapter 5, Theorem 7.5].) Here the constant $C_{V/K}$ depends on V, K, and the choice of the height function H_V used in the definition of $\mathcal{N}_{V(R)}$.

Example 1.7. The above examples are projective, so now we consider an affine variety. Let $V = \mathbf{G}_m = \mathbf{P}^1 \setminus \{0, \infty\}$. V may be defined by taking no f_i's and $g_1 = x_0 x_1$. The set of integral points is then given by

$$V(R) = \{[x_0, x_1] \in \mathbf{P}^1(K) : x_0, x_1 \in R \quad \text{and} \quad x_0 x_1 \in R^*\}.$$

Thus $V(R) \cong R^*$ via $[x_0, x_1] \to x_0$. An easy counting argument ([6, Chapter 3, Theorem 5.2]) shows that

$$\mathcal{N}_{V(R)}(H) \sim C_R(\log H)^{\operatorname{rank} R^*} \qquad \text{as } H \to \infty,$$

where C_R is a certain constant depending on R.

In each of the above examples, we were able to give asymptotic formulas for the counting function $\mathcal{N}_{V(R)}(H)$; but for general varieties it is too much to ask for such precise formulas (at least, it is given our present state of knowledge.) So we will ask only for a general idea of how $\mathcal{N}_{V(R)}(H)$ grows as a function of H. For the above examples, we note that the quantity $\log\log\left(\mathcal{N}_{V(R)}(H)\right)$ grows like a well-behaved function of H, essentially independently of the field K or the ring R. (To be precise, for examples 1.6 and 1.7 we must enlarge R to ensure that $\operatorname{rank} V(R) \geq 1$ and $\operatorname{rank} R^* \geq 1$ respectively.) Thus for the given variety V, we have the following formulas, valid provided the ring R is large enough:

$$V = \mathbf{P}^n \qquad \frac{\log\log \mathcal{N}_{V(R)}(H)}{\log\log H} \sim 1 \qquad \text{as } H \to \infty.$$

$$V = \text{abelian variety} \qquad \frac{\log\log \mathcal{N}_{V(R)}(H)}{\log\log\log H} \sim 1 \qquad \text{as } H \to \infty.$$

$$V = \mathbf{G}_m = \mathbf{P}^1 \setminus \{0,\infty\} \qquad \frac{\log\log \mathcal{N}_{V(R)}(H)}{\log\log\log H} \sim 1 \qquad \text{as } H \to \infty.$$

It thus seems reasonable to describe the growth rate of $\mathcal{N}_{V(R)}(H)$ by comparing $\log\log\left(\mathcal{N}_{V(R)}(H)\right)$ to some multiple iterate $\log\ldots\log H$. This prompts the following preliminary definition. (This definition will work well for curves, but will require some modification for higher dimensional varieties. See Section 3 for details)

Notation. For any integer $k \geq 0$, let $\log^{(k)}$ denote the kth iterate

$$\log^{(k)}(t) = \overbrace{\log \circ \log \circ \cdots \circ \log}^{k \text{ times}}(t).$$

(Preliminary) Definition. Let V/K be a quasi-projective variety given by equations as above, let R be a sufficiently large ring, and let $V(R)$ be the set of R-integral points of V. We will say that V has *arithmetic order* α *(over* R*)* if

$$\lim_{H\to\infty} \frac{\log^{(2)} \mathcal{N}_{V(R)}(H)}{\log^{(2+\alpha)} H} = 1.$$

If the arithmetic order of V over R exists, we denote it by $\alpha(V(R))$. By convention, we will set $\alpha(V(R)) = \infty$ if $\mathcal{N}_{V(R)}(H)$ is bounded (i.e. if $V(R)$ is finite.)

We then define the ***absolute (or stable) arithmetic order of*** V to be α if for every set of equations for V there exists a finitely generated ring $R_0 \subset \overline{K}$ so that for all finitely generated rings $R_0 \subseteq R \subset \overline{K}$,

$$\alpha(V(R)) = \alpha.$$

If the absolute arithmetic order exists, we denote it by $\overline{\alpha}(V)$. (Note that the choice of the ring R_0 is allowed to depend on the particular equations chosen for V.)

Intuition. The arithmetic order of a variety V is large if the integral points on V are sparsely distributed.

Example 1.8. Using the asymptotic formulas given above, we see that

$$\overline{\alpha}(\mathbf{P}^n) = 0, \qquad \overline{\alpha}(\text{abelian variety}) = 1, \qquad \overline{\alpha}(\mathbf{G}_m) = 1.$$

Similarly, Faltings' theorem says that if V is a projective curve of genus at least 2, then $\overline{\alpha}(V) = \infty$.

The above definition of the arithmetic order of a variety appears to depend on the choice of a particular height function H_V, since the counting function $\mathcal{N}_{V(R)}(H)$ depends on this choice. (It also may appear to depend on the equations chosen for V. But from assumption (R4), we see that if we choose a new set of equations, then we will get the same set of integral points for all sufficiently large R.) We now check that the arithmetic order of V over R is independent of the choice of H_V (assuming it exists at all!)

Proposition 1.9. (a) *If the arithmetic order $\alpha(V(R))$ exists for one choice of height function H_V, then it exists for any choice of height function, and is independent of that choice.*
(b) *If the absolute arithmetic order of V exists for one choice of height function H_V, then it exists for any choice of height function, and is independent of that choice.*

Proof. (a) Let H_V and H'_V be height functions corresponding to two different embeddings, and let $\mathcal{N}_{V(R)}$ and $\mathcal{N}'_{V(R)}$ be the corresponding counting functions. Suppose that the arithmetic order $\alpha = \alpha(V(R))$ exists using the counting function $\mathcal{N}_{V(R)}$. Then using lemma 1.1, we obtain

$$\frac{\log^{(2+\alpha)} c_1 H^{c_2}}{\log^{(2+\alpha)} H} \cdot \frac{\log^{(2)} \mathcal{N}_{V(R)}(c_1 H^{c_2})}{\log^{(2+\alpha)} c_1 H^{c_2}} \le \frac{\log^{(2)} \mathcal{N}'_{V(R)}(H)}{\log^{(2+\alpha)} H}$$
$$\le \frac{\log^{(2+\alpha)} c_3 H^{c_4}}{\log^{(2+\alpha)} H} \cdot \frac{\log^{(2)} \mathcal{N}_{V(R)}(c_3 H^{c_4})}{\log^{(2+\alpha)} c_3 H^{c_4}}.$$

Letting $H \to \infty$, the left and righthand sides approach 1, which proves that the arithmetic order computed using $\mathcal{N}'_{V(R)}$ exists and equals α. (We also see why one must take $\log^{(2)} \mathcal{N}_{V(R)}$ in the definition of the arithmetic order, rather than something like $\log \mathcal{N}_{V(R)}$. The point is that $\log^{(2)}(aH^b) \sim \log^{(2)}(H)$, so taking $\log^{(2)}$ is precisely what is needed to eliminate the ambiguity in the choice of height function.)
(b) This is immediate from (a). □

§2. Integral Points on Curves

In this section we will study counting functions and arithmetic orders for curves (i.e smooth, one dimensional varieties.) The arithmetic order is a somewhat coarse measure of the distribution of integral points on a variety, so it is perhaps not surprising that it is completely understood for curves. However, one might be surprised (at least at first) to find that the arithmetic order of a curve is completely determined by the coarsest imaginable topological invariant, namely the sign of the Euler characteristic of the corresponding Riemann surface. We examine the three cases in more detail.

Positive Euler Characteristic.

There are two curves with positive Euler characteristic, namely the projective line (with $\chi(C) = 2$) and the affine line (with $\chi(C) = 1$). In both cases, it is an easy exercise to show that $\mathcal{N}_{C(R)}(H)$ grows polynomially in H (provided that R is chosen sufficiently large and such that $C(R) \neq \emptyset$.) Hence in both cases, $\overline{\alpha}(C) = 0$.

Zero Euler Characteristic.

The curves with zero Euler characteristic are all group varieties. First, there is the multiplicative group $\mathbf{G}_m = \mathbf{P}^1 \setminus \{0, \infty\}$. As noted in the previous section, $\mathbf{G}_m(R) \cong R^*$. Now using Dirichlet's theorem on the finite generation of R^* and elementary properties of heights, one easily shows that $\mathcal{N}_{C(R)}(H)$ grows polynomially in $\log H$. Hence $\overline{\alpha}(\mathbf{G}_m) = 1$.

Second, there are elliptic curves. The Mordell-Weil theorem says that the group of integral (i.e. rational) points on an elliptic curve is finitely generated; and Néron combined this with a height calculation to prove that $\mathcal{N}_{C(R)}(H)$ grows polynomially in $\log H$. (Cf. Example 1.6.) Again we find that $\overline{\alpha}(C) = 1$.

Negative Euler Characteristic.

Here there are three possibilities. First, there is $\mathbf{P}^1$ with (at least) three points removed, say $C = \mathbf{P}^1 \setminus \{0, 1, \infty, P_1, \ldots, P_m\}$ with $m \geq 0$. (Remember that we are free to enlarge the ring R so as to obtain 3 rational points on the projective closure of C.) If $[x, y] \in \mathbf{P}^1(R)$ is chosen with $xR + yR = R$, then $[x, y]$ will be in $C(R)$ if and only if x, y, and $x - y$ are all units. But then $u = \frac{x}{x-y}$ and $v = \frac{-y}{x-y}$ give a solution to the unit equation

$$u + v = 1 \qquad u, v \in R^*.$$

It was proven by Siegel and Lang that there are only finitely many solutions [6, Chapter 8, Section 5], so we find that $\mathcal{N}_{C(R)}(H)$ is bounded, and hence $\overline{\alpha}(C) = \infty$.

Second, there are elliptic curves with (at least) one point removed. Enlarging the ring R, we may write $C = E \setminus \{O, P_1, \ldots, P_m\}$ for some $m \geq 0$, where we take $O \in E(K)$ as the origin of the elliptic curve E. Let $x \in K(E)$ be a rational function on E with a double pole at O. Then (again enlarging R if necessary,) every point $P \in C(R)$ will have $x(R) \in R$. Again, by a theorem of Siegel [6,

Chapter 8, Theorem 1.3], there can be only finitely many such points; and so again we find that $\mathcal{N}_{C(R)}(H)$ is bounded and $\overline{\alpha}(C) = \infty$.

Third, there are curves of genus two or greater with any number (including zero) points removed. From Faltings' theorem [4, Theorem 7], such a curve can have only finitely many rational points, and so $\mathcal{N}_{C(R)}(H)$ is bounded and $\overline{\alpha}(C) = \infty$.

The above discussion is summarized in Table 1 below.

Remark 2.1. There is a theorem of Mumford (cf. [6,Chapter 5, Theorem 8.1]) which we want to mention here, even though the version that we state has been superceded by Faltings' result. Let C/K be a projective curve of genus at least 2. Then Mumford proved that the integral points on C (i.e. points in $C(K)$) are quite sparsely distributed; in our notation, he proved that

$$\mathcal{N}_{C(R)}(H) \ll \log\log H.$$

(As usual, $f(H) \ll g(H)$ means that there is a constant $c > 0$ so that $f(H) \leq cg(H)$ for all sufficiently large H.) In other words, Mumford proved that if the arithmetic order of $\overline{\alpha}(C)$ exists, then it satisfies $\overline{\alpha}(C) \geq 2$. Of course, Faltings' theorem says that $\overline{\alpha}(C) = \infty$; but this brings up a natural problem, to which we will return, of exhibiting even a single example of a variety which has arithmetic order strictly between 1 and ∞. (We remind the reader that we have not yet given the complete definition for $\overline{\alpha}(V)$ in the case that V has dimension greater than 1. See section 3.)

As we see from Table 1, the arithmetic order of a curve, whether affine or projective, is determined by the geometry of the associated Riemann surface. This leads us to our main objective, which will only be partially realized in this paper:

Main Goal. Characterize the arithmetic order of (smooth) varieties in terms of the geometry of their associated complex manifolds.

§3. The Arithmetic Order of Varieties

In section 1 we gave a preliminary definition of the arithmetic order of an algebraic variety. This arithmetic order measures the distribution of integral points on a variety; the larger the arithmetic order, the more sparsely distributed are the points. In the last section we saw that our original definition works quite well for curves; the arithmetic order of a curve is determined by the topology of its corresponding Riemann surface. When we turn to higher dimensional varieties, however, a new phenomenon occurs. The given variety may contain proper subvarieties which have many more integral points than one would expect from the variety as a whole.

$\chi(C)^\dagger$	C	$\log\log \mathcal{N}_C(H)$	$\overline{\alpha}(C)$	
>0	$\mathbf{P}^1$	$\sim \log\log H$	0	easy
	$\mathbf{A}^1 = \mathbf{P}^1 \setminus \{P\}$			easy
$=0$	$\mathbf{G}_m = \mathbf{P}^1 \setminus \{P_1, P_2\}$	$\sim \log\log\log H$	1	Dirichlet
	E (genus $E = 1$)			Néron
<0	$\mathbf{P}^1 \setminus \{P_1, P_2, P_3\}$	$= O(1)$	∞	Siegel
	$E \setminus \{P_1\}$ (genus $E = 1$)			Siegel
	C (genus $C \geq 2$)			Faltings

† $\chi(C)$ = Euler characteristic of $C(\mathbf{C})$

The Arithmetic Order of Curves

(Table 1)

Example 3.1. Let V be the affine surface given by the equation

$$V : x^5 + y^5 + z^5 = 1.$$

(Using the conventions from section 1, V would be the surface in $\mathbf{P}^3$ defined by the two equations $f_1 = x_0^5 + x_1^5 + x_2^5 - x_3^5$ and $g_1 = x_3$.) For a number of reasons, one does not expect V to have very many integral points. (There are conjectures of Bombieri and Vojta which suggest that in fact $V(K)$ should be "small".) However, the variety V contains several affine lines, and these lines contribute a large number of integral points. For example, for any finitely generated ring $R \subset \bar{\mathbf{Q}}$,

$$V(R) \supset \{(1, t, -t) : t \in \mathbf{Z}\}.$$

Hence $\mathcal{N}_{V(R)}(H) \geq 2H$, so according to our original definition, $\alpha(V(R)) = \overline{\alpha}(V) = 0$. But in fact we would like to have a definition which, for this example, yields (at least conjecturally) the value $\overline{\alpha}(V) = \infty$.

It is clear how to proceed. Rather than looking at all of the integral points on the variety V, we instead look at only those points lying in a certain Zariski open subset U of V. (In Vojta's terminology, we discard a *degenerate* subset of $V(K)$.) Then we define the arithmetic order of V by using a counting function which only counts points in U. This prompts the following definition.

Definition. Let V/K be a quasi-projective variety, $U \subseteq V$ a non-empty Zariski open subset of V, and R a sufficiently large ring. The *counting function for V over R relative to U* is defined to be

$$\mathcal{N}^U_{V(R)}(H) = \left|\{P \in V(R) : H_V(P) \leq H \quad \text{and} \quad P \in U\}\right|.$$

This seemingly solves our problem, but there remains the question of how to choose U. As the following (almost trivial) example indicates, the choice of U can be a somewhat delicate matter. In particular, it makes a great deal of difference whether the set U or the ring R is to be fixed first.

Example 3.2. Let C be a (projective) curve of genus at least 2, and let V be the product variety $V = C \times \mathbf{P}^1$. Then we have the following two facts.

(a) For any Zariski open subset $U \subseteq V$ there exists a finitely generated ring R so that $\mathcal{N}^U_{V(R)}(H) \to \infty$ as $H \to \infty$. (In fact, so that

$$\frac{\log\log \mathcal{N}^U_{V(R)}(H)}{\log\log H} \to 1 \qquad \text{as } H \to \infty.)$$

(b) For any finitely generated ring R there exists a non-empty Zariski open subset $U \subseteq V$ so that

$$\mathcal{N}^U_{V(R)}(H) = 0.$$

To check (a), it suffices to take any R for which $V(R) \cap U \neq \emptyset$. For if $(P_0, Q_0) \in V(R) \cap U$, then $\mathcal{N}^U_{V(R)}$ will include all but finitely many points in $\{P_0\} \times \mathbf{P}^1(R)$. Hence $\mathcal{N}^U_{V(R)}(H)$ will be at least as large as $\mathcal{N}_{\mathbf{P}^1(R)}(H)$. To verify (b), we note that for any R, Faltings' theorem implies that $C(R)$ is finite. Hence $C(R) \times \mathbf{P}^1$ is a proper Zariski closed subset of V. Since $V(R) = C(R) \times \mathbf{P}^1(R) \subset C(R) \times \mathbf{P}^1$, it suffices to take $U = V \setminus \left(C(R) \times \mathbf{P}^1\right)$ to obtain $\mathcal{N}^U_{V(R)}(H) = 0$.

If we use the obvious notation $\alpha^U(V(R))$ for the arithmetic order of V over R obtained by counting only points in U, then we can summarize the above example as follows:

Given U, there exists an R so that $\alpha^U(V(R)) = 0$.

Given R, there exists a U so that $\alpha^U(V(R)) = \infty$.

But now how should we define the absolute arithmetic order of V? Which should we fix first, the set U or the ring R? The answer depends on what value we want $\overline{\alpha}(V)$ to have. We will make the choice that $\overline{\alpha}(V)$ should equal 0, which means that U should be picked first. Our reasons for doing so will become clearer in the next section, but let us briefly say that with this definition there seems to be a better correspondence between the arithmetic order of a variety and its geometry. For example, with this definition there is a conjecture of Lang which (essentially) says that a variety has arithmetic order ∞ if and only if it is of general type.

On the other hand, with the definition we have chosen, it is not possible to define the arithmetic order of V over each (large) ring R, and then define the absolute arithmetic order as the limiting value for increasing R. We thus go directly to the definition of the absolute arithmetic order, which we will generally just call the arithmetic order.

Definition. Let V/K be a quasi-projective variety. We will say that V has *(absolute or stable) arithmetic order* α if there is an integer $k \geq 2$ and a non-empty Zariski open subset $U_0 \subseteq V$ so that for each non-empty Zariski open subset $U \subseteq U_0 \subseteq V$ there is a finitely generated ring $R_U \subset \overline{K}$ so that for all finitely generated rings $R_U \subseteq R \subset \overline{K}$,

$$\lim_{H \to \infty} \frac{\log^{(k)} \mathcal{N}^U_{V(R)}(H)}{\log^{(k+\alpha)} H} = 1.$$

If the arithmetic order of V exists, we denote it by $\overline{\alpha}(V)$.

Remark 3.3. As usual, the choice of the ring R_U is allowed to depend on the particular equations chosen for V and U.

Remark 3.4. The reader will notice that we have made one further change from our preliminary definition, namely in the use of $\log^{(k)}$ rather than $\log^{(2)}$. This can be justified on the grounds that the arithmetic order is supposed to be a fairly coarse measure of the distribution of integral points; and even with the weaker definition using k, it is still true that if V_1 and V_2 have arithmetic orders α_1 and α_2 with $\alpha_1 > \alpha_2$, then V_2 has many more integral points than V_1. It may also be justified by examples which we will describe in the next section. (See Examples 4.12, 4.13.) We will show that for certain affine surfaces $V \subset \mathbf{P}^2$ and open subsets $U \subset V$,

$$\log^{(2)} \mathcal{N}^U_{V(R)}(H) \gg \frac{\log\log H}{\log\log\log H}.$$

Further, a heuristic argument (which is far too speculative for the author to dare to publish) suggests that the opposite inequality $\ll$ might also hold. Thus if we only allowed $k = 2$ in the definition of arithmetic order, then it is possible that the arithmetic order of this variety would not exist. On the other hand, taking $k = 3$ gives $\overline{\alpha}(V) = 0$.

Remark 3.5. We note again that the way the definition of $\overline{\alpha}(V)$ is set up, the ring R_U does not need to be chosen until the affine subset U is specified. To summarize, the arithmetic order of V is α if the following holds:

There exists an integer $k \geq 2$

and a non-empty open subset $U_0 \subseteq V$ such that

for all non-empty open subsets $U \subseteq U_0 \subseteq V$,

there exists a ring R_U such that

for all finitely generated rings $R_U \subseteq R \subset \overline{K}$,

$$\log^{(k)} \mathcal{N}^U_{V(R)}(H)/\log^{(k+\alpha)} H \to 1 \qquad \text{as} \qquad H \to \infty.$$

Although it is far from clear that every variety has an arithmetic order, we will at least show that if the arithmetic order does exist, then there is a largest open set U_0 which can be used in the definition of arithmetic order.

Proposition 3.6. *Let V/K be as usual, and assume that the arithmetic order of V exists. Then there exists a Zariski open subset $V^* \subseteq V$ such that the following holds:*

Let $k \geq 2$ be an integer, let $U \subseteq V$ be a Zariski open set, and let $R_U \subset \overline{K}$ be a finitely generated ring. Suppose that for every finitely generated ring $R_U \subseteq R \subset \overline{K}$,

$$\lim_{H\to\infty} \frac{\log^{(k)} \mathcal{N}^U_{V(R)}(H)}{\log^{(k+\overline{\alpha}(V))} H} = 1. \tag{$*$}$$

Then $U \subseteq V^*$.

Proof. Let Σ be the collection of Zariski open subsets U of V for which there exists a k and an R_U such that the limit $(*)$ exists for all $R \supseteq R_U$. Let U_1 and U_2 be in Σ, and let k_1 and k_2 and $R_{U,1}$ and $R_{U,2}$ correspond to U_1 and U_2 respectively. Note that for any ring R, we have the trivial bounds

$$\max\left\{\mathcal{N}^{U_1}_{V(R)}(H), \mathcal{N}^{U_2}_{V(R)}(H)\right\} \leq \mathcal{N}^{U_1\cup U_2}_{V(R)}(H) \leq 2\max\left\{\mathcal{N}^{U_1}_{V(R)}(H), \mathcal{N}^{U_2}_{V(R)}(H)\right\}.$$

Applying $\log^{(k)}(\,\cdot\,)$, dividing by $\log^{(k+\overline{\alpha}(V))}(H)$, and letting $H \to \infty$, it follows immediately that $U = U_1 \cup U_2 \in \Sigma$. (E.g. Take $k = \max\{k_1, k_2\}$ and R_U any ring containing both $R_{U,1}$ and $R_{U,2}$.) This shows that Σ is closed under finite unions. But a quasi-projective variety with its Zariski topology is a Noetherian topological space (cf. [5, exercises I.1.7 and I.2.5].) Further, since we have assumed that the arithmetic order of V exists, the set Σ is not empty. Therefore Σ has a maximal element V^*. Now for any $U \in \Sigma$, we have $V^* \subseteq V^* \cup U \in \Sigma$; hence by the maximality of V^*, $U \subseteq V^*$. □

Notation. If the arithmetic order of V exists, then we will let V^* denote the open set described in Proposition 3.1.

Example 3.7. Continuing with our example $V = C \times \mathbf{P}^1$ from above, we see that

$$\overline{\alpha}(V) = 0 \qquad \text{and} \qquad V^* = V.$$

Example 3.8. On the other hand, it is not hard to construct examples in which $V^* \neq V$. For example, let V be the subset of $\mathbf{P}^2$ defined by $xyz(x+y+z) \neq 0$. Then a theorem of W. Schmidt (cf. Example 4.10) implies that $\overline{\alpha}(V) = \infty$; while the map $R^* \to V$ defined by $u \mapsto [1, -1, u]$ shows that $\mathcal{N}_{V(R)}(H)$ grows at least like a power of $\log(H)$. Hence $V \setminus V^*$ must contain at least the line $x + y = 0$. In fact, for this example one can specify V^* precisely:

$$V \setminus V^* = \{x+y=0\} \cup \{x+z=0\} \cup \{y+z=0\}.$$

We next prove some elementary properties concerning the arithemtic order of varieties connected by various sorts of maps. These will be useful when we discuss surfaces.

Proposition 3.9. *Let V and V' be (quasi-projective) varieties, and let $\phi\colon V' \to V$ be an algebraic map between them.*

(a) *If ϕ is an étale morphism, and one of the arithmetic orders $\overline{\alpha}(V)$ and $\overline{\alpha}(V')$ exists, then the other one exists and they are equal.*

(b) *If V and V' are projective, if ϕ is a birational isomorphism (i.e. ϕ is an isomorphism on some non-empty Zariski open subset of V',) and if one of the arithmetic orders $\overline{\alpha}(V)$ and $\overline{\alpha}(V')$ exists, then the other one exists and they are equal.*

Proof. To ease notation, we write $\overline{\alpha}$ and $\overline{\alpha}'$ for $\overline{\alpha}(V)$ and $\overline{\alpha}(V')$ respectively.

(a) Since ϕ is étale, the inverse image $\phi^{-1}(D)$ of any ample divisor D on V will be ample on V'. It follows from [6, Chapter 4 Proposition 5.4] that there are positive constants c_1, c_2, c_3, c_4 so that

$$c_1 H_{V'}(P)^{c_2} \leq H_V \circ \phi(P) \leq c_3 H_{V'}(P)^{c_4} \qquad \text{for all } P \in V'(\overline{K}).$$

Let us suppose first that $\overline{\alpha}$ exists. Choose an integer k and open set U_0 as in the definition of arithmetic order for V. We will use the same k on V', and will take the open set $U_0' \stackrel{\text{def}}{=} \phi^{-1}(U_0)$ as our open set on V'.

Now let $\emptyset \neq U' \subseteq U_0'$ be any Zariski open set of V'. Then $U \stackrel{\text{def}}{=} \phi(U') \subseteq U_0$ is an open subset of V, so there is a ring R_U corresponding to U in the definition of $\overline{\alpha}$. Since ϕ is étale, the Chevalley-Weil theorem ([6, Chapter 2, Theorem 8.1]) says that the points in $\phi^{-1}(V(R_U))$ are all defined over a finite extension of R_U. Let $R'_{U'}$ be a ring containing R_U such that $\phi^{-1}(V(R_U)) \subseteq V'(R'_{U'})$. We will take $R'_{U'}$ to be the ring corresponding to U' in the definition of $\overline{\alpha}'$, and show that with this choice, $\overline{\alpha}'$ exists.

Let R be any finitely generated ring, $R'_{U'} \subseteq R \subset \overline{K}$. Since also $R \supset R_U$, the definition of $\overline{\alpha}$ gives

$$\log^{(k)}\left(\mathcal{N}^{U}_{V(R)}(H)\right) \sim \log^{(k+\overline{\alpha})}(H).$$

Note that this relation holds in particular for $R = R_U$.

Next, we note that there are set inclusions

$$V(R_U) \cap U \subseteq \phi(V'(R'_{U'})) \cap \phi(U') \subseteq \phi(V'(R) \cap U') \subseteq V(R) \cap U.$$

It follows from these inclusions and the comparison between H_V and $H_{V'}$ given above that

$$\mathcal{N}^{U}_{V(R_U)}(c_1 H^{c_2}) \leq \mathcal{N}^{U'}_{V'(R)}(H) \leq (\deg\phi)\mathcal{N}^{U}_{V(R)}(c_3 H^{c_4}),$$

where c_1, c_2, c_3, c_4 are positive constants. Now apply $\log^{(k)}(\,\cdot\,)$ to this inequality, divide by $\log^{(k+\overline{\alpha})}(H)$, let $H \to \infty$, and use the limit given above. This yields

$$\lim_{H\to\infty} \frac{\log^{(k)} \mathcal{N}^{U'}_{V'(R)}(H)}{\log^{(k+\overline{\alpha})} H} = 1,$$

thereby proving that $\overline{\alpha}'$ exists and that it equals $\overline{\alpha}$.

We now consider the case that $\overline{\alpha}'$ exists. Since the proof is similar, we will omit some details. Choose k' and U'_0 for V', and let $U_0 \stackrel{\text{def}}{=} \phi(U'_0)$. Note that U_0 is an open subset of V, since ϕ is étale. Let $\emptyset \neq U \subseteq U_0$ be an open set, and choose $R'_{U'}$ corresponding to $U' \stackrel{\text{def}}{=} \phi^{-1}(U)$. We will take $R_U = R'_{U'}$ and show that $\overline{\alpha}$ exists.

Let $R \supset R_U$. As above, the Chevalley-Weil theorem says that there is an $R' \supset R$ such that $\phi^{-1}(V(R)) \subseteq V(R')$. It follows that we have inclusions

$$\phi(V'(R) \cap U') \subseteq V(R) \cap U \subseteq \phi(V'(R') \cap U'),$$

which as above lead to the estimate

$$\frac{1}{\deg\phi}\mathcal{N}^{U'}_{V'(R)}\left((c_3^{-1}H)^{1/c_4}\right) \leq \mathcal{N}^{U}_{V(R)}(H) \leq \mathcal{N}^{U'}_{V'(R')}\left((c_1^{-1}H)^{1/c_2}\right).$$

Now applying $\log^{(k')}(\,\cdot\,)$, dividing by $\log^{(k'+\overline{\alpha})}(H)$, and letting $H \to \infty$ gives the desired result.

(b) Since birationality is an equivalence relation, we may assume by symmetry that $\overline{\alpha}$ exists. Choose an open set U_0 and integer k for V as in the definition of $\overline{\alpha}$. Since ϕ is a birational isomorphism, there is a non-empty open subset of U_0 on which ϕ is an isomorphism. Such a subset will also work in the definition of $\overline{\alpha}$, so without loss of generality we may assume that $\phi\colon \phi^{-1}(U_0) \xrightarrow{\sim} U_0$ is an isomorphism. We take $U_0' \stackrel{\text{def}}{=} \phi^{-1}(U_0)$ as our open set on V'.

Now take any open set $\emptyset \neq U' \subseteq U_0'$, and let $U \stackrel{\text{def}}{=} \phi(U')$. Since $\phi\big|_{U'}$ is an isomorphism, it follows that for any ring R there is a one-to-one correspondence

$$\phi\colon V'(R) \cap U' \xrightarrow{\sim} V(R) \cap U.$$

(N.B. This is where we use the fact that V and V' are projective. In general the isomorphism $\phi\colon U_0' \xrightarrow{\sim} U_0$ would not take integral points to integral points. But in our case there is no distinction between integral points and rational points, and this isomorphism does give a one-to-one correspondence between rational points.)

Next we must compare the height functions H_V and $H_{V'}$. For any rational map, such as ϕ, a trivial triangle inequality argument ([6, Chapter 4, Proposition 1.7]) gives a relation

$$H_V \circ \phi(P) \leq c_1 H_{V'}(P)^{c_2} \qquad \text{for all } P \in V'(\overline{K}) \text{ such that } \phi(P) \text{ is defined.}$$

(We have also used [6, Chapter 4, Proposition 5.4].) In particular, this inequality holds for all $P \in V'(R) \cap U'$. But $\phi^{-1}\colon V' \to V$ is also a rational map. Hence we obtain a similar inequality

$$H_{V'} \circ \phi^{-1}(Q) \leq c_3 H_V(Q)^{c_4} \qquad \text{for all } Q \in V(\overline{K}) \text{ such that } \phi^{-1}(Q) \text{ is defined.}$$

Combining these estimates with the correspondence $V'(R) \cap U' \xrightarrow{\sim} V(R) \cap U$, we find that

$$\mathcal{N}_{V(R)}^{U}\left((c_3^{-1}H)^{1/c_4}\right) \leq \mathcal{N}_{V'(R)}^{U'}(H) \leq \mathcal{N}_{V(R)}^{U}\left(c_1 H^{c_2}\right).$$

Now applying $\log^{(k)}(\cdot)$, dividing by $\log^{(k+\overline{\alpha})} H$, and letting $H \to \infty$, we see from the definition of $\overline{\alpha}$ that both sides of this inequality go to 1. Therefore

$$\frac{\log^{(k)} \mathcal{N}_{V'(R)}^{U'}(H)}{\log^{(k+\overline{\alpha})} H} \to 1,$$

which proves that $\overline{\alpha}'$ exists and equals $\overline{\alpha}$. □

§4. Integral Points on Surfaces

The distribution of integral points on curves, as described in section 2, is now well understood. The situation for surfaces is far different. There are a handful of theorems, a number of conjectures, and a large number of open questions (especially for non-projective surfaces.) As we discuss various examples, keep in mind that our ultimate goal is to describe the arithmetic order of varieties in terms of their geometry. Most of our discussion in this section is summarized in Table 2 below.

We will begin with projective surfaces, and will take as our starting point the classification of such surfaces as described, for example, in [5, Chapter V, Section 6] or [2]. (Remember that on a projective variety, the integral points coincide with the rational points.) We briefly describe what is known and conjectured for each type of projective surface, proceeding in order of increasing Kodaira dimension κ.

We briefly recall the definition of the Kodaira dimension of a (smooth) projective variety V, although we will not need to use it. Let $\mathcal{K}_V$ be the canonical bundle on V, and let $\phi_N\colon V \to \mathrm{P}^n$ be the map of V corresponding to the line bundle $\mathcal{K}_V^{\otimes N}$. The Kodaira dimension κ of V is the largest possible dimension of the image $\phi_N(V)$ for $N \geq 1$. (If $\mathcal{K}_V^{\otimes N}$ has no sections for all N, then κ is set equal to -1.)

Example 4.1: Projective Space — $\kappa = -1$.
As explained in Section 1 (cf. Example 1.5), $\mathcal{N}_{\mathbf{P}^2(R)}(H)$ grows like a power of H; and the same is true of $\mathcal{N}^U_{\mathbf{P}^2(R)}(H)$ for any non-empty open subset $U \subseteq \mathbf{P}^2$. Therefore $\overline{\alpha}(\mathbf{P}^2) = 0$.

Example 4.2: Ruled Surfaces — $\kappa = -1$.
A ruled surface is a surface V together with a map $\pi\colon V \to C$ to a smooth curve C such that every fiber of π is isomorphic to $\mathbf{P}^1$. A ruled surface always has a section $\sigma : C \to V$. (I.e. $\pi \circ \sigma = \mathrm{id}_C$. See [5, Chapter V, Section 2].) Let $\emptyset \neq U \subseteq V$ be an open set. Then $\pi(U)$ is an open subset of C, so we can choose a finitely generated ring $R_U \subset \overline{K}$ and a point $t_0 \in \pi(U) \cap C(R_U)$. Then $\pi^{-1}(t_0) \cong \mathbf{P}^1$, and $\pi^{-1}(t_0) \cap U$ is a non-empty open subset of this $\mathbf{P}^1$. It follows that for any $R \supset R_U$, $V(R) \cap U$ contains a subset which looks like $\mathbf{P}^1(R) \setminus \{\text{finite set}\}$. Therefore

$$\mathcal{N}^U_{V(R)}(H) \geq \mathcal{N}_{\mathbf{P}^1(R)}(H) - O(1);$$

and so $\overline{\alpha}(V) = 0$.

Notice that $\overline{\alpha}(V) = 0$ for all ruled surfaces, even though $V(R)$ is never Zariski dense. This generalizes our discussion in Example 3.2, further suggesting that we made the "correct" choice in our definition of $\overline{\alpha}$.

Example 4.3: Abelian Surfaces — $\kappa = 0$.
Let A be an abelian surface, and choose R large enough so that $A(R)$ is Zariski dense in A. From Néron (Example 1.6), we know that $\mathcal{N}_{A(R)}(H)$ grows like a power of $\log H$. In order to show that $\overline{\alpha}(A) = 1$, it suffices to show that the same

is true of $\mathcal{N}^U_{A(R)}(H)$ for any non-empty Zariski open subset U of V. (Notice in this case we can actually choose the ring R independently of the open set U.)

Write $D = A \backslash U$ as a finite union of irreducible curves, say $D = C_1 \cup \cdots \cup C_n$; and let $\mathcal{N}_{C_i(R)}(H)$ be the counting function on C_i obtained by restricting the chosen height function on A to C_i. Then

$$\begin{aligned}\mathcal{N}^U_{A(R)}(H) &\geq \mathcal{N}_{A(R)}(H) - \sum_{i=1}^{n} \mathcal{N}_{C_i(R)}(H) \\ &\geq c\big(\log H\big)^{\frac{1}{2}\operatorname{rank} A(R)} - \sum_{i=1}^{n} \mathcal{N}_{C_i(R)}(H).\end{aligned}$$

(Here $c > 0$ is a constant depending on A and R.) By Faltings' theorem, those C_i with genus at least 2 have only finitely many rational points, so they may be ignored. (In fact, it suffices to use Mumford's weaker estimate $\mathcal{N}_{C_i(R)}(H) \ll \log\log H$. Cf. Remark 2.1.)

Next, since abelian varieties do not contain rational subvarieties, it only remains to deal with those C_i having genus 1. Let E be such a C_i. If $E(R) = \emptyset$, there is no problem, so we may assume there is some $P_0 \in E(R)$. Using P_0 as an origin, we give E the structure of an elliptic curve; and then using Néron's estimate again, we find

$$\mathcal{N}_{E(R)}(H) \gg\ll \big(\log H\big)^{\frac{1}{2}\operatorname{rank} E(R)}.$$

It thus suffices to show that we have a strict inequality $\operatorname{rank} E(R) < \operatorname{rank} A(R)$.

Suppose that $\operatorname{rank} E(R) = \operatorname{rank} A(R)$. Then translating $E(R)$ by $-P_0$, we see that $E(R) - P_0$ would be a subgroup of finite index in $A(R)$. Thus $E(R)$ contains $[N]A(R) + P_0$ for some integer $N \geq 1$. But this would imply that

$$A(R) \subset \bigcup_{T \in [N]^{-1}(P_0)} E + T,$$

contradicting the original assumption that $A(R)$ is Zariski dense in A. Therefore $\operatorname{rank} E(R) < \operatorname{rank} A(R)$, which completes the proof that $\overline{\alpha}(A) = 1$.

Example 4.4: Enriques and K3 Surfaces — $\kappa = 0$.
If V is an Enriques surface, then there is a K3 surface V' and an étale double covering $V' \to V$. (Cf. [2, Proposition VIII.17].) It follows from Proposition 3.9(a) that the arithmetic order of V is determined by that of V'. Thus the study of integral points on Enriques surfaces is reduced to the same problem for K3 surfaces.

In general, little seems to be known about the distribution of integral points on K3 surfaces. For the class of K3 surfaces called *Kummer surfaces*, one can say a bit. These are surfaces which are birational to a quotient $A/\{\pm 1\}$, where A is an abelian surface. The map $A \to A/\{\pm 1\}$ is not étale, so we cannot apply Proposition 3.9 directly. However, a slight variant of Proposition 3.9 shows that if the arithmetic order of a Kummer surface V exists, then its arithmetic order will be less than that of the covering abelian surface. (Intuitively, integral points on A give integral points on V, but not conversely; so one only obtains an inequality.) Hence from above, we see that if a Kummer surface V has an arithmetic order, then $\overline{\alpha}(V) \le 1$.

Example 4.5: Hyperelliptic Surfaces — $\kappa = 0$.
A hyperelliptic surface V is a surface of the form $(E \times C)/\Gamma$, where E and C are elliptic curves, and Γ is a finite set of translations of E which acts on C in such a way that $C/\Gamma \cong \mathbf{P}^1$. (See, e.g., [2, Chapter VI]) The action of Γ on V has no fixed points, so the map $E \times C \to V$ is étale. It follows from Proposition 3.2(a) that $\overline{\alpha}(V) = \overline{\alpha}(E \times C)$. But $E \times C$ is an abelian surface, so it follows from our above discussion that $\overline{\alpha}(V) = 1$.

Example 4.6: Elliptic Surfaces — $\kappa = 1$.
An elliptic surface is a surface V together with a map $\pi: V \to C$ to a smooth projective curve C such that the generic fiber of π is an elliptic curve. (Not all elliptic surfaces have $\kappa = 1$, but those with smaller κ fall into one of the catagories already discussed.)

We first note that if $\overline{\alpha}(V)$ exists, then it satisfies $\overline{\alpha}(V) \le 1$. For if $\emptyset \ne U \subseteq V$ is any open subset of V, then we can choose a ring R_U so that $\pi(U)(R_U) \ne \emptyset$, say $t_0 \in \pi(U)(R_U)$. Further, we can enlarge R_U so that the set of integral points on the fiber $(\pi^{-1}(t_0))(R_U)$ is infinite. (I.e. The Mordell-Weil group of the fiber has positive rank.) Then for any $R \supseteq R_U$, $U(R)$ contains all but finitely many points of $(\pi^{-1}(t_0))(R_U)$, so $\mathcal{N}^U_{V(R)}(H)$ grows at least as fast as a power of $\log(H)$. (I.e. It grows as fast as the number of points on an elliptic curve with positive rank.) Hence $\overline{\alpha}(V) \le 1$.

Next suppose that the base curve C has genus at least 2. Then Faltings' theorem says that $C(R)$ is finite, so $V(R) \subset \pi^{-1}(C(R))$ is contained in finitely many fibers. Let $U_0 \subset V$ be the open subset of V obtained by discarding all of the singular fibers. Then for any $U \subseteq U_0$, $V(R) \cap U$ is contained in a finite union of elliptic curves. It follows that $\mathcal{N}^U_{V(R)}(H)$ grows no faster than the counting functions on elliptic curves. Therefore $\overline{\alpha}(V) \ge 1$. Combining this with the above remark, we obtain

$$\text{genus}(C) \ge 2 \quad \Longrightarrow \quad \overline{\alpha}(V) = 1.$$

If the map π admits a section, then the collection of sections $V(C)$ admits a group structure. (I.e. V is an abelian scheme over C of relative dimension 1.) If $\sigma \in V(C)$ is a section, then one can relate the heights of $t \in C(\overline{K})$, $\sigma(t) \in V(\overline{K})$, and $\sigma \in V(C)$ (cf. [6, Chapter 12, Sections 2 & 5].) Suppose now that there is

$\overline{V}$	D	$\overline{\alpha}(V)$	
$\mathbf{P}^2$	$\deg(D) \leq 2$	0	Trivial
	$\deg(D) = 3$		
	$D = 3$ lines	1	Dirichlet
	$D =$ conic + line	0	
	$D =$ nodal cubic	0	
	$D =$ smooth cubic	≤ 1	
	$\deg(D) \geq 4$		
	D has ≥ 4 components	∞	Schmidt,...
	D arbitrary	∞?	Conjecture
Ruled Surface	$D = 0$	0	Elementary
$X \to C$	$D > 0$	?	
Abelian Surface	$D = 0$	1	Néron
	$D > 0$	≥ 2	JS
		∞?	Conjecture
Enriques or	$D = 0$	?	
K3 Surface	$D > 0$ ample	∞?	Conjecture
Hyperelliptic	$D = 0$	1	Elementary
Surface	$D > 0$ ample	∞?	Conjecture
Elliptic Surface $E \to C$	$D = 0$		
	genus$(C) = 0$		
	rank $E(C) = 0$	≤ 1	
	rank $E(C) \geq 1$	0	G. Call
	genus$(C) = 1$	≤ 1	
	genus$(C) \geq 2$	1	
	$D > 0$	?	
General Type	$D \geq 0$	∞?	Conjecture

The Arithmetic Order of the Surface $V = \overline{V} \setminus D$

(D an effective divisor with normal crossings)

(Table 2)

a $\sigma \in V(C)$ having infinite order. Then using σ, one obtains integral points on V from integral points on C. (I.e. $\sigma\colon C(R) \hookrightarrow V(R)$.) This calculation has been carried out in detail by G. Call ([3]). If C has genus 0, then the integral points on the image of any section grow like the points in $\mathbf{P}^1$, so one obtains $\overline{\alpha}(V) = 0$. If C has genus 1, then there is a balance between the points on any given section and the points on various sections. Thus the number of integral points of V *lying on sections* grows like the number of points on an elliptic curve, so we obtain only the inequality already derived above, $\overline{\alpha}(V) \leq 1$. Note that we do not obtain an equality; there is always the possibility that $V(R)$ might contain many additional points.

On the other hand, if the group of sections $V(C)$ is finite, then little seems to be known. For example, even for $C = \mathbf{P}^1$, it is not known whether or not for every elliptic surface $V \to \mathbf{P}^1$ there is a finitely generated ring R such that $V(R)$ is Zariski dense in V.

Example 4.7: Surfaces of General Type — $\kappa = 2$.
There is a conjecture of Bombieri that the rational points on a surface of general type always lie on a proper subvariety. (This is the analogue for surfaces of Mordell's conjecture.) Vojta ([13, Conjecture 3.4.3]) and Bombieri have suggested that except for finitely many points, the proper subvariety can be chosen independently of the ring. These conjectures are slightly stronger than the assertion that $\overline{\alpha}(V) = \infty$. The only examples which the author is aware of are those surfaces V which admit a surjective map $V \to C$ onto a curve of genus at least 2. For such a surface, all but finitely many of the fibers will also have genus 2 (otherwise V would not be of general type.) The degeneracy of $V(R)$ follows from two applications of Faltings' theorem, and then one easily checks that $\overline{\alpha}(V) = \infty$.

Next we consider surfaces which are not projective. If V is a smooth quasi-projective variety, then it can be embedded in a smooth projective variety $\overline{V}$ in such a way that the complement $D = \overline{V} \setminus V$ is a divisor with normal crossings. (I.e. D is a subvariety of $\overline{V}$ of codimension 1 all of whose singularities consist of transversal intersections. The fact that V admits such an embedding follows from (embedded) resolution of singularities.) Notice we could include the projective case by letting $D = \emptyset$. We will assume for the remainder of this section that $\overline{V}$ has been chosen so that D is a normal crossings divisor. Further, we will let $\mathcal{L}(D)$ be the line bundle corresponding to D, and let $\mathcal{K}_{\overline{V}}$ be the canonical bundle on $\overline{V}$.

Example 4.8. Let $\overline{V} = \mathbf{P}^2$, and let $D = H_1 + \cdots + H_d$ be a sum of d distinct lines. Then D is a normal crossings divisor if and only if $H_1, \ldots, H_d$ are in general position; that is, no three of the lines have a common intersection point. Similarly, if D is an irreducible curve, then it is a normal crossings divisor if and only if it has only nodes as singularities.

There is a beautiful conjecture of Vojta which gives a purely geometric condition in terms of $\overline{V}$ and D under which the integral points on V are degenerate.

(Vojta's conjecture is actually far more precise than what we state. For details, see [13].)

Conjecture 4.9. (Vojta) *Let $\overline{V}$ be a smooth projective variety, let D be an effective normal crossings divisor on $\overline{V}$, and let $V \subseteq \overline{V}$ be the complement of D. Suppose that the line bundle $\mathcal{L}(D) \otimes \mathcal{K}_{\overline{V}}$ is ample on $\overline{V}$. Then there is a Zariski open set $U \subseteq V$ such that for any finitely generated ring $R \subseteq \bar{\mathbf{Q}}$ (over which $\overline{V}$ and D are defined,) the set of integral points $V(R) \cap U$ is finite.*

In other words, using the notions developed above, Vojta's conjecture can be rephrased as

$$\mathcal{L}(D) \otimes \mathcal{K}_{\overline{V}} \text{ ample} \quad \Longrightarrow \quad \overline{\alpha}(V) = \infty.$$

For the special case that $D = \emptyset$ (i.e. V is projective,) Lang has suggested that the converse might also hold. (Cf. [13, Question 4.3.2P]. Actually, one must replace the ampleness condition by *almost ample*, which means that a sufficiently large power of the line bundle gives an embedding outside of a proper Zariski closed subset.) This at least suggests asking the question:

$$\mathcal{L}(D) \otimes \mathcal{K}_{\overline{V}} \text{ almost ample} \quad \overset{?}{\Longleftrightarrow} \quad \overline{\alpha}(V) = \infty.$$

Notice that the $\Rightarrow$ implication in the case $D = 0$ is similar to Bombieri's conjecture, since by definition a variety is of general type if its canonical bundle is almost ample.

A remarkable aspect of Vojta's conjecture is that the main requirement on D, namely that $\mathcal{L}(D) \otimes \mathcal{K}_{\overline{V}}$ be (almost) ample, depends only on the linear equivalence class of D.

Example 4.10. Let $\overline{V} = \mathbf{P}^2$, and let D be a normal crossings divisor of degree d. Then

$$\mathcal{L}(D) \otimes \mathcal{K}_{\overline{V}} \cong \mathcal{O}_{\mathbf{P}^2}(d) \otimes \mathcal{O}_{\mathbf{P}^2}(-3) \cong \mathcal{O}_{\mathbf{P}^2}(d-3)$$

will be ample if and only if $d > 3$. Let us consider the two extreme cases, taking D to be either a collection of lines or an irreducible curve.

First, suppose that $D = H_1 + \cdots + H_d$ is a sum of lines in general position. If we choose linear equations $a_i X + b_i Y + c_i Z = 0$ for each H_i, then $V(R)$ is the set of points $[x, y, z] \in \mathbf{P}^2(R)$ with $x, y, z \in R$ satisfying

$$\prod_{i=1}^{d} (a_i x + b_i y + c_i z) \in R^*.$$

(Remember we always choose a "large" ring R.) It is known that the set of solutions is contained in finitely many *lines* in $\mathbf{P}^2$. (This follows from Schlickewei's p-adic generalization of W. Schmidt's subspace theorem. Cf. [13, Theorem 2.2.4].)

Now if L is a line containing infinitely many solutions, so $(L \cap V)(R)$ is infinite, then from Section 2 we see that L can have at most two points in common with the H_i's. Since $d \geq 4$, there are only finitely many such lines. Taking $U_0 \subset V$ to be the complement of these lines, we see that for any R, $V(R) \cap U_0$ is finite. It follows that $\overline{\alpha}(V) = \infty$, verifying Vojta's conjecture in this case.

Second, suppose that D is an irreducible curve of degree d. Let $F(X,Y,Z) \in K[X,Y,Z]$ be a homogeneous polynomial with $D = \{F = 0\}$. Then

$$V(R) = \left\{[x,y,z] \in \mathbf{P}^2(R) : x,y,z \in R \text{ and } F(x,y,z) \in R^*\right\};$$

and Vojta's conjecture says that there is a homogeneous polynomial $G(X,Y,Z) \in K[X,Y,Z]$ such that for any finitely generated ring $R \subset \overline{K}$, there are only finitely many points $[x,y,z] \in V(R)$ with $G(x,y,z) \neq 0$. For example, we could take a diagonal polynomial $F = aX^d + bY^d + cZ^d$. To illustrate the current state of our knowledge, it does not appear to be known that the solutions to $X^4 + Y^4 - Z^4 = 1$ with $x,y,z \in \mathbf{Z}$ are not Zariski dense in $\mathbf{P}^2$!

In view of Vojta's conjecture, it is reasonable to ask whether the arithmetic order of the quasi-projective variety V depends only on the linear equivalence class of the divisor D. More precisely, fix a smooth projective variety $\overline{V}$, and for each normal crossings divisor D on $\overline{V}$, let $V_D \stackrel{\text{def}}{=} \overline{V} \setminus D$. Then we pose:

Question. If D_1 and D_2 are linearly equivalent, is it true that $\overline{\alpha}(V_{D_1}) = \overline{\alpha}(V_{D_2})$?

The answer to this question, as we will shortly see, is NO. It is natural to start by looking at the simplest case, namely $\overline{V} = \mathbf{P}^2$. Then the linear equivalence class of D is determined by its degree. If $\deg(D) \leq 2$, then it is an easy exercise to check that $\overline{\alpha}(V_D) = 0$. On the other hand, if $\deg(D) \geq 4$, then Vojta's conjecture implies that $\overline{\alpha}(V_D) = \infty$. We thus will concentrate on $\deg(D) = 3$. (Notice this means that $\mathcal{L}(D) \otimes \mathcal{K}_{\mathbf{P}^2}$ is the trivial bundle, so V_D has (logarithmic) Kodaira dimension 0.) Now a normal crossings divisor of degree three in $\mathbf{P}^2$ takes one of four forms: three lines, conic plus line, nodal cubic, or smooth cubic. We will consider each in turn.

Example 4.11: $\mathbf{P}^2 \setminus \{\text{three lines}\}$.

Since the three lines comprising D are in general position, there is a $\overline{K}$-linear change of variables which takes those three lines to the coordinate axes. Take R_0 sufficiently large so that this change of variables has coefficients in R_0 and determinant in R_0^*. Then for any $R \supset R_0$, the linear transformation gives a one-to-one correspondence between R-integral points on the original $\mathbf{P}^2 \setminus D$ and on $\mathbf{P}^2 \setminus \{XYZ = 0\}$. It thus suffices to consider the case that the three lines are $\{XYZ = 0\}$.

Now $V(R)$ consists of points with homogeneous coordinates $x,y,z \in R$ satisfying $xyz \in R^*$. There is thus an isomorphism

$$R^* \times R^* \xrightarrow{\sim} V(R) \qquad\qquad (u,v) \longmapsto [u,v,1];$$

so $\mathcal{N}_{V(R)}(H)$ grows like a power of $\log(H)$. In order to prove that $\overline{\alpha}(V) = 1$, it remains to show that for any open set $\emptyset \neq U \subseteq V$, $\mathcal{N}^U_{V(R)}(H)$ also grows like a power of $\log(H)$.

This is easy. Let $u \in R^*$ be any unit that is not a root of unity. Choose some $n \in \mathbf{Z}$ such that the line $\{X - u^n Z = 0\}$ and the open set U have a point in common. (This will be true for almost all n, since $\mathbf{P}^2 \setminus U$ can contain only finitely many lines.) Then $V(R) \cap U$ will contain all but finitely many of the points $[u^n, u^m, 1]$, $m \in \mathbf{Z}$. It follows that $\mathcal{N}^U_{V(R)}(H) \gg \log(H)$, which completes the proof that $\overline{\alpha}(V) = 1$.

Example 4.12: $\mathbf{P}^2 \setminus \{\text{conic} + \text{line}\}$.
Write $D = C + L$, where C is an irreducible conic and L is a line. As in the previous example, we may make a linear change of variables without affecting the arithmetic order of V. We start by moving L so that it is given by the equation $X = 0$ and moving the two points of $C \cap L$ so that they are $[0,0,1]$ and $[0,1,0]$. (Note that $C \cap L$ consists of two points because D has normal crossings.) This means that C will have an equation of the form $aX^2 + bXY + cXZ + dYZ = 0$. Next we make the substitution $Y \mapsto Y - (c/d)X$, $Z \mapsto Z - (b/d)X$, which leaves L alone and gives C an equation of the form $a'X^2 + dYZ = 0$. Finally, we let $X \mapsto (1/\sqrt{a'})X$, $Y \mapsto (1/d)Y$, which puts C in the form $X^2 + YZ = 0$. We are thus reduced to studying

$$V(R) = \left\{[x,y,z] \in \mathbf{P}^2(R) : x,y,z \in R \text{ and } x(x^2+yz) \in R^*\right\}.$$

Notice that if $u \in R^*$ is any unit, then any factorization $u - 1 = yz$ leads to a point $[1,y,z] \in V(R)$. Thus we may be able to produce many integral points by choosing units for which $u-1$ is highly composite. For example, suppose that $2 \in R^*$. (Since we can always enlarge R, this is no restriction.) Then $2^n \in R^*$ for all n, and we can use the factorization $2^n - 1 = \prod_{d|n} \Phi_d(2)$, where Φ_d is the usual cyclotomic polynomial. It thus appears that if n is highly composite, then $2^n - 1$ will be also. The details of making this intuition quantitative are a bit involved, so we will postpone them until the next section. But the final conclusion is that for any open set $\emptyset \neq U \subseteq V$ there are constants H_0 and $c > 0$ such that

$$\log\log \mathcal{N}^U_{V(R)}(H) \geq c\frac{\log\log H}{\log\log\log H} \qquad \text{for all } H \geq H_0.$$

(See the proof of Theorem 5.6.) Taking one more logarithm, dividing by $\log^{(3)}(H)$, and letting $H \to \infty$, we see that $\overline{\alpha}(V) = 0$.

Comparing the last two examples, we see that the answer to the question that we asked above is no. There are linearly equivalent divisors D_1 and D_2 in $\mathbf{P}^2$ for which $\overline{\alpha}(V_{D_1}) \neq \overline{\alpha}(V_{D_2})$. We now continue by examining the other two cases of degree 3 divisors in $\mathbf{P}^2$.

Example 4.13: $\mathbf{P}^2 \setminus \{\text{nodal cubic}\}$.
Again, we change coordinates so as to put D in a convenient form. As explained, for example, in the proof of [11, Proposition III.2.5], a nodal cubic curve in $\mathbf{P}^2$ can always be put in the form $XYZ - (X-Y)^3 = 0$. (Briefly, one starts by moving the node to $[0,0,1]$, the nodal tangents to the lines $X = 0$ and $Y = 0$, and an inflexional point to $[1,1,0]$ with inflexional line $Z = 0$.) Thus we look at the set

$$V(R) = \left\{[x,y,z] \in \mathbf{P}^2(R) : x,y,z \in R \text{ and } xyz - (x-y)^3 \in R^*\right\}.$$

It is now possible to proceed by a direct argument as in Example 4.12. However, Paul Vojta has shown me the following short proof. There is an étale map

$$\mathbf{P}^2 \setminus \{\text{conic+line}\} \longrightarrow \mathbf{P}^2 \setminus \{\text{nodal cubic}\}$$

$$[u,s,t] \mapsto [su(u^2-st), u^2(u^2-st), (u^2-st)(3u^2-3su+s^2)+tu^3].$$

(Here we are taking the "conic+line" to be $X(X^2-YZ) = 0$, and the nodal cubic to be $XYZ - (X-Y)^3 = 0$.) Now using Proposition 3.9(a) and Example 4.12, we find that

$$\overline{\alpha}(\mathbf{P}^2 \setminus \{\text{nodal cubic}\}) = \overline{\alpha}(\mathbf{P}^2 \setminus \{\text{conic+line}\}) = 0.$$

We make one further remark concerning this example. As one easily checks, $[s,1,s^2-3s+3] \in V(R)$ for every $s \in R$. Hence along with the two nodal tangents, one should also remove from V the curve C with equation $ZY = X^2-3XY+3Y^2$. We note that $C \cong \mathbf{P}^1$, and the reader may check that $C \cap D$ consists of a single point, namely the node $[0,0,1]$. This explains why $C \setminus D$ contains so many integral points. There is also a symmtric curve obtained by interchanging X and Y and changing sign.

Example 4.14: $\mathbf{P}^2 \setminus \{\text{smooth cubic}\}$.
Again making a $\overline{K}$-linear change of variables, we may assume that D is given by an equation of the form

$$F(X,Y,Z) \stackrel{\text{def}}{=} Y^2Z - X^3 - aXZ^2 - bZ^3 = 0$$

for some $a,b \in R$ with $4a^3 + 27b^2 \neq 0$. (Cf. [11,Chapter III, Section 1].) How should we choose the open set $U_0 \subset V$? First, there are nine inflexional lines to D which should be discarded. (If L is an inflexional line, then $L \cap D$ consists of a single point, so $L \setminus D$ is an affine line.) But there may be other rational curves C intersecting D in a single point.

For example, let $F(X,Y,Z) = Y^2Z - X^3 + Z^3$. Then along with the inflexional lines, the curves

$$C_1 : Y^2Z = X^3 \qquad \text{and} \qquad C_2 : 3X^2 - 3XZ - Y^2 = 0$$

intersect D in single points. Hence for any $t \in R$, there are integral points

$$[t^2, t^3, 1], [-3t^2, 3t, 1 - 3t^2] \in V(R)$$

lying on C_1 and C_2 respectively. This example illustrates that even for degree 3 divisors in $\mathbf{P}^2$, before studying the arithmetic one should solve the not entirely trivial problem of determining all rational curves intersecting the divisor in a single point.

We will now verify that the arithmetic order of V is either 0 or 1 (assuming it exists at all.) We will use the fact that if L is a line tangent to D, then $L \setminus D$ has a group structure, namely

$$L \setminus D = L \setminus \{2 \text{ points}\} \cong \mathbf{P}^1 \setminus \{0, \infty\} = \mathbf{G}_m.$$

For any point $P \in D$, we use L_P to denote the tangent line to D at P, and P' to denote the third intersection point of L_P with D. (If P is a flex point, we let $P' = P$.) Now let U be a non-empty open subset of V, and choose a finitely generated ring R_0 so that there exists a point $P \in D(R_0)$ and a point $Q \in L_P(R_0) \cap U$. (Note that P' is also in $D(R_0)$.)

We have a canonical isomorphism $L_P \longrightarrow \mathbf{P}^1$ defined over the quotient field of R_0, specified by the conditions

$$P \mapsto 0, \quad Q \mapsto 1, \quad P' \mapsto \infty.$$

Further, one easily sees that if R_0 is sufficiently large, then we even have an isomorphism

$$(L_P \setminus \{P, P'\})(R) \cong \mathbf{G}_m(R).$$

Precisely, this holds if for every prime ideal π of R_0, the points P, P', and Q remain distinct modulo π. We let R_U be a finitely generated ring containing R_0 which has this property.

It only remains to point out that

$$(L_P \setminus \{P, P'\})(R) \subset (\mathbf{P}^2 \setminus D)(R)$$

for any ring $R \supseteq R_U$. Since all but finitely many of the points on L_P are in U, it follows that $V(R) \cap U$ essentially contains a copy of R^*. Therefore V has arithmetic order at most 1.

Note that although we have shown that $\overline{\alpha}(V) \leq 1$, we have not been able to show that $V(R)$ is Zariski dense in V. (The previous three examples, as well as a numeric example given below, suggest that it should be dense.) We pose the following two problems.

Problem 4.15. Find a finitely generated ring $R \subset \bar{\mathbf{Q}}$ and a smooth cubic polynomial $F(X,Y,Z) \in R[X,Y,Z]$ such that the set of points $[x,y,z] \in \mathbf{P}^2(R)$ satisfying $F(x,y,z) = 1$ is Zariski dense in $\mathbf{P}^2$.

Problem 4.16. Find a smooth cubic polynomial $F(X,Y,Z) \in \mathbf{Z}[X,Y,Z]$ and an open subset $\emptyset \neq U \subseteq \mathbf{P}^2$ such that for all finitely generated rings $R \subset \mathbf{Q}$ and all $\epsilon > 0$,

$$\mathcal{N}^U_{V(R)}(H) \ll H^\epsilon.$$

(Here V is the complement of $\{F = 0\}$, and the $\ll$ constant will depend on F, U, R, and ϵ.)

We will conclude our discussion of $\mathbf{P}^2 \setminus D$ with a numerical example, taking for D the curve

$$F(X,Y,Z) \stackrel{\text{def}}{=} Y^2Z - X^3 + 2XZ^2 - 3Z^3 = 0$$

and for R the ring $\mathbf{Z}[1/2]$. It is clear that the inflexional line $\{Z = 0\}$ should be discarded. The other inflexional lines are not defined over $\mathbf{Q}$, and there do not appear to be other rational curves intersecting D in only one point. We thus count triples $x,y,z \in \mathbf{Z}$ with $\gcd(x,y,z) = 1$, $z \neq 0$, and

$$y^2z - x^3 + 2xz^2 - 3z^3 = 2^n \qquad \text{for some integer } n \geq 0.$$

(Note that by taking 2^n rather than $\pm 2^n$, we prevent double counting the points $[x,y,z] = [-x,-y,-z]$.) Our results are given in Table 3.

It is not known that there are infinitely many such triples, although Table 3 makes this a plausible guess. On the other hand, the numerical data is far too scanty to allow a conjecture on the growth rate of $\mathcal{N}^U_{V(R)}(H)$. The largest solution found was $F(209,500,35) = 2^{12}$; the largest power of 2 was $F(17,414,-273) = 2^{24}$.

We continue with our study of integral points on quasi-projective varieties. In view of the fragmentary state of our knowledge, we will be content to look at several examples.

Example 4.17: Open Subvarieties of Abelian Surfaces.
Let A be an abelian surface, $D > 0$ an effective divisor on A, and $V = A \setminus D$. The divisor D in ample, while the canonical bundle on A is trivial. It follows from Vojta's conjecture that $\overline{\alpha}(V) = \infty$. In fact, there is an earlier conjecture of Lang which says that for any finitely generated ring R, $V(R)$ is finite. (Lang's stronger

H	$\mathcal{N} = \mathcal{N}^U_{V(R)}(H)$	$\dfrac{\log \mathcal{N}}{\log H}$	$\dfrac{\log \mathcal{N}}{\log \log H}$
50	152	1.284	3.683
100	228	1.179	3.555
150	268	1.116	3.469
200	302	1.078	3.425
250	326	1.048	3.387
300	346	1.025	3.358
350	362	1.006	3.333
400	388	0.995	3.330
450	400	0.981	3.311
500	410	0.968	3.293

Integral Points on $V = \mathbf{P}^2 \setminus \{Y^2Z - X^3 + 2XZ^2 - 3Z^3\}$ with $R = \mathbf{Z}[1/2]$ and $U = \{Z \neq 0\}$

(Table 3)

conjecture actually follows from Vojta's conjecture [13, Chapter 4, Section 2].) We briefly review what is known.

(a) Vojta [13, Corollary 2.4.5] has proven that if the divisor D has sufficiently many irreducible components, then $V(R)$ is not Zariski dense in V. Unfortunately, the required number of components depends on the ring R.

(b) As an elementary consequence of Faltings' theorem [4], the author [10] has shown that for each A there exists an irreducible divisor D so that $V(R)$ is finite for all R. In particular, for abelian surfaces, which are automatically jacobian varieties, one can take D to be the θ-divisor. Thus for this very special class of V's, one has $\overline{\alpha}(V) = \infty$.

(c) The author [12] has also given a general upper bound for $\mathcal{N}_{V(R)}(H)$ which is analogous to Mumford's estimate for rational points on curves (Remark 2.1). Precisely, for any finitely generated ring $R \subset \overline{K}$ there is a constant $c = c(A, D, R)$ so that

$$\mathcal{N}_{V(R)}(H) \leq c \log\log H \qquad \text{for all } H \geq e^e.$$

Thus if the arithmetic order of V exists, then it satisfies $\overline{\alpha}(V) \geq 2$.

Example 4.18: *Open Subsets of Surfaces with Kodaira Dimension* 0.
We have already considered one such family of surfaces, namely abelian surfaces. Recall that the others are Enriques, hyperelliptic, and K3 surfaces. These surfaces have the property that some power of their canonical bundle is trivial. It follows

from Vojta's conjecture that if D is an ample, normal crossings divisor, then the complement $V = \overline{V} \setminus D$ has arithmetic order $\overline{\alpha}(V) = \infty$. (I.e. These are the *affine* open subsets of surfaces with Kodaira dimension 0.) No general results appear to be known. For example, the equation $aX^4 + bY^4 + cZ^4 = 1$ defines an open affine subset of a K3 surface; it is not known whether the Zariski closure of the integral solutions to such an equation always form a proper subset of the surface. The case of non-projective, non-affine subsets of such surfaces do not appear to have been considered from an arithmetic point of view.

Example 4.19: *Open Subsets of Elliptic Surfaces.*
Let $\pi: E \to C$ be an elliptic surface. The canonical bundle on E is given by $\mathcal{K}_E \cong \pi^*\left(\mathcal{K}_C \otimes \mathcal{D}_{E/C}\right)$, where $\mathcal{D}_{E/C}$ is the minimal discriminant of E over C. One can combine this formula with Vojta's conjecture to give conditions under which certain open subsets of E should have arithmetic order ∞. Beyond this, little of significance seems to be known.

We conclude with an example in which the divisor D is non-trivial but not ample, so the variety V is neither projective nor affine.

Example 4.20: $\mathbf{P}^2 \setminus \{P_1, \ldots, P_r\}$.
If $V = \mathbf{P}^2 \setminus \{P_1, \ldots, P_r\}$, then we can find a $\overline{V}$ by blowing up $\mathbf{P}^2$ at the points $P_1, \ldots, P_r$. Choose coordinates for the P_i's, say $P_i = [a_i, b_i, 1]$. (If need be, change coordinates so no P_i lies on the line $Z = 0$.) Then one easily checks that

$$V(R) = \big\{[x, y, z] \in \mathbf{P}^2(R) : x, y, z \in R \text{ and } \\ (x - a_i z)R + (y - b_i z)R = R \quad \forall 1 \le i \le r\big\}.$$

Let $F(Y)$ be the polynomial $F(Y) = \prod_{1 \le i \le r}(Y - b_i)$; and let $\mathcal{I}_F$ be the fractional ideal of R generated by $F(y)$ as y ranges over R. Since we can always enlarge the ring R, we may assume that $\mathcal{I}_F = R$.

Now we take some value x_0 for x (other than one of the a_i's,) and look at possible values for y which give points $[x_0, y, 1]$ in $V(R)$. Let

$$A_0 = \prod_{1 \le i \le r} (x_0 - a_i).$$

Since $\mathcal{I}_F = R$, the Chinese remainder theorem implies that there is some $y_0 \in R$ such that $A_0 R + F(y_0)R = 1$. (Remember we assume that R is a PID.) Hence $[x_0, y_0, 1] \in V(R)$. More generally, for any $t \in R$, $[x_0, y_0 + tA_0, 1] \in V(R)$. It follows that $\mathcal{N}_{V(R)}(H)$ grows essentially like the number of points of R with height less than H. (I.e. like a power of H.) Since the choice of x_0 was arbitrary, it is clear that $\mathcal{N}^U_{V(R)}(H)$ grows similarly for any open subset $\emptyset \ne U \subseteq V$. Therefore

$$\overline{\alpha}\left(\mathbf{P}^2 \setminus \{P_1, \ldots, P_r\}\right) = 0.$$

To recapitulate, the (stable) distribution of integral points on projective surfaces is reasonably well understood, at least conjecturally. The major open questions (in which even conjectures are lacking) concern K3 surfaces and certain sorts of elliptic surfaces. For affine open subsets of projective surfaces there are a few general theorems, some conjectures, and many cases in which there are currently no conjectures. And for non-affine, non-projective surfaces, little work seems to have been done. Our goal of describing the arithmetic order of a variety in terms of its geometry, which worked so well in the case of curves, is still quite incomplete for surfaces. However, it is at least clear that the arithmetic order of a surface depends on more than the surface's Kodaira dimension, and on more than the linear equivalence class of the divisor removed from a projective surface. There is clearly still much work left to be done. Table 2 above contains a summary of our discussion of surfaces.

§5 A Lower Bound for the Counting Function of Certain Varieties

In this section we give detailed proofs of the arguments sketched in Example 4.12. Specifically, we prove that if $D \in \mathrm{Div}(\mathbf{P}^2)$ is a "conic + line," then the arithmetic order of $\mathbf{P}^2 \setminus D$ is 0. The underlying reason for this fact is that numbers of the form $\alpha^n - 1$ tend to have more divisors that one would expect. (I.e. On average, an integer x will have $O(\log x)$ divisors; but for highly composite n, an integer of the form $x = \alpha^n - 1$ will have at least $\exp(\exp(c \log\log x / \log\log\log x))$ divisors. The reader will note that $\log x = \exp(\exp(\log\log\log x))$ grows less rapidly.

In order to obtain these lower bounds, we will use the following version of a "folklore" theorem first shown to us by Charles Matthews. For any rational number $\alpha \in \mathbf{Q}^*$, $\alpha \neq \pm 1$, and any rational prime $p \in \mathrm{Spec}\,\mathbf{Z}$, we define

$$\begin{aligned} n_p(\alpha) &= \inf\{n \in \mathbf{N} : \alpha^n \equiv 1 \pmod{p}\}\,; \\ \Omega(\alpha) &= \{n_p(\alpha) : p \in \mathrm{Spec}\,\mathbf{Z}\}\,; \\ \Omega^*(\alpha) &= \mathbf{N} \setminus \Omega(\alpha) = \text{complement of } \Omega(\alpha)\ . \end{aligned}$$

(If α is not a p-adic unit, then we set $n_p(\alpha) = 0$.) We are interested in the size of the set $\Omega(\alpha)$.

Theorem 5.1. (a) *For all $\alpha \in \mathbf{Q}^*$, $\alpha \neq \pm 1$, $\Omega^*(\alpha)$ is finite.*

(b) *In fact, the union* $\bigcup_{\alpha \neq \pm 1} \Omega^*(\alpha)$ *is finite.*

Remark 5.2. It is undoubtedly true that for most α, $\Omega(\alpha) = \mathbf{N}$. For example, one can check that if $|\alpha| > 64$, $\alpha \in \mathbf{Z}$, then $\Omega(\alpha) = \mathbf{N}$. On the other hand, $6 \notin \Omega(2)$, and $2 \notin \Omega(\frac{a}{a-1})$ for all integers $a \neq 0, \pm 1$, so the precise behavior of $\Omega(\alpha)$ is not entirely clear.

We will not actually use Theorem 5.1, but rather the following corollary which gives a lower bound for the number of divisors of $2^n - 1$. Since this will suffice for our purposes, we will leave for the reader the obvious generalization to $\alpha^n - 1$. For any non-zero integer N, we let $d(N)$ denote the number of divisors of N.

Corollary 5.3. *There are constants $a > 1$ and $b > 0$ so that*

(i) $$d(2^n - 1) \geq a^{d(n)};$$

(ii) $$\sup_{n \leq N} d(2^n - 1) \geq a^{N^{\frac{b}{\log\log N}}}$$

We start with a lemma describing the factorization of cyclotomic polynomials. For $n \geq 1$, let $\Phi_n(X) \in \mathbf{Z}[X]$ denote the nth cyclotomic polynomial. Since $X^n - 1$ has the canonical factorization

$$X^n - 1 = \prod_{d|n} \Phi_d(X), \qquad (*)$$

we note that $n_p(\alpha)$ could also have been defined as the smallest $n \geq 1$ such that $\Phi_n(\alpha) \equiv 0 \pmod{p}$.

Lemma 5.4. *Let $\alpha \in \mathbf{Q}^*$, $\alpha \neq \pm 1$, let $p \in \operatorname{Spec} \mathbf{Z}$, $p > 2$, and assume that α is a p-adic unit. Let $n \geq 1$ be an integer.*

(a) *If* $\Phi_n(\alpha) \equiv 0 \pmod{p}$ *and* $n \neq n_p(\alpha)$, *then* $n \equiv 0 \pmod{p}$.

(b) *If* $n \equiv \Phi_n(\alpha) \equiv 0 \pmod{p}$, *then* $\operatorname{ord}_p \Phi_n(\alpha) \leq \operatorname{ord}_p n$.

(c) *There is an (absolute) constant $c > 1$ such that for all $n > e^e$,*

$$|\text{Numerator } \Phi_n(\alpha)| \geq c^{\frac{n}{\log\log n}}.$$

Proof. Since α is fixed, we will ease notation by writing n_p for $n_p(\alpha)$.

(a) By definition, $\Phi_{n_p}(\alpha) \equiv 0 \pmod{p}$. It follows from $(*)$ that α is a double root of $X^n - 1 \equiv 0 \pmod{p}$. Therefore $n \equiv 0 \pmod{p}$.

(b) First we note that the given assumptions together with $(*)$ imply

$$0 \equiv \alpha^n - 1 \equiv \left(\alpha^{n/p} - 1\right)^p \pmod{p}.$$

In particular, $n_p \leq \frac{n}{p} < n$. Also, n_p divides n. Second, we remark that for any integer $m \geq 1$,

$$\begin{aligned}\operatorname{ord}_p(\alpha^{n_p m} - 1) &= \operatorname{ord}_p(\alpha^{n_p} - 1) + \operatorname{ord}_p(m) \\ &= \operatorname{ord}_p(\Phi_{n_p}(\alpha)) + \operatorname{ord}_p(m).\end{aligned}$$

Here the first equality is true because α^{n_p} is a 1-unit in $\mathbf{Z}_p^*$, and the 1-units in $\mathbf{Z}_p^*$ are isomorphic (via the logarithm map) to $\mathbf{Z}_p^+$. The second equality then follows from $(*)$ and the definition of n_p.

We now apply this equality with $m = n/n_p$, and use $(*)$ to compute

$$\begin{aligned}\mathrm{ord}_p\left(\Phi_{n_p}(\alpha)\right) + \mathrm{ord}_p(m) &= \mathrm{ord}_p\left(\alpha^{n_p m} - 1\right) \\ &= \mathrm{ord}_p\left(\Phi_n(\alpha)\right) + \mathrm{ord}_p\left(\Phi_{n_p}(\alpha)\right) + \text{ non-negative stuff.}\end{aligned}$$

Subtracting $\mathrm{ord}_p(\Phi_{n_p}(\alpha))$ from both sides gives the desired result.

(c) Let N_n be the numerator of $\Phi_n(\alpha)$. Write $\alpha = a/b$ as a fraction in lowest terms, and let μ_n^* denote the set of primitive n^{th} roots of unity. Then factoring $\Phi_n(X)$, we obtain

$$|N_n| = \left|b^{\varphi(n)}\Phi_n\left(\frac{a}{b}\right)\right| = \prod_{\zeta\in\mu_n^*} |a - b\zeta|.$$

By assumption, $|a| \neq |b|$, so every term in the product satisfies

$$\left|a - b\zeta\right| \geq \left||a| - |b|\right| \geq 1.$$

In fact, if $\left||a| - |b|\right| \geq 2$, then every term contributes at least 2, so we immediately obtain an estimate of the desired form,

$$|N_n| \geq 2^{\frac{1}{2}\varphi(n)}.$$

Finally, if $\left||a| - |b|\right| = 1$, then we look at only those $\zeta \in \mu_n^*$ for which a and the real part of $b\zeta$ have opposite signs. For such ζ,

$$\left|a - b\zeta\right| \geq \sqrt{a^2 + b^2} \geq \sqrt{5}.$$

Now for large n, μ_n^* is (essentially) uniformly distributed on the unit circle, so about half of its elements will have this property. This gives

$$|N_n| \geq \sqrt{5}^{(\frac{1}{2}-\epsilon)\varphi(n)},$$

valid for all sufficiently large n. But for any fixed n, we certainly have $|N_n| \geq c(n) > 1$ for a constant $c(n)$ independent of α. (Note that $\alpha \in \mathbf{Q}$, while $\Phi_n(X)$ has no real roots.) Hence we can find an absolute constant $c > 1$ so that $|N_n| > c^{\varphi(n)}$. To complete the proof, we use the well-known lower bound

$$\varphi(n) \gg \frac{n}{\log\log n}$$

(cf. [1, Theorem 13.14]) to obtain the desired lower bound for $|N_n|$. □

We are now ready to prove Theorem 5.1.

Proof of Theorem 5.1. Clearly (b) implies (a), so we will concentrate on proving (b). Let $n \in \mathbf{N}$, and consider a prime p dividing the numerator N_n of $\Phi_n(\alpha)$. From Lemma 5.4(a), we will have $n = n_p(\alpha) \in \Omega$ provided that p does not divide n. We are thus reduced to showing that for all sufficiently large n, there exists a prime p dividing N_n which does not divide n. But using Lemma 5.4(b) and 5.4(c), we see that

$$\frac{|N_n|}{\prod\limits_{p|n} p^{\mathrm{ord}_p(N_n)}} \geq \frac{c^{\frac{n}{\log\log n}}}{n} \xrightarrow[n\to\infty]{} \infty.$$

This gives the desired result. □

Remark 5.5. One could define more generally

$$\Omega_k(\alpha) = \{n \in \mathbf{N} : \text{there exist } k \text{ distinct primes } p_1, \ldots, p_k \text{ such that } n_{p_1}(\alpha) = \cdots = n_{p_k}(\alpha) = n\}.$$

It seems to be a difficult problem to say anything significant about the set $\Omega_k(\alpha)$ for any $k \geq 2$.

Proof of Corollary 5.3. (i) Let $m_0 = \max \Omega^*(2)$. Then for all $m > m_0$ we have $m \in \Omega(2)$, so there exists a prime p with $n_p(2) = m$; equivalently, there exists a prime p dividing $\Phi_m(2)$ such that p does not divide $\Phi_{m'}(2)$ for all $m' < m$. Hence most of the terms in the factorization $2^n - 1 = \prod_{m|n} \Phi_m(2)$ will be divisible by some prime not dividing any of the previous terms. Precisely, if we let $\nu(N)$ denote the number of prime divisors of the natural number N, then we obtain the lower bound

$$\nu(2^n - 1) \geq d(n) - m_0.$$

Combining this with the trivial estimate $d(N) \geq 2^{\nu(N)}$ gives the desired result.

(ii) Applying (i), it suffices to prove that

$$\sup_{n \leq N} d(n) \geq N^{\frac{b}{\log\log N}}.$$

This is an easy exercise using the prime number theorem. (Take n to be the product of the first r primes, choosing the largest possible r for which $n \leq N$. For a more precise estimate, see [9, Theorem 13.12].) □

Finally, we use Corollary 5.3 to count the integral points on $\mathbf{P}^2 \setminus D$.

Theorem 5.6. *Let $V \stackrel{\text{def}}{=} \mathbf{P}^2 \setminus D$, where $D = C + L$ is the union of a smooth conic C and a line L which intersect transversally. Then*

$$\overline{\alpha}(V) = 0.$$

Proof. As explained in Example 4.12, after extending the base ring, we may assume that D is given by the equation $X(X^2 - YZ) = 0$. Further, we may assume that 2 is invertible in our ring. Then, for any factorization $2^n - 1 = yz$, we obtain a point of $V(R)$:

$$\{(y,z) \in \mathbf{Z}^2 : yz = 2^n - 1\} \hookrightarrow V(R) \qquad (y,z) \mapsto [1, y, z].$$

In particular, using Corollary 5.3, we obtain a lower bound

$$\mathcal{N}_{V(R)}(H) \geq \sum_{n=1}^{\log_2(H)} \sum_{y|2^n-1} 1 = \sum_{n=1}^{\log_2(H)} d(2^n - 1)$$

$$\geq \sup_{1 \leq n \leq \log_2(H)} d(2^n - 1) \geq a^{(\log H)^{\frac{b}{\log\log\log H}}}.$$

Hence $\log^{(3)} \mathcal{N}_{V(R)}(H) / \log^{(3)}(H) \to 1$ as $H \to \infty$. However, $\overline{\alpha}(V)$ is defined using the smaller counting function $\mathcal{N}^U_{V(R)}(H)$, so it remains to check that the integral points we have produced are reasonably well dispersed relative to the Zariski topology.

Thus let $U \subseteq V$ be a non-empty Zariski open set; and write $\mathbf{P}^2 \setminus U$ as a union of distinct irreducible curves, say $C_1 \cup \cdots \cup C_m$. We distinguish between two types of curves. First, there are curves with equations of the form $YZ = cX^2$ for some constant c. If c happens to have the form $2^n - 1$, then we will lose all of the points corresponding to the factorization of $2^n - 1$; but since there are only finitely many such curves, we may assume that H is chosen sufficiently large so that for the $n_0 \leq \log_2(H)$ which gives the supremum of $d(2^n - 1)$, the curve $YZ = (2^{n_0} - 1)X^2$ is not one of the C_i's. Second, any curve not of the form $YZ = cX^2$ will intersect the curve $YZ = (2^{n_0} - 1)X^2$ in only finitely many points, independent of n_0. (Precisely, C_i will intersect it in at most $2\deg(C_i)$ points.) Combining this with the above estimate for $\mathcal{N}_{V(R)}(H)$, it follows that there is a constant $c = c(U)$, *independent of* H, so that for all sufficiently large H,

$$\mathcal{N}^U_{V(R)}(H) \geq a^{(\log H)^{\frac{b}{\log\log\log H}}} - c.$$

Hence

$$\frac{\log^{(3)} \mathcal{N}^U_{V(R)}(H)}{\log^{(3)} H} \longrightarrow 1 \qquad \text{as } H \to \infty.$$

Since U is an arbitrary non-empty open subset of V, this completes the proof that $\overline{\alpha}(V) = 0$. □

§6 Some Open Questions

In this section we collect some of questions mentioned earlier and pose some new ones.

General Questions.

Question 6.1. Do there exist any varieties V with arithmetic order

$$2 \leq \overline{\alpha}(V) < \infty?$$

Question 6.2. More specifically, does there exist a variety V and a finitely generated ring $R \subset \overline{\mathbf{Q}}$ such that $V(R)$ is Zariski dense in V and

$$\mathcal{N}_V(R)(H) \ll \log\log H \qquad \text{as } H \to \infty?$$

(Or even satisfying $\mathcal{N}_V(R)(H) \ll (\log H)^\epsilon$ for every $\epsilon > 0$?)

Question 6.3. Let $V/\overline{\mathbf{Q}}$ be a projective variety, and suppose that the canonical bundle $\mathcal{K}_V$ has finite order in $\operatorname{Pic}(V)$. (I.e. There is an integer $n \geq 1$ such that $\mathcal{K}_V^{\otimes n}$ is the trivial bundle.) Does there always exist a number field K such that $V(K)$ is Zariski dense in V?

Projective Surfaces.

Question 6.4. What is the arithmetic order of a K3-surface? In particular, what is the arithmetic order of the Fermat hypersurface $X^4+Y^4+Z^4+W^4=0$?

Question 6.5. What is the arithmetic order of an elliptic surface $E \to \mathbf{P}^1$ having Kodaira dimension $\kappa_E = 1$ in the case that the set of sections $E(\mathbf{P}^1)$ is finite? (If $\kappa_E < 1$, then one would expect $\overline{\alpha}(E)$ to behave differently. For example, let E be the elliptic surface defined by desingularizing the subset of $\mathbf{P}^2 \times \mathbf{P}^1$ given by the equation $SY^2Z = SX^3 + TZ^3$. The map $E \to \mathbf{P}^1$ is projection onto the second factor. Then $\overline{\alpha}(E) = 0$, since E is birationally isomorphic to $\mathbf{P}^2$.)

Question 6.6. What is the arithmetic order of an elliptic surface $E \to C$ with base curve C having genus 1? (This would appear to involve fairly delicate questions concerning the variation of the elliptic regulator for families of elliptic curves.)

Quasi-Projective Surfaces.

We confine ourselves to questions concerning surfaces of the form $\mathbf{P}^2 \setminus D$.

Question 6.7. Let $V(R)$ be the set

$$V(R) = \left\{[x,y,z] \in \mathbf{P}^2(R) : x,y,z \in R \text{ and } x(x^2+yz) \in R^*\right\}.$$

We proved the $\log\log \mathcal{N}_{V(R)}(H)$ grows faster than $\log\log H/\log\log\log H$. (Cf. Example 4.12.) Let $U \subset V$ be given by $U = \{yz \neq 0\}$. (Notice that the lines

$y = 0$ and $z = 0$ are the tangent lines to the conic $x^2 + yz = 0$ at the points where it intersects the line $x = 0$.) Is it true that $\mathcal{N}^U_{V(R)}(H) \ll H^\epsilon$ for every $\epsilon > 0$? (For the ring $\mathbf{Z}[\frac{1}{2}]$, this would mean showing that the numbers $\Phi_n(2)$ tend not to have too many divisors, where the Φ_n's are the cyclotomic polynomials. This might be feasible, although the author was unable to do so. On the other hand, for a ring such as $\mathbf{Z}[\frac{1}{2}, \frac{1}{3}]$, the underlying problems seem considerably more difficult.)

Question 6.8. Same as the previous question, but for $V = \mathbf{P}^2 \setminus \{\text{nodal cubic}\}$,

$$V(R) = \left\{[x, y, z] \in \mathbf{P}^2(R) : x, y, z \in R \text{ and } xyz - (x - y)^3 \in R^*\right\}.$$

The set U is formed by discarding from V the tangents to the nodes and the two conics $ZY = X^2 - 3XY + 3Y^2$ and $ZX = -Y^2 + 3XY - 3X^2$. (Cf. Example 4.13.)

Question 6.9. Let $V = \mathbf{P}^2 \setminus \{\text{smooth cubic}\}$. What is the arithmetic order of V? Since this seems to be a difficult question, we pose the following two preliminary problems.

Question 6.10. Find a finitely generated ring $R \subset \overline{\mathbf{Q}}$ and a smooth cubic polynomial $F(X, Y, Z) \in R[X, Y, Z]$ such that the set of points $[x, y, z] \in \mathbf{P}^2(R)$ satisfying $F(x, y, z) \in R^*$ are Zariski dense in $\mathbf{P}^2$.

Question 6.11. Find a smooth cubic polynomial $F(X, Y, Z) \in \mathbf{Z}[X, Y, Z]$ and an open subset $\emptyset \neq U \subseteq \mathbf{P}^2$ such that for all finitely generated rings $R \subset \mathbf{Q}$ and all $\epsilon > 0$,

$$\mathcal{N}^U_{V(R)}(H) \ll H^\epsilon.$$

(The $\ll$ constant will depend on F, U, R, and ϵ. Note we are only asking for this to hold for subrings of $\mathbf{Q}$, not $\overline{\mathbf{Q}}$.)

Question 6.12. Let $F(X, Y, Z) \in R[X, Y, Z]$ be a homogeneous polynomial of degree $d \geq 4$ such that $\{F = 0\}$ is a normal crossings divisor in $\mathbf{P}^2$. As usual, let

$$V(R) = \left\{[x, y, z] \in \mathbf{P}^2(R) \colon F(x, y, z) \in R^*\right\}.$$

From Vojta's conjecture, $V(R)$ should never be Zariski dense in $\mathbf{P}^2$. Prove at least that there is an open subset $\emptyset \neq U \subset \mathbf{P}^2$ such that $\mathcal{N}^U_{V(R)}(H) \ll H^\epsilon$ for all $\epsilon > 0$. (Even for very special cases, such as $F = X^d + Y^d - Z^d$ and $R = \mathbf{Z}$, this seems quite difficult. For some work in this direction, see [9].)

References

[1] Apostel, T, *Introduction to Analytic Number Theory,* Springer-Verlag, N.Y., 1976

[2] Beauville, *Complex Algebraic Surfaces,* LMS Lecture Notes **68**, Cambridge University Press, 1983

[3] Call, G., *Local Heights on Families of Abelian Varieties,* thesis, Harvard, 1986

[4] Faltings, G., Endlichkeitsätz für abelsche Varietäten über Zahlkörpern, *Invent. Math.* **73** (1983), 349–366

[5] Hartshorne, R., *Algebraic Geometry,* GTM **52**, Springer-Verlag, 1977

[6] Lang, S., *Fundamentals of Diophantine Geometry,* Springer-Verlag, N.Y., 1983

[7] Mumford, D., A remark on Mordell's conjecture, *Amer. J. of Math.* **87** (1965), 1007–1016

[8] Schanuel, S., Heights in number fields, *Bull. Soc. Math. France* **107** (1979), 433–449

[9] Schmidt, W., Integer points on curves and surfaces, *Monatsh. Math.* **99** (1985), 45–72

[10] Silverman, J., Integral points on Abelian varieties, *Invent. Math.* **81** (1985), 341–346

[11] Silverman, J., *The Arithmetic of Elliptic Curves,* GTM **106**, Springer-Verlag, N.Y., 1986

[12] Silverman, J., Integral points on abelian surfaces are widely spaced, *Compositio Math.* **61** (1987), 253–266

[13] Vojta, Paul, Diophantine approximations and value distribution theory, *Lecture Notes in Math.* **1239**, Springer-Verlag, 1987

submitted: October 9, 1987
revised: February 11, 1988

Joseph H. Silverman
Mathematics Department
Brown University
Providence, RI 02912, U.S.A.

WEAK UNIFORM DISTRIBUTION OF SECOND-ORDER LINEAR RECURRING SEQUENCES

Gerhard Turnwald

Introduction. A sequence (u_n) of integers is said to be weakly uniformly distributed modulo m (WUD mod m), if $(u_n, m) = 1$ for infinitely many n and

$$\lim_{n\to\infty} \frac{|\{0 \le k < n : u_k \equiv a\ (m)\}|}{|\{0 \le k < n : (u_k, m) = 1\}|} = \frac{1}{\varphi(m)} \quad \text{for all } a \text{ with } (a,m) = 1$$

(where φ denotes Euler's function). For several important multiplicative functions f (like d (number of divisors), σ (sum of divisors), φ , or Ramanujan's τ-function) the set of all m such that the sequence defined by $u_n = f(n)$ is WUD mod m is explicitly known (cf.[4]); e.g., φ is WUD mod m iff (if and only if) $(m, 6) = 1$. In this paper we investigate linear recurring sequences defined by $u_{n+2} = c_1 u_{n+1} + c_0 u_n$ with fixed integers c_0, c_1 and initial values u_0, u_1. Since (u_n) is periodic mod m (possibly with a preperiod), (u_n) is WUD mod m iff in a fixed period all invertible residues occur with the same multiplicity. (By period we do not necessarily mean a period of minimal length.) For sequences with $u_{n+1} = au_n + b$ (hence $u_{n+2} = (a+1)u_{n+1} - au_n$) a complete description of all m such that (u_n) is WUD mod m has been given recently in [6] for Dedekind domains (thus including rings of algebraic integers and p-adic integers). In §1 we study WUD with respect to prime power moduli. It turns out that (u_n) is WUD mod p^h for all h if (u_n) is WUD mod p^3 and $p \neq 2$. (In this paper p always means a prime number.) In §2 a complete solution is given for composite moduli m such that the characteristic polynomial $c(x) = x^2 - c_1 x - c_0$ is reducible mod p for all $p|m$. (Here it is assumed that one knows how to see whether (u_n) is WUD mod p, p^2 if $c(x)$ splits into different linear factors mod p.). In §3 it is proved that the Fibonacci sequence is WUD mod m iff m is the product of a divisor of 6 with a power of 5.

A sequence (u_n) is said to be uniformly distributed modulo m (UD mod m) if all residues mod m appear with the same asymptotic frequency. A characterization of all moduli m such that a given second-order linear recurring sequence is UD mod m was found by Bumby ([2]; in [7] the solution for Dedekind domains is given). While it is obvious that a sequence is UD mod d for all $d|m$ if it is UD mod m, this property does not hold for WUD. Since, as is not hard to see, every invertible residue class mod d comprises $\varphi(m)/\varphi(d)$ invertible residue classes mod m , it is clear that the sequence is WUD mod d if d is divisible by all prime factors of m. Nothing more can be said in general (cf. no. 6 in the addenda of [4]). In order to overcome some of the resulting difficulties, we introduce the following notion:

Definition. The sequence (u_n) is said to cover m if $u_n \equiv a\,(m)$ for infinitely many n whenever $(a, m) = 1$.

By what we have noted above, it is clear that (u_n) covers d if it covers m and $d|m$. Obviously, (u_n) covers m if it is WUD mod m. In §2 we prove that a second-order linear recurring sequence is WUD mod m if it covers m and the characteristic polynomial is reducible mod p for all $p|m$. Hence in this case WUD mod d follows from WUD mod m for $d|m$.

The following simple observation is a powerful tool to investigate WUD (for linear recurring sequences) and is used throughout the paper. Assume that (u_n) has period l mod m and $u_{n+kl} \equiv u_n + k(u_{n+l} - u_n)$ (mp) for some prime p. If $u_{n+l} - u_n \not\equiv 0(mp)$ for all $n \geq n_0$ such that u_n is invertible mod m, then every invertible residue mod mp appears with the same multiplicity in a period of length pl as the corresponding (invertible) residue mod m. Hence (u_n) is WUD mod mp iff it is WUD mod m. (Recall that every invertible residue mod m corresponds to some invertible residue mod mp.)

We will make extensive use of the methods and results from [7]; for the convenience of the reader we state the following special case of Theorem 1:

Theorem 0. If $x^{n_0}(x^l - 1) \equiv 0$ $(c(x), p)$ then for every linear recurring sequence (u_n) with characteristic polynomial $c(x)$ we have :

(a) (u_n) has period $p^h l$ mod p^{h+1} for all $h \geq 0$.

(b) $u_{n+kp^h l} \equiv u_n + kp^h(u_{n+l} - u_n)$ (p^{h+2}) for $h \geq 0$, $n \geq max\{3 \cdot 2^{h-1}, 2\}n_0$ if $p \neq 2$.

(c) $u_{n+k2^h l} \equiv u_n + k2^{h-1}(u_{n+2l} - u_n)$ (2^{h+2}) for $h > 0$, $n \geq 3 \cdot 2^{h-1} n_0$.

§1

Proposition 1. Assume that $c(x) = x^2 - c_1 x - c_0$ splits into different linear factors mod p. Then for every sequence (u_n) with characteristic polynomial $c(x)$ we have:

(a) (u_n) has period $(p-1)p^{h-1}$ mod p^h; hence (u_n) is WUD mod p^h iff it covers p^h $(h \geq 1)$.

(b) If $p \neq 2$ and (u_n) is WUD mod p^2, then it is WUD mod p^h for all h.

(c) For $p = 2$, (u_n) is WUD mod p iff u_1 is odd; (u_n) is WUD mod p^2 iff $u_1 \equiv 1\,(2)$ and $c_1 \equiv c_0 - 1\,(4)$; (u_n) is not WUD mod p^h for $h \geq 3$.

Proof. Since $x(x^{p-1} - 1) \equiv 0$ $(c(x), p)$, (u_n) has period $(p-1)p^{h-1}$ mod p^h by Theorem 0. Since this number coincides with the number of invertible residue classes mod p^h, (a) is proved. If (u_n) is WUD mod p^2 then we must have $u_{n+p-1} - u_n \not\equiv 0\,(p^2)$ for all $n \geq n_0$. Hence for $p \neq 2$ from Theorem 0(b) we obtain

$$u_{n+kp^h(p-1)} \equiv u_n + k(u_{n+p^h(p-1)} - u_n)\ (p^{h+2}) \quad \text{and} \quad u_{n+p^h(p-1)} - u_n \not\equiv 0\ (p^{h+2})$$

for all $h \geq 0$ and $n \geq n_0(h)$. Hence (u_n) is WUD mod p^{h+2} with period $p^{h+1}(p-1)$ if this holds with h replaced by $h-1$. Inductively, (b) follows.

Now let us assume $p = 2$; then $c(x) \equiv x(x-1)$ (2), i.e. $c_0 \equiv 0\,(2)$ and $c_1 \equiv 1\,(2)$. From $u_{n+2} \equiv u_{n+1}\,(2)$ (for $n \geq 0$) it is clear that (u_n) is WUD mod 2 iff u_1 is odd. If u_1 is odd (hence $u_n \equiv 1\,(2)$ for all $n \geq 1$) then from $u_{n+3} - u_{n+1} = (c_1^2 + c_0 - 1)u_{n+1} + c_1 c_0 u_n$ we obtain $u_{n+3} - u_{n+1} \equiv c_0 u_{n+1} + c_0 u_n \equiv 0\,(4)$ for $n \geq 1$; hence (u_n) is WUD mod 4 iff $u_3 \not\equiv u_2\,(4)$, i.e. $1 + c_0 - c_1 \equiv 2\,(4)$ since $u_3 - u_2 \equiv (1 + c_0 - c_1)u_1\,(4)$. Assume that (u_n) is WUD mod 8. Then we must have $u_1 \equiv 1\,(2)$ and $c_1 \equiv c_0 - 1\,(4)$. Thus from $u_{n+4} - u_{n+2} = (c_1^2 + c_0 - 1)u_{n+2} + c_1 c_0 u_{n+1} \equiv c_1 c_0 (u_{n+2} + u_{n+1})\,(8)$ we see that (u_n) has period 2 mod 8, since $u_{n+2} + u_{n+1} = (1 + c_1)u_{n+1} + c_0 u_n \equiv c_0(u_{n+1} + u_n) \equiv 0\,(4)$ for

$n \geq 1$. As there are 4 invertible residue classes mod 8, this is impossible, thus proving (c).

Lemma 1. Let (u_n) be a linear recurring sequence with characteristic polynomial $c(x)$ (of arbitrary degree). If l is a period of (u_n) mod p^h and every linear recurring sequence of integers with characteristic polynomial $c(x)$ has period l mod p, then $u_{n+kl} - u_n \equiv k(u_{n+l} - u_n)\ (p^{h+1})$.

Proof. Set $v_n = (u_{n+l} - u_n)/p^h$. Then (v_n) is a linear recurring sequence of integers with characteristic polynomial $c(x)$. Hence $v_{n+l} \equiv v_n\,(p)$ and $u_{n+kl} - u_n = p^h(v_n + v_{n+l} + \ldots + v_{n+(k-1)l}) \equiv p^h k v_n \equiv k(u_{n+l} - u_n)\ (p^{h+1})$.

Proposition 2. Let $p \neq 2$; assume that $c(x) \equiv (x-\gamma)^2\,(p)$ and $u_1 \equiv \gamma u_0\,(p)$ for some integer $\gamma \not\equiv 0\,(p)$. Then we have:
(a) (u_n) has period $(p-1)p^{h-1}$ mod p^h; hence (u_n) is WUD mod p^h iff it covers p^h.
(b) (u_n) is WUD mod p iff γ is a primitive root mod p and $u_0 \not\equiv 0\,(p)$.
(c) (u_n) is WUD mod p^2 iff it is WUD mod p and $c_1^2 + 4c_0 \equiv 0\,(p^2)$, $2^p u_1 \not\equiv c_1^p u_0\,(p^2)$.
(d) If (u_n) is WUD mod p^2 then (u_n) is WUD mod p^h for all h.

(Note that the hypotheses are equivalent to $c_1^2 + 4c_0 \equiv 0\,(p)$, $c_0 \not\equiv 0\,(p)$, and $2u_1 \equiv c_1 u_0\,(p)$.)

Proof. From $c(x) \equiv (x-\gamma)^2\,(p)$ and $u_1 \equiv \gamma u_0\,(p)$ we obtain $u_n \equiv \gamma^n u_0\,(p)$. Since $x^{p(p-1)} - 1 \equiv (x^{p-1} - 1)^p \equiv 0\,(c(x), p)$, from Theorem 0 we get $u_{n+p^{h+1}(p-1)} - u_n \equiv p^h(u_{n+p(p-1)} - u_n)\ (p^{h+2})$ for all $h \geq 0$. Thus it remains to prove $u_{n+p(p-1)} \equiv u_n\,(p^2)$. For $p \geq 5$ this follows from [7] Lemma 3(c), since then $u_{n+p(p-1)} \equiv u_n + p(u_{n+p-1} - u_n)\ (p^2)$. For $p = 3$ put $v_n = (u_{n+2} - u_n)/3$. Since (v_n) is a linear recurring sequence with characteristic polynomial $c(x)$, we have $v_{n+2} \equiv 2\gamma v_{n+1} - v_n\ (3)$ and $v_{n+4} + v_{n+2} + v_n \equiv 2\gamma v_{n+3} + v_n \equiv v_{n+2} - 2\gamma v_{n+1} + v_n \equiv 0\,(3)$. Hence $u_{n+6} - u_n = 3(v_{n+4} + v_{n+2} + v_n) \equiv 0\,(9)$, thus finishing the proof of (a).

Since (b) is obvious, we proceed with (c) and assume that (u_n) is WUD mod p, $p \neq 2$. Set $v_n = (u_{n+p-1} - u_n)/p$. Since (v_n) has characteristic polynomial $c(x)$, we have $\gamma v_n \equiv ((v_1 - \gamma v_0)n + \gamma v_0)\gamma^n\ (p)$. If $v_1 \not\equiv \gamma v_0\ (p)$ or $v_0 \equiv 0\,(p)$, then for suitable n we have $v_n \equiv 0\,(p)$, i.e. $u_{n+p-1} \equiv u_n\,(p^2)$. Hence in this case (u_n) cannot cover p^2 (since (u_n) has period $(p-1)p$ mod p^2). If, however, $v_1 \equiv \gamma v_0\ (p)$ and $v_0 \not\equiv 0\,(p)$, then $v_n \equiv \gamma^n v_0 \not\equiv 0(p)$ and $v_{n+p-1} \equiv v_n(p)$ for all n. Then from

$$u_{n+k(p-1)} - u_n = p\sum_{j=0}^{k-1} v_{n+j(p-1)} \equiv pkv_n \equiv k(u_{n+p-1} - u_n)\ (p^2)$$

we conclude that (u_n) is WUD mod p^2, since $u_{n+p-1} - u_n \not\equiv 0\,(p^2)$. Hence (u_n) is WUD mod p^2 iff $v_1 \equiv \gamma v_0\ (p)$ and $v_0 \not\equiv 0\,(p)$.

From $(u_{j+2} - \gamma u_{j+1}) - \gamma(u_{j+1} - \gamma u_j) = (c_1 - 2\gamma)u_{j+1} + (c_0 + \gamma^2)u_j \equiv (c_1 - 2\gamma)\gamma^{j+1}u_0 + (c_0 + \gamma^2)\gamma^j u_0 \equiv -c(\gamma)\gamma^j u_0\ (p^2)$ by multiplying with γ^{n-2-j} and summing up for $j = 0, \ldots, n-2$ we obtain

$$(u_n - \gamma u_{n-1}) - \gamma^{n-1}(u_1 - \gamma u_0) \equiv -(n-1)\,c(\gamma)\gamma^{n-2}u_0\ (p^2)$$

for $n \geq 2$. Hence from

$$p(v_1 - \gamma v_0) = (u_p - \gamma u_{p-1}) - (u_1 - \gamma u_0) \equiv (u_p - \gamma u_{p-1}) - \gamma^{p-1}(u_1 - \gamma u_0)\ (p^2)$$

we see that $v_1 \equiv \gamma v_0\,(p)$ is equivalent to $c(\gamma) \equiv 0\,(p^2)$, since $\gamma, u_0 \not\equiv 0\,(p)$. If $c(\gamma) \equiv 0\,(p^2)$ then for $j \geq 1$ we obtain $u_j - \gamma u_{j-1} \equiv \gamma^{j-1}(u_1 - \gamma u_0)\ (p^2)$ and

$$pv_0 = \sum_{j=0}^{p-2}(u_{j+1} - \gamma u_j)\gamma^{p-2-j} + (\gamma^{p-1} - 1)u_0$$

$$\equiv (p-1)(u_1 - \gamma u_0)\gamma^{p-2} + (\gamma^{p-1} - 1)u_0 \equiv -\gamma^{p-2}(u_1 - \gamma^p u_0)\ (p^2).$$

Hence, for $c(\gamma) \equiv 0\,(p^2)$, $v_0 \not\equiv 0\,(p)$ is equivalent to $u_1 \not\equiv \gamma^p u_0\ (p^2)$, which is equivalent to $2^p u_1 \not\equiv c_1^p u_0\ (p^2)$, since $c_1 \equiv 2\gamma\,(p)$. Finally, the class of $c(\gamma)$ mod p^2 only depends on the residue class of γ mod p, since $c'(\gamma) \equiv 0\,(p)$. Choosing $c_1 \equiv 2\gamma\,(p^2)$, we get $4c(\gamma) \equiv -c_1^2 - 4c_0\ (p^2)$, i.e. $c(\gamma) \equiv 0\,(p^2)$ holds iff $c_1^2 + 4c_0 \equiv 0\,(p^2)$. Thus (c) is proved.

From Lemma 1 we get

$$u_{n+kp^{h-1}(p-1)} - u_n \equiv k(u_{n+p^{h-1}(p-1)} - u_n)\ (p^{h+1})$$

for $h \geq 2$. Hence (u_n) is WUD mod p^{h+1} if (u_n) is WUD mod p^h $(h \geq 2)$ and $u_{n+p^{h-1}(p-1)} - u_n \not\equiv 0\,(p^{h+1})$ for all $n \geq n_0(h)$. By Theorem 0, every linear recurring sequence with characteristic polynomial c(x) has period $p^h(p-1)$ mod p^h, i.e. $x^{p^h(p-1)} - 1$ is a characteristic polynomial for these sequences mod p^h. Since the same polynomial is a characteristic polynomial for (u_n) mod p^{h+1}, we conclude that the j-th power of it is a characteristic polynomial of (u_n) mod p^{jh+1}. Thus from

$$x^{p^{h+1}(p-1)} - 1 = \sum_{j=1}^{p}\binom{p}{j}(x^{p^h(p-1)} - 1)^j$$

we obtain (cf. [7], Lemma 1)

$$u_{n+p^{h+1}(p-1)} - u_n \equiv p(u_{n+p^h(p-1)} - u_n)\ (p^{h+3})$$

provided $jh + 1 \geq h + 2$ for $j \geq 2$ and $ph + 1 \geq h + 3$, i.e. $h \geq 1$ (since $p > 2$). Inductively this yields

$$u_{n+p^{h-1}(p-1)} - u_n \equiv p^{h-2}(u_{n+p(p-1)} - u_n)\ (p^{h+1}) \text{ for } h \geq 2$$

(note that the case $h = 2$ is trivial). Hence, for $h \geq 2$, (u_n) is WUD mod p^{h+1} if it is WUD mod p^h and $u_{n+p(p-1)} - u_n \not\equiv 0\,(p^3)$. It remains to prove that $u_{n+p(p-1)} \not\equiv u_n\ (p^3)$ if (u_n) is WUD mod p^2. In the proof of (c) we have noted that $u_{n+k(p-1)} - u_n \equiv k(u_{n+p-1} - u_n)\,(p^2)$. Taking $k = 2$ we see that $(x^{p-1}-1)^2$ is a characteristic polynomial of (u_n) mod p^2. This is also a characteristic polynomial mod p for all sequences with characteristic polynomial $c(x)$, since $(x^{p-1} - 1)^2 \equiv 0\,(c(x), p)$. Hence $(x^{p-1} - 1)^4$ is a characteristic polynomial of (u_n) mod p^3. Thus from $x^{p(p-1)} - 1 = \sum_{j=1}^{p}\binom{p}{j}(x^{p-1} - 1)^j$

we obtain $u_{n+p(p-1)} \equiv u_n + p(u_{n+p-1} - u_n)\,(p^3)$ for $p \geq 5$. Since $u_{n+p-1} - u_n \not\equiv 0\,(p^2)$, this proves (d) for $p \geq 5$.

Now assume that (u_n) is WUD mod p^2 for $p = 3$. In the proof of (c) we have seen that $c_1^2 + 4c_0 \equiv 0\,(3^2)$, $v_{n+1} \equiv \gamma v_n\,(3)$, and $v_n \not\equiv 0\,(p)$ (where $v_n = (u_{n+2} - u_n)/3$). Hence $(u_{n+6} - u_n)/3 = v_{n+4} + v_{n+2} + v_n = (c_1^2 + c_0)v_{n+2} + (c_1 c_0 + c_1)v_{n+1} + (c_0 + 1)v_n \equiv (c_1^2 + c_0)v_{n+2} + (c_0 + 1)(c_1\gamma + 1)v_n \equiv (c_1^2 + c_0)v_{n+2} \equiv -3c_0 v_{n+2} \not\equiv 0\,(9)$.

Proposition 3. Assume that $c(x) \equiv (x-1)^2\,(2)$ and $u_1 \equiv u_0\,(2)$. Then we have:

(a) (u_n) has period 2^{h-1} mod 2^h $(h \geq 1)$ except for $h = 2$ and $c_0 \equiv c_1 - 1\,(4), u_0 \equiv 1\,(2)$, in which case the period is 4 and $u_{n+2} \equiv u_n + 2\,(4)$; (u_n) is WUD mod 2^h iff it covers 2^h.

(b) (u_n) is WUD mod 2 iff $u_0 \equiv 1\,(2)$. If (u_n) is WUD mod 2^4 then (u_n) is WUD mod 2^h for all h.

(c) Let $c_0 \equiv c_1 - 1\,(4)$. Then (u_n) is WUD mod 4 iff it is WUD mod 2; (u_n) is WUD mod 8 iff it is WUD mod 4, $c_1 \equiv 0\,(4)$, and $u_1 \equiv u_0 + 4\,(8)$ or $u_2 \equiv u_1 + 4\,(8)$; (u_n) is WUD mod 16 iff it is WUD mod 8 and $c_0 \not\equiv c_1 - 1\,(8)$.

(d) Let $c_0 \equiv c_1 + 1\,(4)$. Then (u_n) is WUD mod 4 iff it is WUD mod 2 and $u_0 \not\equiv u_1\,(4)$; (u_n) is WUD mod 8 iff it is WUD mod 4 and $c_0 \not\equiv c_1 + 1\,(8)$; (u_n) is WUD mod 16 iff it is WUD mod 8.

Proof. Note that $c_0 \equiv 1\,(2), c_1 \equiv 0\,(2)$. Obviously, $u_n \equiv u_0\,(2)$ for all n, i.e. (u_n) has period 1 mod 2. From $u_{n+2} - u_n = c_1 u_{n+1} + (c_0 - 1)u_n \equiv (c_1 + c_0 - 1)u_0\,(4)$ we see that (u_n) has period 2 mod 4 unless $c_0 + c_1 - 1 \equiv 2\,(4)$ and $u_0 \equiv 1\,(2)$; in this case the period is 4 and $u_{n+2} \equiv u_n + 2\,(4)$. Note that in the exceptional case (u_n) is WUD mod 4. From

$$u_{n+4} - u_n = c_1(c_1^2 + 2c_0)u_{n+1} + (c_0 c_1^2 + c_0^2 - 1)u_n$$

we get $u_{n+4} - u_n \equiv (c_1(c_1^2 + 2c_0) + c_0 c_1^2 + c_0^2 - 1)u_0 \equiv (2c_0 c_1 + c_0 c_1^2)u_0 \equiv c_0 c_1(c_1 + 2)u_0 \equiv 0\,(8)$. By Theorem 0 for $h \geq 1$ we have $u_{n+2^{h+1}} - u_n \equiv 2^{h-1}(u_{n+4} - u_n)\,(2^{h+2})$. Hence 2^{h+1} is a period of (u_n) mod 2^{h+2} for $h \geq 1$. Thus the proof of (a) is complete.

The first part of (b) is trivial. For the second part, assume that (u_n) is WUD mod 2^4. Then $u_0 \equiv 1\,(2)$ and $u_{n+4} - u_n \equiv 8\,(16)$, since $u_{n+4} \equiv u_n\,(8)$ and (u_n) has period 8 mod 16. For $h \geq 3$, $v_n = (u_{n+2^{h-1}} - u_n)/2^h$ is integral for all n. Since (by Theorem 0) every sequence with characteristic polynomial $c(x)$ has period 2^{h-1} mod 2^{h-1}, from $u_{n+2^h} - u_n = 2(u_{n+2^{h-1}} - u_n) + 2^h(v_{n+2^{h-1}} - v_n)$ we obtain $u_{n+2^h} - u_n \equiv 2(u_{n+2^{h-1}} - u_n)\,(2^{h+2})$ (since $2h - 1 \geq h + 2$). Inductively this yields $u_{n+2^h} - u_n \equiv 2^{h-2}(u_{n+4} - u_n) \equiv 2^{h+1}\,(2^{h+2})$. Thus, for $h \geq 3$, (u_n) is WUD mod 2^{h+2} (with period 2^{h+1}) if (u_n) is WUD mod 2^{h+1} (with period 2^h). Hence (b) is proved.

Let $c_0 \equiv c_1 - 1\,(4)$ and $u_0 \equiv 1\,(2)$. Then (u_n) is WUD mod 4, since $u_{n+2} - u_n \equiv c_1 + c_0 - 1 \equiv 2\,(4)$. Assume $u_1 \equiv u_0\,(4)$ first. Then (u_n) is WUD mod 8 iff $u_1 \not\equiv u_0\,(8)$ and $u_3 \not\equiv u_2\,(8)$. Note that $(u_3 - u_2) - (u_1 - u_0) = (u_3 - u_1) - (u_2 - u_0) \equiv c_1(u_2 - u_1) \equiv 2c_1\,(8)$. Hence (u_n) is WUD mod 8 iff $u_1 \not\equiv u_0\,(8)$ and $c_1 \equiv 0\,(4)$. If $u_1 \not\equiv u_0\,(4)$ then $u_2 \equiv u_1\,(4)$ and we may apply the above arguments to the sequence (u_{n+1}), i.e. (u_n) is WUD mod 8 iff $u_2 \not\equiv u_1\,(8)$ and $c_1 \equiv 0\,(4)$. Now assume that (u_n) is WUD mod 8; hence $c_0 \equiv 3\,(4)$ and $c_1 \equiv 0\,(4)$. Then from

$$u_{n+4} - u_n \equiv 2c_1 u_{n+1} + (c_0^2 - 1)u_n \equiv 2c_1 + 2(c_0 + 1)\ (16)$$

we conclude that (u_n) is WUD mod 16 iff $c_1+c_0+1 \not\equiv 0\,(8)$. Since for $c_1 \equiv 0\,(4)$ this means $c_0 \not\equiv c_1 - 1$ (8), the proof of (c) is complete.

Now let $c_0 \equiv c_1+1\,(4)$ (and $u_0 \equiv 1\,(2)$). Since (u_n) has period 2 mod 4, (u_n) is WUD mod 4 iff $u_1 \not\equiv u_0\,(4)$. Since (u_n) has period 4 mod 8, (u_n) is WUD mod 8 iff (u_n) is WUD mod 4 and $u_2 \not\equiv u_0\,(8)\,, u_3 \not\equiv u_1\,(8)$. From $u_{n+1} \equiv u_n+2\,(4)$ we obtain $u_{n+2}-u_n = c_1u_{n+1}+(c_0-1)u_n = c_1(u_n+2)+(c_0-1)u_n \equiv (c_1+c_0-1)+2c_1 \equiv c_0-1-c_1$ (8). Hence (u_n) is WUD mod 8 iff $c_0 \not\equiv c_1+1$ (8) (and $u_1 \not\equiv u_0\,(4)$). If this holds, then $c_0 \equiv c_1+5$ (8) and $u_{n+4}-u_n \equiv c_1(c_1^2+10+2c_1)u_{n+1}+(c_1^3+6c_1^2+10c_1+24)u_n \equiv 2c_1u_{n+1}+(2c_1+8)u_n \equiv 8$ (16). Hence (u_n) is WUD mod 16.

Proposition 4. Assume that $c(x) \equiv (x-\gamma)^2$ (p) and $u_1 \not\equiv \gamma u_0\,(p)$ for some integer $\gamma \not\equiv 0\,(p)$. Then we have:

(a) (u_n) is WUD mod p and has period $(p-1)p^h$ mod p^h; if $p \geq 5$ then (u_n) is WUD mod p^h for all h.
(b) Let $p=3$. If (u_n) covers 9 then $c_1^2+c_0 \not\equiv 0\,(9)$; if $c_1^2+c_0 \not\equiv 0\,(9)$ then (u_n) is WUD mod 3^h for all h. If (u_n) covers 9 then $u_{n+6}-u_n \not\equiv 0\,(9)$ for all n.
(c) Let $p=2$. If (u_n) covers 4 then $c_0 \equiv 3\,(4)$; if $c_0 \equiv 3\,(4)$ then (u_n) is WUD mod 4. If (u_n) covers 8 then $c_1 \equiv 2\,(4)$; if $c_0 \equiv 3\,(4)$ and $c_1 \equiv 2\,(4)$ then (u_n) is WUD mod 2^h for all h.

Proof. By Theorem 0, (u_n) has period $(p-1)p^h$ mod p^h, since $x^{p(p-1)}-1 \equiv (x^{p-1}-1)^p \equiv 0\,(c(x),p)$. By [2] (or [7], Theorem 3), (u_n) is UD mod p and, for $p \geq 5$, (u_n) is UD mod p^h for all h. This proves (a), since UD implies WUD.

Assume $p=3$. After some calculation we obtain (cf.[7], p.198) $u_{n+6}-u_n \equiv 2\gamma^3(3\gamma^2-c(\gamma))(u_{n+1}-\gamma u_n)$ (9) and $3\gamma^2-c(\gamma) \equiv c_1^2+c_0$ (9). Hence (u_n) has period 6 mod 9 if $c_1^2+c_0 \equiv 0\,(9)$. Since not all u_n are invertible mod 3, (u_n) cannot cover 9. If $c_1^2+c_0 \not\equiv 0\,(9)$ then, by [2] (or [7], Theorem 3), (u_n) is UD (hence WUD) mod p^h for all h; moreover, $u_{n+6}-u_n \not\equiv 0\,(9)$, since $u_{n+1}-\gamma u_n \equiv \gamma^n(u_1-\gamma u_0) \not\equiv 0\,(3)$.

Now assume $p=2$. Then $u_{n+1} \equiv u_n+1$ (2) and $u_{n+2}-u_n \equiv c_1u_{n+1}+(c_0-1)u_n \equiv (c_1+c_0-1)u_n+c_1$ (4); note that $c_1 \equiv 0\,(2)$ and $c_0 \equiv 1\,(2)$. Hence $u_{n+2}-u_n \equiv c_0-1$ (4) if $u_n \equiv 1\,(2)$. Thus (u_n) cannot cover 4 if $c_0 \equiv 1\,(4)$, since (u_n) has period 4 mod 4 and not all u_n are odd. Consequently, (u_n) is WUD mod 4 if $c_0 \equiv 3\,(4)$. Note that $u_{n+4}-u_n \equiv c_1(c_1^2+2c_0)u_{n+1}+(c_0c_1^2+c_0^2-1)u_n \equiv 2c_1(u_n+1)+c_1^2u_n \equiv 2c_1$ (8). Hence, if $c_1 \equiv 0\,(4)$, (u_n) has period 4 mod 8 and thus cannot cover 8 (since not all the u_n are odd). By Theorem 0, $u_{n+2^{h+1}} \equiv u_n+2^{h-1}(u_{n+4}-u_n)$ (2^{h+2}) for $h>0$. Hence $u_{n+2^{h+1}} \equiv u_n+2^{h+1}$ (2^{h+2}) if $c_1 \equiv 2\,(4)$ and $h>0$; thus (u_n) is WUD mod 2^h for all h if it is WUD mod 4. This completes the proof.

Lemma 2. Assume that $c(x)$ is irreducible mod p (and $\deg c(x)=2$). Then $u_{n+p+1} \equiv -c_0u_n\,(p)$ for all n.

Proof. For $p=2$ we have $c(x)=x^2+x+1$. Hence $u_{n+2} \equiv u_{n+1}+u_n$ (2) and $u_{n+3} \equiv u_n$ (2). Now assume $p \neq 2$. It is sufficient to prove $x^{p+1}+c_0 \equiv 0\,(c(x),p)$. Let $c(x)=(x-\alpha)(x-\beta)$ be the factorization of $c(x)$ over the finite field with p^2 elements. Since $\alpha=(c_1+w)/2$ where $w^2=c_1^2+4c_0 \neq 0$, we have $\alpha^{p+1}=(c_1+w)(c_1^p+w^p)/4=(c_1+w)(c_1-w)/4=-c_0$ (note that $w^{p-1}=-1$, since w^2 lies in the ground field while $w \neq 0$ does not) ; for the same reason also $\beta^{p+1}=-c_0$. Hence $c(x)$ divides $x^{p+1}+c_0$, if the coefficients are interpreted as elements of the finite field with p elements (note that $\alpha \neq \beta$).

Remark. If $-c_0$ is a primitive root mod p then from $u_{n+p+1} \equiv -c_0 u_n\ (p)$ it is obvious that (u_n) is WUD mod p unless $u_n \equiv 0\,(p)$ for all n. Artin conjectured that every integer $a \neq -1$ is a primitive root mod p for infinitely many primes p, unless a is a square. Recently, Heath-Brown gave a partial solution in [3]; one of his results states that the conjecture holds for all primes a with at most two exceptions. Hooley has proved that Artin's conjecture follows from the extended Riemann hypothesis for certain Dedekind zeta functions. Hence it seems likely that "most" sequences (u_n) are WUD mod p for infinitely many p. (This would, however, not follow immediately from the truth of Artin's conjecture, since one also has to require that $c_1^2 + 4c_0$ is not a square mod p.) Note that (u_n) can only be UD for infinitely many p if $c(x)$ is a square (since $c_1^2 + 4c_0$ must be divisible by p).

Proposition 5. Let $p \neq 2$ and assume that $c(x)$ is irreducible mod p (and $\deg c(x) = 2$). Then (u_n) has period $p^h(p^2-1)$ mod p^{h+1} (for $h \geq 0$) and we have:

(a) If (u_n) does not have period $p^2 - 1$ mod p^2, then (u_n) is WUD mod p^2 iff it is WUD mod p and $u_{n+p^2-1} \not\equiv u_n\ (p^2)$ for all n such that $u_n \not\equiv 0\,(p)$; if (u_n) is WUD mod p^2 then it is WUD mod p^h for all h.

(b) If (u_n) is WUD mod p^2 and $u_{n+p^2-1} \equiv u_n\ (p^2)$ for some n with $u_n \not\equiv 0\,(p)$, then (u_n) has period p^2-1 mod p^2 and $-c_0$ is a primitive root mod p; if (u_n) is WUD mod p^3 then it is WUD mod p^h for all h.

Proof. Since $x^{p^2-1} - 1 \equiv 0\,(c(x), p)$, by Theorem 0 (u_n) has period $p^h(p^2-1)$ mod p^{h+1} and $u_{n+kp^h(p^2-1)} \equiv u_n + kp^h(u_{n+p^2-1} - u_n)\ (p^{h+2})$ for $h \geq 0$. Hence (u_n) is WUD mod p^{h+2} if (u_n) is WUD mod p^{h+1} and $u_{n+p^2-1} - u_n \not\equiv 0\,(p^2)$ for all n with $u_n \not\equiv 0\,(p)$. Thus (a) is proved except for the "only if" part of the first statement.

Now assume that (u_n) is WUD mod p^2 and $u_{n+p^2-1} - u_n \equiv 0\,(p^2)$ for some $n = n_0$ with $u_n \not\equiv 0\,(p)$. Let m be the number of indices n, $0 \leq n \leq p$, such that $u_n \not\equiv 0\,(p)$. Then, by Lemma 2, the number of invertible residues in a period of length $p(p^2-1)$ mod p^2 is $pm(p-1)$; hence every invertible residue occurs m times. From $u_{n+k(p^2-1)} \equiv u_n + k(u_{n+p^2-1} - u_n)\ (p^2)$ we see that $u_{n_0+k(p^2-1)} \equiv u_{n_0}\ (p^2)$ for $k = 0, \ldots, p-1$. Hence $m \geq p$. Let f be the order of $-c_0$ mod p. Then, by Lemma 2, (u_n) has period $f(p+1)$ mod p. The number of invertible residues in this period is mf. Hence $p-1$ divides mf. For $m = p$ we obtain $f = p-1$; for $m = p+1$ we obtain $f = p-1$ or $f = (p-1)/2$. Let $v_n = (u_{n+p^2-1} - u_n)/p$; note that, by Lemma 2, $v_{n+p+1} \equiv -c_0 v_n\ (p)$.

We treat the case $f = (p-1)/2$, $m = p+1$ first. If $v_n \not\equiv 0\,(p)$ for some n with $u_n \equiv u_{n_0}\,(p)$ then $u_{n+k(p^2-1)} \equiv u_n + kpv_n\ (p^2)$ yields $u_{n+k(p^2-1)} \equiv u_{n_0}\ (p^2)$ for suitable k. Since n can be replaced by $n+f(p+1)$, we conclude that u_{n_0} appears at least $p+2$ times in a period of length $p(p^2-1)$ mod p^2, contradiction. Hence $u_n \equiv u_{n_0}\,(p)$ implies $v_n \equiv 0\,(p)$. Among the $p+1$ indices n, $1 \leq n < p^2$, with $u_n \equiv u_{n_0}\,(p)$ we may find n_1, n_2 with $u_{n_1} \equiv u_{n_2}\,(p^2)$. Hence the $2p$ indices $n_i + k(p^2-1)$ $(i = 1, 2;\ k = 0, \ldots, p-1)$ yield the same residue mod p^2, contradiction.

Thus we must have $f = p-1$, i.e. $-c_0$ is a primitive root mod p. Then $u_{n+k(p+1)} \equiv u_n\,(p)$ implies $k \equiv 0\,(p-1)$ if $u_n \not\equiv 0\,(p)$, since $u_{n+k(p+1)} \equiv (-c_0)^k u_n\ (p)$. Hence in a period of length p^2-1 mod p the indices corresponding to the same invertible residue belong to different residue classes mod $p+1$. Assume that not all v_n are divisible by p. Since $v_{n+2}=c_1v_{n+1}+c_0v_n$ and $v_{n+p+1}\equiv -c_0v_n\ (p)$, we have $v_n\not\equiv 0\,(p)$ for all n in at least one half of the residue classes mod $p+1$. If n belongs to one of these residue classes, we have $u_{n+k(p^2-1)}\equiv u_{n_0}\,(p^2)$ for suitable k provided that $u_n\equiv u_{n_0}\,(p)$. Since the last condition holds for m residue classes, from $(p+1)/2\geq 2$ we conclude that the residue u_{n_0} appears at least $p+2$ times or $p+1$ times in a period of length $p(p^2-1)$ mod p^2 for $m=p+1$ or $m=p$, respectively. From this contradiction we conclude $v_n\equiv 0\,(p)$ for all n, i.e. (u_n) has period p^2-1 mod p^2; this completes the proof of (a) and of the first part of (b).

Now from $u_{n+p^h(p^2-1)}-u_n\equiv p^h(u_{n+p^2-1}-u_n)\equiv 0\,(p^{h+2})$ (for $h\geq 0$) and Lemma 1 we conclude

$$u_{n+kp^h(p^2-1)}\equiv u_n+k(u_{n+p^h(p^2-1)}-u_n)\ (p^{h+3})\quad\text{for}\quad h\geq 0\,.$$

Hence (u_n) is WUD mod p^{h+3} if it is WUD mod p^{h+2} and $u_{n+p^h(p^2-1)}-u_n\not\equiv 0\ (p^{h+3})$ for all n with $u_n\not\equiv 0\,(p)$. From $x^{p^h(p^2-1)}-1=\sum_{j=1}^{p}\binom{p}{j}(x^{p^{h-1}(p^2-1)}-1)^j$ for $h\geq 1$ we obtain

$$u_{n+p^h(p^2-1)}\equiv u_n+p(u_{n+p^{h-1}(p^2-1)}-u_n)\ (p^{h+3})\,,$$

since $jh+1\geq h+2$ for $j\geq 2$ and $ph+1\geq h+3$; note that $x^{p^{h-1}(p^2-1)}-1$ is a characteristic polynomial of (u_n) mod p^{h+1} and a characteristic polynomial mod p^h for every linear recurring sequence with characteristic polynomial $c(x)$. Hence $u_{n+p^h(p^2-1)}-u_n\equiv p^h(u_{n+p^2-1}-u_n)\ (p^{h+3})$ for all $h\geq 0$ (this being trivial for $h=0$). Note that $m=p$ since the number $m(p-1)$ of invertible residues in a period of length p^2-1 mod p^2 must be divisible by $p(p-1)$. Hence a period of length $p(p^2-1)$ mod p^3 contains only $p^2(p-1)$ invertible residues. Hence, if (u_n) is WUD mod p^3, we must have $u_{n+p^2-1}\not\equiv u_n\ (p^3)$ for all n with $u_n\not\equiv 0\,(p)$. Then also $u_{n+p^h(p^2-1)}\not\equiv u_n\ (p^{h+3})$ and inductively we conclude that (u_n) is WUD mod p^h for all h.

Remark. It is not clear whether the hypotheses of (b) can be satisfied. In the special case of the Fibonacci sequence it is a well known open problem whether p^2-1 can be a period mod p^2.

Proposition 6. Assume that $c(x)$ is irreducible mod 2 (and $\deg c(x)=2$). Then $3\cdot 2^h$ is a period mod 2^{h+1} and (u_n) is WUD mod 2^h for all h if it is WUD mod 2^4.

Proof. We have $c(x)\equiv x^2+x+1\ (2)$. Hence $x^3-1\equiv 0\,(c(x),p)$ and (u_n) has period $3\cdot 2^h$ mod 2^{h+1} for $h\geq 0$ by Theorem 0. Unless $u_0\equiv u_1\equiv 0\,(2)$, we obtain the sequence $\ldots,0,1,1,\ldots$ mod 2. A short calculation yields $u_{n+3}-u_n=(c_1^2+c_0)u_{n+1}+(c_1c_0-1)u_n$ and

$$u_{n+6}-u_n=(c_1^2+c_0)^2u_{n+2}+2c_1c_0(c_1^2+c_0)u_{n+1}+(c_1c_0-1)(c_1c_0+1)u_n\,.$$

Assume that (u_n) is WUD mod 4. Then, since (u_n) has period 6 mod 4, $u_{n+5}-u_{n+2}\equiv u_{n+4}-u_{n+1}$ (4) if $u_n\equiv 0\,(2)$. Hence from $(u_{n+5}-u_{n+2})-(u_{n+4}-u_{n+1})=(c_1^2+c_0)(u_{n+3}-u_{n+2})+(c_1c_0-1)(u_{n+2}-u_{n+1})\equiv c_1^2+c_0$ (4) we obtain $c_1^2+c_0\equiv 0\,(4)$, which yields $u_{n+6}-u_n\equiv 0\,(8)$. Hence, by Theorem 0, (u_n) has period $3\cdot 2^h$ mod 2^{h+2}

for $h > 0$. Then (for $h > 0$) from $x^{3\cdot 2^{h+1}} - 1 = 2(x^{3\cdot 2^h} - 1) + (x^{3\cdot 2^h} - 1)^2$ we conclude $u_{n+3\cdot 2^{h+1}} - u_n \equiv 2(u_{n+3\cdot 2^h} - u_n)\ (2^{h+4})$, since $x^{3\cdot 2^h} - 1$ is a characteristic polynomial mod 2^{h+1} for every linear recurring sequence with characteristic polynomial $c(x)$ and $2h+3 \geq h+4$. Thus we obtain $u_{n+3\cdot 2^{h+1}} - u_n \equiv 2^h(u_{n+6} - u_n)\ (2^{h+4})$ for $h > 0$. If (u_n) is WUD mod 16 then $u_{n+6} - u_n \not\equiv 0\,(16)$, since there are only 8 invertible residues in a period of length 12. Hence $u_{n+6} - u_n \equiv 8\,(16)$ and $u_{n+3\cdot 2^{h+1}} - u_n \equiv 2^{h+3}\ (2^{h+4})$ for $h \geq 0$. Inductively we conclude that (u_n) is WUD mod 2^h for all h.

Remark. If (u_n) is WUD mod 2^3 then (u_n) need not be WUD mod 2^4, as follows from the example $u_{n+2} = u_{n+1} - u_n$, $u_0 = 1$, $u_1 = 5$; the sequence given by $u_{n+2} = 3u_{n+1} - u_n$, $u_0 = 0$, $u_1 = 1$, is WUD mod 2^4.

§2

Theorem1. Let (u_n) be a linear recurring sequence with characteristic polynomial $c(x) = x^2 - c_1 x - c_0$. If $c(x)$ is reducible mod p for all $p|m$ then we have:

(1) If (u_n) covers m then:

- (i) (u_n) is WUD mod p^h if $p^h|m$.
- (ii) There is at most one prime $p \neq 2$ with $p|m$ such that $c_1^2 + 4c_0 \not\equiv 0\,(p)$ or $2u_1 \equiv c_1 u_0\,(p)$; if such p exists then $p \not\equiv 1\,(q)$ for all primes $q \neq 2$ with $q|m$.
- (iii) If $m \equiv 0\,(2)$ and p satisfies the hypotheses of (ii) then $u_2 \equiv u_1\,(2)$; if $m \equiv 0\,(4)$ then $m \not\equiv 0\,(8)$, $p \equiv 3\,(4)$, $c_1 \equiv 0\,(2)$ and $c_0 \equiv c_1 - 1\ (4)$.

(2) If (i), (ii), (iii) hold, then (u_n) is WUD mod m.

Proof. Assume that (u_n) covers m. Then (u_n) covers p^h if $p^h|m$, and from Propositions 1,2,3,4 we conclude that (i) holds. If p satisfies the hypotheses of (ii) (i.e. $c(x)$ splits into different linear factors mod p or $c(x) \equiv (x-\gamma)^2\,(p)$ and $u_1 \equiv \gamma u_0\,(p)$) then (u_n) has period $p-1$ mod p for $p|m$. If p_1, p_2 are distinct odd divisors of m such that $p_i - 1$ is a period of (u_n) mod p_i then (u_n) has period $(p_1-1)(p_2-1)/2$ mod p_1p_2; hence (u_n) cannot cover p_1p_2, contradiction. Thus there is at most one $p \neq 2$ with this property. Similarly, if $p \equiv 1\,(q)$ for some $q \neq 2$ with $q|m$, then (u_n) has period $(p-1)(q-1)/2$ mod pq (note that $q(q-1)$ is a period mod q), hence cannot cover pq, contradiction.

If m is even then all u_n in a period mod $2p$ must be invertible mod $2p$, since (u_n) covers $2p$ and has period $p-1$ mod $2p$. Thus all u_n in a period mod 2 are odd, which implies $u_n \equiv 1\,(2)$ for all $n \geq 1$; hence $u_2 \equiv u_1\,(2)$. If $c(x)$ is not a square mod 2 then, by Proposition 1(a), (u_n) has period 2 mod 4. Hence (u_n) has period $p-1$ mod $4p$. Thus (u_n) cannot cover $4p$ and, consequently, m is not dividible by 4. Assume that m is divisible by 4. Then $c(x)$ is a square mod 2 (hence $c_1 \equiv 0\,(2)$) and $u_0 \equiv u_1 \equiv 1\,(2)$. If $c_0 \not\equiv c_1 - 1\ (4)$ then, by Proposition 3(a), (u_n) has period 2 mod 4. As noted before, this is impossible; hence $c_0 \equiv c_1 - 1\ (4)$. Similarly, $p \equiv 3\,(4)$ since otherwise $p-1$ is a period mod $4p$ (4 is a period mod 4). By Proposition 3(a), (u_n) has period 4 mod 8. Hence (u_n) has period $2(p-1)$ mod $8p$, which implies that (u_n) cannot cover $8p$. Thus $m \not\equiv 0\,(8)$ and the first part of the theorem is proved.

Now let us assume that (i), (ii), (iii) hold. Let q be an odd prime divisor of m such that $c(x) \equiv (x-\gamma)^2\,(q)$ and $u_1 \not\equiv \gamma u_0\,(q)$; note that $\gamma \not\equiv 0\,(q)$, since (u_n) covers q.

From $x^{q(q-1)} - 1 \equiv (x^{q-1}-1)^q \equiv 0\,(c(x),q)$ we conclude

$$u_{n+kq^h(q-1)} \equiv u_n + kq^{h-1}(u_{n+q(q-1)} - u_n)\ (q^{h+1}) \quad \text{for} \quad h \geq 1,$$

by Theorem 0. As we have already observed in the proof of Proposition 4, (u_n) is UD mod q. Hence, by [7], Lemma 3, we obtain $u_{n+k(q-1)} \equiv u_n + k(u_{n+q-1} - u_n)\ (q)$, $u_{n+q-1} \not\equiv u_n\,(q)$, and $u_{n+q(q-1)} - u_n \equiv q(u_{n+q-1} - u_n) \not\equiv 0\,(q^2)$ for $q \geq 5$; if (u_n) covers 9 then, by Proposition 4(b), $u_{n+6} - u_n \not\equiv 0\,(9)$.

Assume we already know that (u_n) is WUD mod $m'q^h$, where $(m',q)=1$, $q^{h+1}|m$, and $h \geq 0$. If l is a period of (u_n) mod m' then

$$u_{n+klq^h(q-1)} \equiv u_n + k(u_{n+lq^h(q-1)} - u_n)\ (m'q^{h+1})$$

(since this congruence clearly holds mod m' and mod q^{h+1}). If $(l,q)=1$ then, by what we have noted above, we obtain

$$u_{n+lq^h(q-1)} - u_n \equiv l(u_{n+q^h(q-1)} - u_n) \not\equiv 0\,(q^{h+1});$$

hence (u_n) is WUD mod $m'q^{h+1}$ with period $lq^{h+1}(q-1)$.

Since (u_n) has period 2^h mod 2^h and (u_n) is WUD mod 2^h if $2^h|m$ (by (i)), inductively we conclude that (u_n) is WUD mod m if there is no p that satisfies the hypotheses of (ii) (starting with $m' = 2^h$ and then taking successively the remaining prime divisors q in increasing order). If, however, there exists such p, then it remains to prove that (u_n) is WUD mod the product of the prime powers belonging to 2 and p; then again inductively we conclude that (u_n) is WUD mod m, since $(p-1,q)=1$ by (ii). (Recall that (u_n) has period $p^{h-1}(p-1)$ mod p^h.) If m is odd this holds by (i); so let us assume $m \equiv 0\,(2)$. Then, by (iii), $u_2 \equiv u_1\,(2)$, which implies $u_n \equiv 1\,(2)$ for all $n \geq 1$ (since $c(x) \equiv x^2 + x\ (2)$ or $c(x) \equiv x^2 + 1\ (2)$). Then, obviously, (u_n) is WUD mod $2p^h$ if (u_n) is WUD mod p^h. This concludes the proof if $m \not\equiv 0\,(4)$. Since $m \not\equiv 0\,(8)$ (by (iii)), it remains to prove that (u_n) is WUD mod $4p^h$ if $m \equiv 0\,(4)$. By (iii) we have $p^{h-1}(p-1) \equiv 2\,(4)$ (since $p \equiv 3\,(4)$) and (u_n) is WUD mod 2 with $c(x) \equiv (x-1)^2\,(2)$, $c_0 \equiv c_1 - 1\ (4)$. Hence $u_1 \equiv u_0\,(2)$ and thus $u_{n+2} \equiv u_n + 2\ (4)$ by Proposition 3(a). Hence $u_{n+p^{h-1}(p-1)} \equiv u_n + 2\ (4)$, and from this we conclude that (u_n) is WUD mod $4p^h$ with period $2p^{h-1}(p-1)$.

Corollary. If $c(x)$ is reducible mod p for all $p|m$, then (u_n) is WUD mod m iff (u_n) covers m; if (u_n) is WUD mod m then (u_n) is WUD mod d for all $d|m$.

Proof. The first part follows immediately from the theorem; the second part follows from the first.

Remark 1. If $c(x)$ is irreducible, then (u_n) may cover p although (u_n) is not WUD mod p. As an example, take $p=7$ and $u_{n+2} = u_{n+1} + u_n$, $u_0 = 0$, $u_1 = 1$. The second part of the Corollary could hold in general (for second-order linear recurring sequences; it certainly fails for order three, since the sequence given by $u_{n+3} = u_n$, $u_0 = 1$, $u_1 = 2$, $u_2 = 5$ is WUD mod 6 but not WUD mod 3).

Remark 2. The conditions for WUD mod p^h can be seen from Propositions 1,...,6. Note that it is always sufficient to check the cases $h \leq 4$ for $p=2$ and $h \leq 3$ for

$p \neq 2$. ($h = 3$ is needed only for the exceptional (perhaps even impossible) case (b) of Proposition 5 if $c(x)$ is irreducible mod p.)

§3

In the following we study the Fibonacci sequence defined by $u_0 = 0$, $u_1 = 1$, and $u_{n+2} = u_{n+1} + u_n$.

Lemma 3. If $p \neq 5$ then $25(u_0^4 + \ldots + u_{p^2-2}^4) \equiv -6\,(p)$.

Proof. Let α, β be the roots of $x^2 - x - 1$ in the finite field with p^2 elements. Then $\alpha^{p^2-1} = \beta^{p^2-1} = 1$ and $(\alpha - \beta)^2 = (\alpha + \beta)^2 - 4\alpha\beta = 5$. Note that $u_n = (\alpha^n - \beta^n)/(\alpha - \beta)$, since both sides are linear recurring sequences with the same characteristic polynomial and the same initial values. Hence $u_n^4 = (\alpha^{4n} - 4\alpha^{3n}\beta^n + 6\alpha^{2n}\beta^{2n} - 4\alpha^n\beta^{3n} + \beta^{4n})/25 = (\alpha^{4n} - 4(-\alpha^2)^n + 6 - 4(-\beta^2)^n + \beta^{4n})/25$. Now $\alpha^4, -\alpha^2, -\beta^2, \beta^4$ are different from 1, since otherwise $\alpha^2 = \pm 1$ and this contradicts $\alpha^2 = \alpha + 1$. Hence summing up for $0 \leq n < p^2 - 1$, the corresponding geometric series have sum zero. Hence $u_0^4 + \ldots + u_{p^2-2}^4 = 6(p^2 - 1)/25 = -6/25$.

Lemma 4. The Fibonacci sequence does not cover p if $p > 7$.

Proof. Assume that (u_n) covers $p > 7$. Note that (u_n) is purely periodic mod p; hence the residue 0 occurs in each period. Thus $c(x) = x^2 - x - 1$ is irreducible mod p, since (by Proposition 1) (u_n) has period $p - 1$ if $c(x)$ splits into different linear factors mod p (and $c(x)$ is not a square for $p \neq 5$). Now Lemma 2 implies $u_{n+p+1} \equiv -u_n\ (p)$. From this we conclude $u_{p+1} \equiv 0\,(p)$ and $u_p \equiv -1\,(p)$ so that $u_{p+1-k} \equiv (-1)^k u_k\ (p)$ holds for $k = 0, 1$ and inductively we obtain $u_{p+1-k} = u_{p+3-k} - u_{p+2-k} \equiv (-1)^k u_{k-2} - (-1)^{k-1} u_{k-1} \equiv (-1)^k u_k\ (p)$ for $k \leq p + 1$. Hence, for $0 \leq k \leq p + 1$, u_{p+1-k} or $u_{2(p+1)-k} \equiv -u_{p+1-k}\ (p)$ is congruent to u_k, which (together with $u_{n+p+1} \equiv -u_n\ (p)$) implies that, with the possible exception of $\pm u_{(p+1)/2}$, each residue appears at least twice in a period of length $2(p + 1)$ mod p. Note that 1 appears at least four times (for $k = 1, 2, p-1, 2p+1$); hence -1 also appears at least four times. Since we assumed that all residues occur (in a period of length $2(p + 1)$) we easily conclude that $e := u_{(p+1)/2} \not\equiv 0\,(p)$ and $1, -1$ appear exactly four times, $e, -e$ appear exactly once, and all the others appear exactly twice (note that $u_{p+1} \equiv u_0 \equiv 0\,(p)$). Hence for any $r \geq 1, r \not\equiv 0\,(\frac{p-1}{2})$, we get $u_0^{2r} + \ldots + u_{2p+1}^{2r} \equiv 2(0^{2r} + \ldots + (p - 1)^{2r}) + 2(2 \cdot 1^{2r} - e^{2r}) \equiv 2(2 - e^{2r})\ (p)$. For $r = 1$ we obtain $e^2 \equiv 2\,(p)$, since $\sum_{k=0}^n u_k^2 = \sum_{k=1}^n u_k(u_{k+1} - u_{k-1}) = u_n u_{n+1}$. For $r = 2$ Lemma 3 yields $25(p - 1)(2 - e^4) \equiv -6\,(p)$, since $u_0^4 + \ldots + u_{p^2-2}^4 \equiv \frac{p-1}{2}(u_0^4 + \ldots + u_{2p+1}^4)\ (p)$. Thus we conclude $56 \equiv 0\,(p)$, which contradicts $p > 7$.

Remark. A different proof of Lemma 4 was given by Shah and Bruckner ([5] for $p \not\equiv 3, 7\,(20)$; [1] for $p \equiv 3(4)$).

Theorem 2. The Fibonacci sequence is WUD mod m iff $m = 2^\alpha 3^\beta 5^\gamma$ with $0 \leq \alpha, \beta \leq 1$ and $\gamma \geq 0$.

Proof. If (u_n) is WUD mod m then, by Lemma 4, m has no prime factor greater than 7. From Theorem 0 (with $p=5$ and $l=4\cdot 5$) and [7], Lemma 3, we get $u_{n+4k5^h} \equiv u_n + k5^h(u_{n+4}-u_n)\,(5^{h+1})$ for $h \geq 0$ and $u_{n+4}-u_n \not\equiv 0\,(5)$. (We have already used this result in the proof of Theorem 1 with q instead of 5. Note that (u_n) is WUD mod 5.) If m' has no prime factor different from $2,3,7$, then Proposition 5 implies that $l = 48m'$ is a period of (u_n) mod m'. Then from $u_{n+4kl5^h} \equiv u_n + k(u_{n+4l5^h} - u_n)\ (m'5^{h+1})$ and $u_{n+4l5^h} - u_n \equiv l5^h(u_{n+4}-u_n) \not\equiv 0\,(5^{h+1})$ we conclude that (u_n) is WUD mod $5^{h+1}m'$ iff (u_n) is WUD mod $5^h m'$ (for $h \geq 0$). Thus it remains to prove that (u_n) is WUD mod m' iff m' divides 6.

It is obvious that (u_n) is WUD mod 2. Note that (by Lemma 2) $u_{n+p+1} \equiv -u_n\,(p)$ for $p = 2,3,7$. From $u_{n+4} \equiv -u_n\,(3)$ and $u_{n+12} \equiv -u_n\,(6)$ we see that the residues $1,-1$ appear with the same frequency. Hence (u_n) is WUD mod 3 and mod 6, thus proving one part of the assertion. Now assume that (u_n) is WUD mod m' (where m' has no prime factor different from $2,3,7$). A short calculation shows that $u_n \equiv 1\,(3)$ implies $u_n \equiv \pm 1\,(7)$. Hence m' is not divisible by 21, since (u_n) does not cover 21. Since there are 6 invertible residues mod 14 and 28 invertible residues in a period of length 48, (u_n) is not WUD mod 14. Hence m' cannot be the product of positive powers of 2 and 7. Thus m' is a power of 7 if it is divisible by 7. Since, as is shown by a short calculation, (u_n) is not WUD mod 7, we conclude that $m' \not\equiv 0\,(7)$. (Note, however, that (u_n) covers 7.) It is easily verified that $u_{n+6} \equiv u_n\,(4)$ and $u_{n+12} \equiv -u_n\,(9)$. Hence (u_n) has period 24 mod $4\cdot 9$; 12 of the u_n in a period are invertible. Since there are 4 invertible residues mod 12 and 6 invertible residues mod 18, from $u_1 \equiv u_2 \equiv u_7 \equiv u_{17} \equiv 1\,(12)$ and $u_1 \equiv u_2 \equiv u_{10} \equiv 1\ (18)$ we conclude that (u_n) is not WUD mod 12 or 18. Hence m' is not divisible by $12 = 2^2\cdot 3$ or $18 = 2\cdot 3^2$. It is easily seen that (u_n) is not WUD mod 4 and not WUD mod 9. Hence m' is neither of the form 2^k nor 3^k for $k \geq 2$. Thus $m' = 1,2,3$, or 6.

References:

[1] G. Bruckner: Fibonacci sequence modulo a prime $p \equiv 3\,(\mathrm{mod}\ 4)$, Fibonacci Quart. 8(1970), 217-220.

[2] R. T. Bumby: A distribution property for linear recurrence of the second order, Proc. Amer. Math. Soc. 50(1975), 101-106.

[3] D. R. Heath-Brown: Artin's conjecture for primitive roots, Quart. J. Math. Oxford Ser.(2) 37(1986), 27-38.

[4] W. Narkiewicz: Uniform distribution of sequences of integers in residue classes, Lecture Notes in Math., vol.1087, Springer-Verlag, Berlin and New York, 1984.

[5] A. P. Shah: Fibonacci sequence modulo m, Fibonacci Quart. 6(1968), 139-141.

[6] R. F. Tichy and G. Turnwald: Weak uniform distribution of $u_{n+1} = au_n + b$ in Dedekind domains, Manuscripta Math. (to appear).

[7] G. Turnwald: Uniform distribution of second-order linear recurring sequences, Proc. Amer. Math. Soc. 96(1986), 189-198.

Mathematisches Institut der Universität, Auf der Morgenstelle 10,
D-7400 Tübingen, Federeal Republic of Germany.

Correspondance modulaire galois - quaternions pour un corps p-adique.

Marie-France Vignéras

Soient F soit une extension finie de $\mathbb{Q}_p$ ou de $F_p((T))$, et C un corps algébriquement clos de caractéristique ℓ. On fixe une clôture algébrique F' de F, qui contiendra toutes les extensions de F que l'on introduira. On note par W le groupe de Weil de F' sur F, et par H le groupe des éléments non nuls d'un corps de quaternions M sur F.

Nous allons (§II) pour tout $\ell \neq p$,

- décrire les représentations irréductibles de H sur C, et irréductibles de dimension 2 pour W, (la théorie complexe s'étend).
- montrer que la réduction modulo ℓ d'une représentation irréductible (de dimension 2 pour W) est toujours irréductible, sauf dans un cas exceptionnel, où la ramification est modérée. Dans ce cas, la représentation est de dimension 2, et sa réduction est somme de deux caractères.
- vérifier que la bijection de Langlands complexe galois-quaternions passe au quotient modulo ℓ

Le cas $\ell = p$ est exceptionnel. Toutes les représentations irréductibles sont alors modérément ramifiées.

Comme application, nous donnons (I) une démonstration pour $\ell \neq p$ d'une conjecture de Serre ([2], 3.2.6? p.196).

- I -

Soient K une clôture algébrique de $\mathbb{Q}$, $p \neq \ell$ deux nombres premiers, K_ℓ une clôture algébrique de $\mathbb{Q}_\ell$, A l'anneau des entiers de K_ℓ (il a un unique idéal maximal Λ, mais qui n'est pas principal), $C = A/\Lambda$ le corps résiduel qui est une clôture algébrique de F_ℓ. Si $a = bu$, a,b, $u \in A$, mais $u \notin \Lambda$, on notera : $a =' b$

Soit G l'un des groupes : $Gal_p = Gal(K_p/\mathbb{Q}_p)$, $W_p \subset Gal_p$ le groupe de Weil, $I_p \subset W_p$ le groupe d'inertie, H_p le groupe des unités d'un corps de quaternions sur F.

J'appelle "réduction modulo ℓ" la surjection canonique de A sur C. Je dis qu' une représentation irréductible (π,V) de G sur K_ℓ est A-admissible s'il existe un A-module $L \subset E$, G_p-stable, qui engendre E, et tel que pour tout sous-groupe ouvert compact $\Gamma \subset G_p$, le sous-module des éléments de L invariants par Γ est libre de type fini (un A-modèle). La "réduction modulo ℓ de L" est la représentation canonique de G_p sur $L/\Lambda L$. Elle est de longueur finie (voir Serre [1], p.138 pour un groupe fini, facilement généralisable à un groupe profini) et (Vignéras) pour G_p) et dépend du choix de L (il est facile de donner des exemples). Son image dans le groupe de Grothendieck des représentations de G sur C de longueur finie, indépendante du choix du modèle (Serre, p.138). C'est par définition la "réduction modulo ℓ" de la représentation.

On dit qu'une représentation de G sur C se relève à la caractéristique 0, si elle est la réduction modulo ℓ d'un réseau A-admissible d'une représentation de G sur K_ℓ.

Notons qu'un caractère (représentation de dimension 1) de G se relève à la caractéristique 0.

La conjecture 3.2.6? (Serre,[2] p.196) qui est formulée au n°6 ci-dessous se démontre ainsi :

1. Les représentations (ρ,E) de Gal_p sur C qui sont de dimension 2, ramifiées, et telles qu'il existe une droite $D \subset E$ stable, telle que $\rho(I_p)$ opère trivialement sur E/D" sont :
les représentations réductibles (D stable existe), donc de semi-simplifiées de la forme $\mu + \mu'$, où μ, μ' sont deux caractères continus : $Gal_p \to C^*$ (ρ de dimension 2), l'un deux est non ramifié (condition sur E/D), et si l'autre est aussi non ramifié, l'action de I_p est unipotente, non triviale (ρ ramifiée).

2. Si $p \neq \ell$, les représentations irréductibles A-admissibles (π,V) de G_p sont (Vignéras) :
- soit principale $i(\chi,\chi')$ induite "unitairement" à partir de deux caractères $\chi, \chi' : (Q_p)^* \to A^*$,
- soit de la forme Steinberg χSt, χ comme ci-dessus,
- soit cuspidale, de caractère central à valeurs dans A^*.

Pour tout caractère $\chi : (Q_p)^* \to A^*$, on pose $\alpha(\chi) = 0$ si χ est ramifié, et $\alpha(\chi) = \chi(p)$ sinon.
Soit $\alpha(\pi) \in A$ tel que : $\alpha(\pi) =' \alpha(\chi) + \alpha(\chi')$ si π est de la série principale $i(\chi,\chi')$,
$=' \alpha(\chi)$ si π est de la forme Steinberg χSt,
$= 0$ si π est cuspidale.

Pour ainsi dire par définition, pour une représentation irréductible A-admissible, ramifiée de G_p, on a les équivalences :

$$\underline{\alpha(\pi) \neq 0 \text{ modulo } \ell \quad \leftrightarrow \quad \pi = i(\chi,\chi') \text{ ou } \chi \, St, \text{ avec } \chi \text{ non ramifié, } \chi' \text{ ramifiée.}}$$

3. Les représentations ℓ-adiques (σ,W) de Gal_p sur K_ℓ (Deligne, §8) admettant un A-modèle $L \subset W$ dont la réduction ρ modulo ℓ est réductible, de même conducteur d'Artin (Serre [3]) que σ, et comme dans 1 sont :
les représentations ℓ-adiques réductibles de dimension 2 de Gal_p
- indécomposables $(\xi \, sp)(Fr_p^n u) = \xi(Fr_p^n u) \exp(t_\ell(u))$, ou
- semisimples $\xi + \xi'$,
où $\xi, \xi' : Gal_p \to A^*$ caractères continus, $u \in I_p$, Fr_p le Frobenius, t_ℓ est un homomorphisme du groupe d'inertie I_p dans $\mathbf{Z}_\ell$, où ξ est non ramifié et la réduction modulo ℓ de ξ' est ramifiée.
Preuve : voir l'appendice (III).

4. Soit Φ le caractère non ramifié de W_p tel que $\Phi(Fr_p) = p^{\pm 1}$ (le signe n'est pas important)..
Soit π la représentation de G_p correspondant à σ' par la correspondance de Hecke (Deligne).
On peut supposer l'isomorphisme du corps de classes local τ (d'où une identification $\xi \to \tau(\xi)$ des caractères de Gal_p sur ceux de G_p) choisi de sorte que :
a) Si $\sigma = \xi + \xi'$, $\xi, \xi' : Gal_p \to A^*$ caractères continus, alors π est induite "unitairement" à partir des deux caractères $\Phi^{\varepsilon/2} \tau(\xi)$, $\Phi^{\varepsilon/2} \tau(\xi')$,
b) Si $\sigma = \xi \, sp$, alors $\pi = \Phi^{\varepsilon/2} \tau(\xi) St$,
c) Si σ est irréductible, alors π est cuspidale.
où ε est un signe dépendant des choix de τ et Φ.
On a, comme $p^{1/2} \in A^*$:
$\alpha(\pi) = 0$ si σ est irréductible,

$\alpha(\pi)$ =' $\alpha(\tau(\xi))$, si $\sigma = \xi$ sp ,
$\alpha(\pi)$ =' $\alpha(\tau(\xi)) + \alpha(\tau(\xi'))$, si $\sigma = \xi + \xi'$.
La correspondance de Langlands ($\varepsilon = 0$) aurait donné le même résultat.

5. PROPOSITION (Conjecture de Serre 3.2.6?, pour $p \neq \ell$, en termes de représentations). Si la représentation irréductible ramifiée π de G_p correspond par la correspondance de Hecke (ou de Langlands) à une représentation ℓ-adique σ de Gal_p de dimension 2, de réduction modulo ℓ (notée ρ) ayant même conducteur d'Artin, alors (i) et (ii) sont équivalents,

(i) $\alpha(\pi) \neq 0$ modulo ℓ,

(ii) ρ réductible, fixe une droite, et $\rho(I_p)$ est triviale sur le quotient.

Preuve : Utilisant 2,3,4 on voit que $\pi = \Phi^{\varepsilon/2}\tau(\xi)$ St ou induite de $\Phi^{\varepsilon/2}\tau(\xi)$, $\Phi^{\varepsilon/2}\tau(\xi')$, avec ξ non ramifié et ξ' de réduction modulo ℓ ramifiée (notée μ'). Ceci est équivalent à :

$$\alpha(\pi) \neq 0 \text{ modulo } \ell \text{ et } \mu' \text{ ramifié.}$$

Pour ξ non ramifié, σ et ρ ramifiées de même conducteur d'Artin, ξ' est ramifié si et seulement si μ' est ramifié.

6. Retour à la formulation originale: soit ρ une représentation continue de Gal(K/Q) sur C de dimension 2. Soit f la forme parabolique f normalisée, propre pour tous les opérateurs de Hecke , de poids $k \geq 2$, de niveau N premier à ℓ , de caractère ε qui lui est associée conjecturalement ([2], 3.2.4?, p.196).

LEMME. f est primitive.

Preuve. C'est un résultat local, en $p \neq \ell$, car N est premier à ℓ. Soit p|N, et N_p la plus grande puissance de p divsqnt N. Soient :

- π la composante en p de la représentation automorphe définie par f,
- $N(\pi) \leq N_p$ l'exposant de son conducteur (la plus petite puissance de p telle que π ait un vecteur non nul invariant par le groupe de congruence habituel de $GL(2,\mathbf{Z}_p)$ (c = 0 modulo ...),
- σ l'image de π par la correspondance de Hecke (on sait qu'elle est galoisienne), ρ la réduction modulo ℓ de σ,
- $a(\sigma) = N(\pi)$ le conducteur d'Artin de σ, $a(\rho) \leq a(\sigma)$ celui de ρ.

Par la conjecture (3.2.4? Serre [2]) : $N_p = a(\rho)$. C'est équivalent à

$$N_p = N(\pi) = a(\sigma) = a(\rho).$$

L'égalité $N_p = N(\pi)$ pour tout p|N est équivalente à : f est primitive. Le lemme est démontré.

On fixe un plongement de K dans K_ℓ. Le p-ième coefficient de Fourier a_p de f modulo ℓ est égal à $\alpha(\pi)$ modulo ℓ . On en déduit la conjecture 3.2.6? (loc.cit.):

a_p <u>est une unité p-adique, si et seulement si ρ a un quotient étale de dimension 1.</u>

(la conjecture précise aussi qu'alors a_p est la valeur propre de Frobenius sur l'espace quotient, je ne l'ai pas vérifié).

II - Soit $G = W_p$ (noté W) ou H_p (noté H). Une représentation (π,V) de G sur C est un espace vectoriel V de dimension finie sur C, muni d'un homomorphisme $\pi : G \to GL(V)$ tel que le stabilisateur dans G de tout élément de V soit ouvert. Une représentation de G qui est induite d'un sous-groupe $U \neq G$ est dite imprimitive. Sinon, elle est dite primitive.
Soient $g \in G$ et (σ,V) une représentation d'un sous-groupe Q de G, on note (σ^g,V) la conjuguée de (σ,V), qui est la représentation de gQg^{-1} sur V telle que $\sigma^g(x) = \sigma(g^{-1}xg)$, $x \in gQg^{-1}$. Le normalisateur de σ dans G est le groupe des $g \in G$, tels que σ^g soit isomorphe à σ. On dit que σ est régulier, si son normalisateur est égal à Q.
Si χ est un caractère de F^*, et π est une représentation de G, le produit tensoriel de π par χ (identifié à un caractère de G, via τ (I.3) ou la norme réduite) est appelée la tordue de π par χ et notée par $\pi\chi$.

O sera l'anneau des entiers du corps des quaternions M, P son idéal premier, ω un générateur de P (on ajoutera un indice k pour un autre corps k), G_o sera le

- groupe des unités O^* si $G = H$,
- groupe d'inertie si $G = W$,

et G_n, $n \geq 1$, les groupes sauvages, seront les

- groupes $1+P^n$, si $G = H$,
- n-ième groupes de ramification si $G = W$,

Les groupes profinis G_n sont distingués dans G.
Le plus petit entier $n \geq 0$ tel que la représentation (π,V) soit triviale sur le sous-groupe G_n s'appelle l'indice de ramification de la représentation, et noté $f(\pi)$. La représentation est dite

- non ramifiée, si $f(\pi) = 0$. Si elle est irréductible, c'est alors un caractère, puisque G/G_o est cyclique.
- modérément ramifiée, si $f(\pi) \leq 1$,
- sauvagement ramifiée, si $f(\pi) > 1$.

Une représentation (π,V) est dite minimale si son indice de ramification ne peut pas être rendu plus petit en tordant la représentation par un caractère :

$$f(\pi) \leq f(\pi\chi) \text{, pour tout caractère } \chi \text{ de } G.$$

Les caractères non ramifiés sont les représentations de dimension 1 minimales.
Le lemme suivant permet d'appliquer la théorie des groupes finis à l'étude des représentations irréductibles de G, comme dans le cas complexe.

LEMME 1. Toute représentation irréductible de G est produit tensoriel d'un caractère non ramifé et d'une représentation d'image finie.

COROLLAIRE 2. Toute représentation irréductible de G est produit tensoriel d'un caractère et d'une représentation minimale d'image finie.

Donc à torsion près, les représentations irréductibles de G s'identifient aux représentations irréductibles des groupes quotients $G' = G/U$, pour les sous-groupes ouverts distingués d'indice fini de G. La filtration de G par les groupes G_n, $n \geq 0$ fournit par passage au quotient une suite

de composition G'_n sur G' dont les quotients Q'_n ont les propriétés :
- Q'_o est cyclique,
- Q'_1 est cyclique d'ordre premier à p,
- Q'_i, $i \geq 2$ est un p-groupe, trivial si i est assez grand, abélien si G= H (non pour G=W),
- G' / G_1' est hyper-résoluble (admet une suite de composition par des sous-groupes distingués dans le groupe entier, à quotients cycliques),
- G_1' est hyper-résoluble si G = W (non si G=H).

Les quotients sont soient cycliques, soient des p-groupes. Par définition de "p'-résoluble" (Serre p.155), on a donc :

LEMME 3. Pour tout sous-groupe ouvert distingué d'indice fini $U \subset G$ et tout nombre premier p' le groupe G/U est p'-résoluble.

Il est facile de voir qu'un caractère de G se relève à la caractéristique 0.
En appliquant le théorème de Fong-Swan (Serre, p.147), on déduit des deux lemmes le résultat suivant.

THEOREME 4. Toute représentation irréductible de dimension finie de G se relève à la caractéristique 0.

La "correspondance de Langlands galois-quaternions" est une certaine bijection entre les classes d'équivalence des représentations irréductibles complexes de H et celles de dimension ≤ 2 de W. On dit qu'elle se "réduit modulo ℓ" si
1) l'image de la classe d'équivalence d'une représentation A-admissible est la classe d'équivalence d'une représentation A-admissible,
2) elle induit par passage au quotient une bijection entre le groupe de Grothendieck de H sur C et le sous-groupe du groupe de Grothendieck de W sur C, engendré par les représentations iréductibles de dimension ≤ 2 de W.

Remarques. 1) Il est clair que la bijection entre les caractères de W et H (rappelée plus haut) se réduit modulo λ.
2) On plonge K_ℓ dans $\mathbb{C}$. Comme les représentations sont de dimension finie, une représentation irréductible complexe isomorphe à ses conjuguées par les K_ℓ-automorphismes de $\mathbb{C}$ est isomorphe à une représentation définie sur K_ℓ.
3) Si la correspondance de Langlands se réduit modulo ℓ, il en est de même pour la correspondance de Hecke, et inversement.

THEOREME 5. Si $\ell \neq p$, la correspondance de Langlands galois-quaternions se réduit modulo λ.

Remarques. 1) Comme le groupe H modulo son centre est isomorphe à O* qui est un groupe profini de cardinal $(q-1)\, q^\infty\, (q+1)$, la théorie de ses représentations de caractère central donné est la même au sens de (Serre [1],p.141) qu'en caractéristique 0 si ℓ ne divise pas $(q-1)q(q+1)$. Par le théorème, il en est de même pour les représentations irréductibles de dimension 2 de W.
2) Comme la correspondance de Langlands est compatible avec la torsion par un caractère, et à la théorie du corps de classes local, il suffit de montrer le théorème 5 pour les représentations

minimales.

La classification des représentations irréductibles de H , ou de dimension 2 de W repose sur les résultats ci-dessous sur la restriction d'une représentation irréductible à un sous-groupe distingué d'indice fini (la théorie de Clifford, (Curtis-Reiner,§11, et spécialement p.278 (11.20)) Ils sont vrais pour les représentations de dimension quelconque.

LEMME 6. 1) La restriction d'une représentation irréductible π d'un groupe B à un sous-groupe distingué d'indice fini A est semi-simple.
2) Soit σ une représentation irréductible de A. Les représentations irréductibles π de B dont la restriction à A contient σ sont les induites $\pi = i(B,Z,\rho)$ à B des représentations irréductibles ρ du normalisateur Z de σ dans B, telles que $\rho|_A \approx m\sigma$, m entier, soit σ-isotypique.
3) Si Z/A est hyper-résoluble, ρ est induite, $\rho = i(Z,Z',\sigma')$, d'une représentation σ' prolongeant σ et $|Z/Z'|$ est premier à ℓ ; si Z/A est abélien, le quotient $|Z/A| / |Z/Z'|^2$ est un entier, dont la partie première à ℓ est égale au nombre de classes d'équivalence de ρ .

Cas particulier : restriction à un sous-groupe d'indice 2.
Supposons que $A \subset B$ soit d'indice 2 dans B. Soit $g \in B$ mais non dans A. Si $\ell \neq 2$, il existe un caractère non trivial de B, trivial sur A noté μ .

LEMME 7. 1) Si σ est régulier, $i(\sigma) = i(B,A,\sigma)$ est irréductible et sa restriction à B est $\sigma+\sigma^g$.
Sinon, σ admet
- deux prolongements distincts π et $\pi\mu$ à B, et $i(\sigma) = \pi + \pi\mu$, si $\ell \neq 2$,
- un seul prolongement π , et $i(\sigma)$ est indécomposable admettant π comme quotient et sous-module si $\ell = 2$.

2) Soit σ' une représentation irréductible d'un sous-groupe $A' \subset B$, d'indice 2 dans B.
Pour que $i(\sigma') = i(\sigma)$ il faut et il suffit que
- $\sigma' = \sigma$ ou σ^g si $A = A'$,
- $\sigma' = \sigma$ sur $A \cap A'$ si $A \neq A'$.

3) Pour toute représentation irréductible π de B, on a : la restriction de π à A n'est pas irréductible $\Leftrightarrow \pi = i(\sigma)$. Si $\ell \neq 2$, c'est aussi équivalent à : $\pi = \pi\mu$.

COROLLAIRE 8. 1) Si $\ell = p$, toutes les représentations irréductibles de dimension >1 de G sont modérément ramifiées.
2) Si $\ell \neq p$, la réduction modulo λ d'une représentation irréductible sauvagement ramifée de G conserve l'indice de ramification et la propriété d'être minimale.
3) La restriction au groupe sauvage G_1 d'une représentation irréductible primitive de W est irréductible.
4) Une représentation irréductible imprimitive de dimension 2 de W est de la forme $i(\chi)$ pour un caractère régulier χ d'une extension quadratique E .
Pour que $i(\chi) = i(\chi')$, il faut et il suffit que
- si $E = E'$, χ' est conjuguée à χ,
- si $E \neq E'$, $\chi' N_{EE'/E'} = \chi N_{EE'/E}$ sur $(EE')^*$, où $N_{EE'/E}$, resp. $N_{EE'/E'}$, est la norme de EE'

sur E, resp. sur E'.

Par la théorie du corps de classes, la dernière égalité signifie que la restriction de χ à $W_{EE'}$ est isomorphe à ses conjugués par $W = W_E W_{E'}$, et égale à la restriction de χ'. Nous donnerons les démonstrations de 6, 7, 8 dans l'appendice.

Représentations modérément ramifiées.
Soit E l'extension quadratique non ramifée de F, et Q le sous-groupe d'indice 2 de G égal à

$$Q = W_E \quad \text{si } G = W, \qquad Q = T_E O^* \quad \text{si } G = H,$$

où T_E est un sous-groupe fixé de H isomorphe à E^*. Un caractère modérément ramifié χ de E^* s'identifie à un caractère modérément ramifié de Q (trivial sur G_1) qui est régulier, si et seulement si χ n'est pas trivial sur le sous-groupe d'ordre $q+1$ du groupe cyclique d'ordre q^2-1 des racines de l'unité de E^*.
On note $i(\chi)$ la représentation induite à G.

PROPOSITION 9 . a) Les représentations irréductibles de dimension finie modérément ramifiées de H sont des caractères ou de dimension 2.
b) Une représentation irréductible modérément ramifée de dimension 2 de G est de la forme $i(\chi)$ pour un caractère modérément ramifié régulier χ de l'extension quadratique non ramifiée.
c) La réduction d'une représentation irréductible modérément ramifiée $i(\chi)$ est irréductible, sauf si $q = -1$ modulo ℓ, et χ est trivial sur les racines de l'unité d'ordre premier à ℓ divisant $q+1$. Alors la réduction d'une tordue de $i(\chi)$ est égale dans le groupe de Grothendieck à $1 + (-1)^{val}$.

Preuve. a) et b) pour $G = H$. Il suffit de remarquer que G/G_1 est isomorphe au produit semi-direct $\mathbb{Z}/2\mathbb{Z} \times E^*/E_1$ ou encore à $\{\pm 1\} \times (F_{q^2})^*$, où E_1 est le groupe des unités congrues à 1 modulo une uniformisante, et -1 agit par l'élévation à la puissance q. On applique le lemme 7.
b) pour $G = W$. Une représentation modérément ramifiée est induite par un caractère d'un sous-groupe, car G/G_1 est hyper-résoluble (lemme 6). Si elle est de dimension 2, c'est l'induite d'un caractère modérément ramifié régulier χ' de $W_{E'}$ pour une extension quadratique modérément ramifiée E'/F. L'extension E'/F est non ramifiée, si $p=2$. Si $p \neq 2$, on peut se ramener à E'/F non ramifiée, par le lemme 7. Si E'/F est ramifiée, et E/F non ramifiée, $\chi' N_{EE'/E'}$ est un caractère de $(EE')^*$ invariant par $Gal(EE'/E)$.
c) résulte du lemme 7.

La correspondance de Langlands complexe a la propriété (voir Gérardin-Kutzko, p.365) :
Pour les représentations complexes modérément ramifiées qui ne sont pas des caractères, elle coincide avec la paramétrisation par χ du théorème 2, b), avec un décalage par multiplication par le caractère $(-1)^{val}$ de E^*. La réduction de $i(\chi)$ est irréductible, si et seulement si la réduction de $i(\chi\,(-1)^{val})$ est irréductible. Lorsque ces réductions sont réductibles, le c) du théorème 2 montre que ces réductions sont égales.

COROLLAIRE 10. La correspondance de Langlands galois-quaternions pour les représentations modérément ramifiées se réduit modulo λ.

Remarque. On n'a fait aucune restriction sur ℓ (ℓ=p convient).

Représentations irréductibles minimales sauvagement ramifiées de H.

On suppose $\ell \neq p$, jusqu'au corollaire 12 inclus.

Soit π une telle représentation, de conducteur $f \geq 2$ fixé. Le sous-groupe $F^*G_{f'} \subset H$, où $f' = [(f+1)/2]$, est distingué d'indice fini dans H, commutatif modulo G_f. On applique le lemme 6. Soit χ un caractère contenu dans la restriction de π à $F^*G_{f'}$. Soit Z le centralisateur de χ dans H. Il existe une représentation irréductible ρ de Z dont la restriction à $F^*G_{f'}$ est χ -isotypique telle que

$$\pi \sim \text{ind}(H, Z, \rho).$$

A conjugaison près dans H, le couple (Z, ρ) est unique.

On a la description suivante de Z et de ρ (voir par exemple, Carayol p.216).

- Si f est pair, $Z = E^*G_{f/2}$ où E/F est quadratique ramifiée (mais non unique). Comme $Z/\text{Ker}\chi$ est abélien, ρ est un caractère. On note $\rho = \rho(\theta, \chi)$, où θ est sa restriction à E^* et

$$\pi = \pi(E, \theta, \chi).$$

La dimension de π est égale à l'indice de Z dans G et donc de la forme

$$\dim \pi = (q+1) \times \text{une puissance de } p.$$

- Si f est impair, $Z = E^* G_{(f-1)/2}$, où E/F est quadratique non ramifiée. Soit $Z' = F^*E_1G_{(f-1)/2}$ le sous-groupe distingué de Z, d'indice q+1. La restriction de ρ à Z' est irréductible. Le groupe $H' = Z'/\text{Ker}\chi$ est un groupe d'Heisenberg, extension centrale non dégénérée par un groupe abélien $A = F^*E_1G_{(f+1)/2}/\text{Ker}\chi$ d'un groupe fini $B = Z'/A$ d'ordre q^2

$$1 \to A \to H' \to B \to 1.$$

Toute représentation irréductible de H' qui prolonge χ est de dimension q, caractérisée par son caractère central que l'on note $\alpha\chi$, où α est sa restriction à F^*E^1. Comme A est central dans $Z/\text{Ker}\chi$, elle se prolonge en une représentation ρ de Z. Ses prolongements sont canoniquement en bijection avec les prolongements θ de α à E^*. On note $\rho = \rho(\theta, \chi)$ et $\pi = \pi(E, \theta, \chi)$.

La dimension de π est égale à $q \times [G:Z]$, elle est de la forme

$$\dim \pi = 2 \times \text{une puissance de } p.$$

Pour que deux représentations π, π' de H, irréductibles, minimales, de conducteur $f \geq 2$ soient isomorphes, il faut et il suffit qu'il existe E, θ, θ', χ comme ci-dessus tel que

$$\pi \sim \pi(E, \theta, \chi) \quad , \quad \pi' \sim \pi(E, \theta', \chi),$$

où θ est conjugué à θ'.

Remarque. Toute représentation irréductible sur K du groupe d'Heisenberg H' se réduit irréductiblement car $(q, \ell) = 1$. En effet, H' est le produit direct $H' = A'H''$, où H'' est un p-groupe d'Heisenberg

$$1 \to A_p \to H'' \to B \to 1,$$

où A_p est le groupe fini des éléments de A d'ordre une puissance de p, A' celui des éléments d'ordre ∞ ou premier à p. On sait que la théorie des représentations d'un p-groupe en caractéristique différente de p est la même qu'en caractéristique 0.

PROPOSITION 11 ($\ell \neq p$). 1) La réduction modulo ℓ d'une représentation de G, irréductible, minimale, de conducteur $f \geq 2$, définie sur K, est irréductible.

2) Pour que π et π' comme en 1) aient des réductions isomorphes, il faut et il suffit que:

$$\pi \sim \pi(E, \theta, \chi) \ , \quad \pi' \sim \pi(E, \theta' \mu , \chi), \quad \text{si } G = H,$$
$$\pi \sim i(E,\theta) \quad , \quad \pi' \sim i(E, \theta' \mu) \quad , \quad \text{si } G = W,$$

où θ, θ' sont conjugués, et μ est un caractère de E^* sur K qui se réduit trivialement.
Si $p=2$ et f est pair, ceci équivalent à $\pi' \sim \pi\mu'$, où μ' est un caractère de F^* sur K qui se réduit trivialement .

COROLLAIRE 12 ($\ell \neq p$). La correspondance de Langlands entre représentations irréductibles minimales sauvagement ramifiées de W et de H se réduit modulo λ .

Si π , π' sont deux représentations A-admissibles de H comme ci-dessus, σ , σ' leurs images par la correspondance galois-quaternions complexe, le corollaire signifie

- σ est A-admissible,
- π est réductible modulo ℓ si et seulement si σ est iréductible modulo ℓ , et alors $\dim\pi = 2$, et les réductions modulo ℓ de π et de σ sont égales (via l'identification des caractères),
- π, π' ont des réductions isomorphes, si et seulement si σ, σ' ont des réductions isomorphes.

Le corollaire se déduit de la proposition 11, et de la description de la correspondance $\pi \to \sigma$ par Gérardin et Kutzko (p.365 et 370), en termes de la paramétrisation de la proposition 11, si E/F est non ramifiée, ou si $p \neq 2$.
L'image de $\pi(E, \theta, \chi)$ est égale à $i(E,\theta')$ où $\theta' = \theta \beta$,

- Si E/F n'est pas ramifié, ß est égal à $(-1)^{val}$.
- Si E/F est ramifié et $p \neq 2$, ß ne dépend de la restriction de θ à $1+P_E$. En particulier, ß est le même pour θ et $\theta \mu$ comme ci-dessus.

Il n'est pas nécessaire de connaître explicitement la correspondance, si E/F est ramifié et $p = 2$, grâçe à la proposition 11, 2).

III - APPENDICE (Démonstrations).

<u>Preuve de I.3.</u>
Une représentation ℓ-adique de dimension finie n du groupe compact Gal_p a toujours un modèle L sur A. On distingue deux cas.
a) σ est irréductible. Par II-9, et II-11,1) le seul cas où σ se réduit "réductiblement" modulo ℓ se produit quand $\ell \mid p+1$, et $\sigma = \sigma' \xi$ est le produit tensoriel d'un caractère $\xi : Gal_p \to A^*$, et de la représentation irréductible σ' induite du caractère modérément ramifié régulier (qui correspond à l'un des deux caractères fondamentaux de F_{p^2}) de l'extension quadratique non ramifiée de $\mathbb{Q}_p$. Soit μ la réduction modulo ℓ de ξ, ρ' celle d'un modèle de σ'.
La semi-simplifiée de ρ' est $1 + (-1)^{val}$.
Pour que ρ soit comme en I.1, il faut que μ soit non ramifié, et $\rho' (I_p)$ unipotent non trivial.
On fait l'exercice du calcul des conducteurs d'Artin :

Soient $a(\sigma)$, $a(\xi)$, $a(\mu)$,les exposants des conducteurs d'Artin de σ, ξ, . μ. La formule d'induction implique :

$$a(\sigma) = 2 \quad \text{si } \xi \text{ n'est pas ramifié}, \qquad a(\sigma) = 2\, a(\xi) > 2, \quad \text{si } \xi \text{ est ramifié.}$$

On vérifie facilement que :

$a(\rho) = 2\, a(\mu)$, si $\rho'(I_p) = id.$, ou μ ramifiée, et $a(\rho) = 1$, si $\rho'(I_p) \neq id.$ et μ non ramifié.

On compare $a(\sigma)$ et $a(\rho)$: μ non ramifié implique $a(\rho) < a(\sigma)$ (car $a(\rho) \leq 1$); c'est même équivalent. Il n'y a donc aucune σ irréductible, telle que $a(\sigma) = a(\rho)$ et ρ comme dans I.1.

b) σ est réductible. Les représentations ℓ-adiques réductibles de dimension 2 de Gal_p sont (c'est bien connu) :

- indécomposables $(\xi\, sp)(Fr_p{}^n u) = \xi(Fr_p{}^n u) \exp(t_\ell(u))$,
- semisimples $\xi + \xi'$,

où $\xi, \xi' : Gal_p \to A^*$ caractères continus, $u \in I_p$, Fr_p le Frobenius, t_ℓ est un homomorphisme du groupe d'inertie I_p dans $\mathbf{Z}_\ell$.

Soient μ, μ', ρ les réductions modulo ℓ de ξ, ξ', σ. Si ρ est comme dans I.1, on peut supposer μ non ramifié, μ' ramifié (donc ξ' ramifié). Les exposants des conducteurs d'Artin sont :

$$a(\xi\, sp) = \sup(1, 2\, a(\xi)) \ , \quad a(\xi + \xi') = a(\xi) + a(\xi') \, ,$$

même résultat pour les réductions modulo ℓ en remplaçant ξ, ξ' par μ, μ'.

On compare: $a(\sigma) = a(\rho)$ si $\sigma = \xi\, sp$, ou $\xi + \xi'$ avec $a(\xi) = a(\mu)$, $a(\xi') = a(\mu')$. C'est équivalent à : ξ non ramifiée, et μ' ramifiée.

<u>Preuve du lemme 1.</u> Si (π, V) est irréductible, on veut montrer qu'il existe χ non ramifié tel que le noyau de $\pi\chi$ dans G soit d'indice fini.

- Le groupe G = H est extension d'un groupe profini, par son centre. Il suffit de prendre χ non ramifié tel que $\pi\chi$ soit trivial sur l'uniformisante de F.

- Pour W, la restriction de π au groupe d'inertie G_0 a un noyau J d'indice fini, normalisé par un Frobenius Fr de W; il existe donc un entier n tel que Fr^n opère trivialement sur G_0/J, ce qui implique $\pi(Fr^n)$ appartient au centre de $\pi(W)$. La dimension de π est finie, puisque W est profini.Par le lemme de Schur, le centre opère par homothéties, et il suffit de prendre un caractère non ramifié χ de G tel que $\pi\chi$ soit trivial sur Fr^n .

<u>Démonstration du lemme 6.</u>

1) Soit $X \subset B$ un ensemble de représentants de B/A dans B . Si (π, V) est une représentation irréductible de B, et $v \in V$ non nul, V est un A-module engendré par les $\pi(g)v$, $g \in X$. Si $\pi(X)v$ est fini, $\pi_{|A}$ est à engendrement fini. Elle possède alors un quotient irréductible (σ, V); soit p la projection canonique de π sur σ. La projection canonique : $v \in V \to \{p(\pi(g)v)\}$ est un A-homomorphisme non nul de π dans $\oplus\, \sigma^g$, $g \in X$, de noyau un sous-espace stable par B. Il est injectif, puisque π est irréductible. Un sous-module d'un somme directe est semi-simple, donc $\pi_{|A}$ est semi-simple.

2) Soit (σ, V) une représentation irréductible de A, isomorphe à ses conjuguées par B, et qui satisfait le lemme de Schur. On définit une extension centrale :

$$1 \to C^* \to B' \to B/A \to 1,$$

où B' est le quotient du sous-groupe de $B \times GL(V)$

$\{ (g,r(g)) \in B \times GL(V)$, $g \in B$, $r(g) \in GL(V)$ vérifie $\sigma(g^{-1}ag) = r(g)^{-1} \sigma(a) r(g)$, $a \in A$ $\}$,

par le sous-groupe $\{ (a, \sigma(a)$, $a \in A$ $\}$.

Si π est une représentation irréductible de B de restriction à A isomorphe à $m\sigma$, où $m \geq 1$ est un entier, alors π se réalise sur l'espace $W \otimes_C V$, où W est un espace vectoriel sur C de dimension m, et

$$\pi(g) = M(g) \otimes r(g)$$

où $M(g) \in GL(W)$ et $r(g) \in GL(V)$ est comme ci-dessus. L'application

$$(g,r(g)) \to M(g)$$

définit une représentation irréductible (M,W) de B' triviale sur C*. Inversement, les classes d'isomorphie des représentations irréductibles de B, dont la restriction à A est isomorphe à un multiple de σ sont paramétrées canoniquement par les classes d'isomorphie des représentations irréductibles de B', triviales sur C*.

3) Si B/A est abélien, le groupe B' est un groupe d'Heisenberg, i.e. abélien modulo son centre Z'. La théorie de ses représentations est bien connue. L'application commutateur induit sur B' induit une forme bilinéaire $[\, ,\,] : B/A \times B/A \to C^*$, de radical le sous-groupe distingué

$$A' = \{ g' \in B/A , [g', g''] = 1 \text{ , pour tout } g'' \in B' \}$$

de B/A. Son image inverse Z' est le centre de B'. Le nombre de caractères de Z' prolongeant l'identité sur C* est égal à l'entier m' premier à ℓ, tel que $|A'| = m' \times$ une puissance de ℓ. Un caractère χ de Z' induisant l'identité sur C* définit une extension centrale de B/A' par C* non dégénérée (le radical est trivial). Il existe une unique représentation irréductible de B' de caractère central χ. Sa dimension est l'entier n premier à ℓ tel que $|B/A'|^{1/2} = n$. Elle est induite par un prolongement de χ à un sous-groupe commutatif maximal.

Si B/A est cyclique, alors le groupe B' est abélien.

Si B/A est hyper-résoluble non abélien, alors B' contient un sous-groupe D' abélien, distingué, non contenu dans le centre (Serre,p.82 : l'image inverse dans B' d'un sous-groupe abélien distingué abélien D de B/A , contenant strictement le centre de B/A, convient). Si π n'est pas induite, la restriction de π à A' est isotypique. Elle définit une extension centrale B'' de B'/D' par C* et π correpond canoniquement à une représentation irréductible π' de B'' triviale sur C*. On raisonne par récurrence sur l'ordre de B/A . Alors π' est induite par un caractère. Il en est de même de π.

Démonstration du lemme 7 .

Le théorème d'induction-restriction de Mackey (Serre p.175) et la formule de réciprocité de Frobenius (Serre p.173) sont valables en toute caractéristique.

a) Par l'induction-restriction de Mackey : $res(A',B)ind(B,A)(\sigma) = \sigma + \sigma^g$ si $A = A'$, et égale à $ind(A', A \cap A')res(A \cap A',A)(\sigma)$ sinon.

b) Par Frobenius : $Hom_B(\rho, i(\sigma')) = Hom_A(res(A,B)(\rho) , \sigma')$, si ρ est une représentation de B. Par le lemme précédent, σ est régulier si et seulement s'il ne se prolonge pas à B. Si $i(\sigma)$ est réductible, un sous-module irréductible $\pi \subset i(\sigma)$ prolonge σ (Frobenius), et inversement, s'il existe un prolongement de σ à G, il est contenu dans $i(\sigma)$. L'espace de $i(\sigma)$ contient strictement celui de σ, donc $i(\sigma)$ est réductible. Donc σ est régulier, si et seulement si $i(\sigma)$ est irréductible.

Si $i(\sigma)$ est réductible, par a) et b), $Hom_B(i(\sigma), i(\sigma))$ est de dimension 2. Il existe un quotient non trivial égal à un sous-module non trivial. Par Frobenius, tout prolongement de σ se plonge dans $i(\sigma)$. Si $\ell \neq 2$, il existe deux prolongements et $i(\sigma)$ sera leur somme, sinon il existe un seul prolongement et $i(\sigma)$ sera indécomposable. Le lemme se déduit de ceci facilement. La théorie est la même qu'en caractéristique 0 si $\ell \neq 2$.

Démonstration du corollaire 8.

Il est assez évident avec les lemmes, sauf peut-être 1). On peut supposer que π est une représentation irréductible d'image finie. Elle s'identifie à une représentation π' du groupe fini $G' = G/\mathrm{Ker}\pi$. L'image G'_1 de G_1 dans G' est un p-sous-groupe distingué, donc la restriction de π' au p-groupe G'_1 est semi-simple. La représentation triviale est la seule représentation irréductible de G'_1. Donc G_1 opére trivialement dans π.

Remarque. Le conducteur d'un caractère $\chi = \mu$ Nrd de H, sauvagement ramifié, est impair, car

$$\mathrm{Nrd}(1+P^i) = F^* \cap (1+P^i) = (1+P_F^{f(\mu)}) \quad , \text{ si } i = 2f(\mu), 2f(\mu)-1.$$

Une représentation π de H irréductible, sauvagement ramifiée avec f pair, est donc toujours minimale. Si f est impair, l'hypothèse que π est minimale est équivalente à :

(*) $$\chi(1+x) = \psi_H(\omega_F^{-(f+1)/2} ax),$$

où ψ_H est un caractère de M de conducteur P^{-1}, et $a \notin F$ appartient au groupe cyclique d'ordre q^2-1 des racines de l'unité de E.

Preuve de la proposition 11.

G = H. 1) Comme F^* est le centre de H, et G_f est un p-groupe profini, Z est encore le centralisateur de la réduction de χ. La réduction commute avec l'induction, et la réduction de ρ est irréductible.

2) Si π et π' ont des réductions isomorphes, leur restrictions au p-groupe G_f sont isomorphes : on peut choisir le même χ, donc le même E. On en déduit que

$$\pi \sim \pi(E, \theta, \chi) \quad , \quad \pi' \sim \pi(E, \theta'\mu, \chi)$$

où θ, θ' sont conjugués, et μ est un caractère de E sur K qui se réduit trivialement. Le caractère μ est un caractère trivial sur les unités de E^* qui ne sont pas dans le groupe X des racines de l'unité d'ordre un puissance ℓ.

– Si p=2, et f pair, alors ℓ est impair et E/F est ramifiée. Le groupe des racines de l'unité de E^* d'ordre q-1 est isomorphe à celui de F^*, par la norme qui est l'élévation au carré $x \to x^2$. On en déduit que μ se factorise par la norme : $\mu = \mu' N_{E/F}$, et $\pi' \sim \pi\mu'$.

G=W. - Si π est imprimitive, minimale et $f \geq 2$, alors $\pi = i(E, \theta)$, où $f(\theta/\theta') \geq 2$, si θ' est le conjugué de θ sur F (Gérardin-Kutzko, p.358 et 367). Par réduction, θ et θ' restent différents, et π reste irréductible, d'où 1). La restriction commute avec la réduction. Donc la réduction de π est imprimitive.

- Si π est primitive, sa restriction au p-groupe G_1 est irréductible, et le reste par réduction, d'où 1). La réduction conserve la propriété d'être primitive ou non. Deux représentations irréductibles primitives, isomorphes après réduction, prolongent la même représentation irréductible du p-groupe G_1, et sont déduites l'une de l'autre par multiplication par un caractère modérément ramifié, dont la réduction est triviale.D'où 2) pour les primitives.

Pour les imprimitives, minimales, sauvagementramifiées, on va vérifier que l'égalité

$$\pi = \pi' \text{ modulo } \ell ,$$

implique que $\pi = i(E, \theta)$, $\pi' = i(E', \theta')$, avec $E = E'$. Alors, on a $\theta = \theta' \mu$, où μ est un caractère de E^* trivial modulo ℓ , d'où 2).

Supposons que $\pi = i(E, \theta)$, $\pi' = i(E', \theta')$ et que l'on ait $i(E, \theta) = i(E', \theta')$ modulo ℓ .

Le caractère θ est minimal, dans le sens suivant : son conducteur est ne peut pas être abaisser en le multipliant par un caractère $\chi N_{E/F}$ de E^*, où χ est un caractère de F^*, et $N_{E/F}$ la norme de E sur F. Pour que l'on ait $E' \neq E$, il faut et il suffit qu'il existe un caractère χ de F^*, tel que

$$\theta = \chi N_{E/F} \text{ modulo } \lambda,$$

sur le sous-groupe $N_{EE'/E}(EE')^* \subset E^*$ d'indice 2 dans E^*. Autrement dit, la restriction de θ à $N_{EE'/E}(EE')^*$ "se factorise par la norme".

Si $p \neq 2$, F admet seulement trois extensions quadratiques, toutes contenues dans l'unique extension bi-quadratique EE'. On peut supposer que E/F est ramifiée (quitte à remplacer E par E'), donc EE'/E est non ramifiée, et par la théorie du corps de classes,

$$1+P_E \subset N_{EE'/E}(EE')^*.$$

Il existe un caractère χ de F^*, tel que $\theta \chi N_{E/F}$ est trivial sur $1+P_E$ modulo ℓ . C'est absurde puisque θ est minimal, et sauvagement ramifié.

Si $p = 2$, l'argument ne marche pas, puisqu'il est possible que les trois sous-extensions quadratiques de F, contenues dans EE' soient ramifiées. Mais dans ce cas, les extensions quadratiques EE'/E, et E/F étant ramifiées, l'on remarque que θ "se factorise par le norme" sur $N_{EE'/E}(EE')^*$ est équivalente à θ "se factorise par la norme" sur le 2-pro-goupe $N_{EE'/E}(1+P_{EE'})$. Comme 2 ne divise pas λ, si θ "se factorise par la norme" modulo ℓ sur $N_{EE'/E}(EE')^*$, alors θ "se factorise par la norme" sur $N_{EE'/E}(EE')^*$.

BIBLIOGRAPHIE.

Henri Carayol. Représentations cuspidales du groupe linéaire. Ann. Scient. Ec. Norm. Sup. 4°série, t.17, 1984, p.191à 225.

C. Curtis, I. Reiner.Methods of representation theory. Pure and Applied Mathematics. John Wiley&Sons. New-York.

Pierre Deligne.Les constantes des équations fonctionnelles des fonctions L, Antwerp II, Lecture Notes in Math., vol.349, Springer-Verlag, 1973, PP. 501-595.

Paul Gérardin et Philip Kutzko. Facteurs locaux pour GL(2) . Ann. Scient. Ec. Norm. Sup. 4°série, t.13, 1980, p.349 à 384.

Jean-Pierre Serre. [1] Représentations linéaires des groupes finis. Hermann. Deuxième édition.
[2] Sur les représentations modulaires de degré 2 de Gal(Q/Q). Duke Math. J., 1987, vol.54, n°1, 179-230.
[3] Corps locaux. 3ème édition. Hermann, Paris, 1980.

Marie-France Vignéras. Représentations modulaires de GL(2,F) en caractéristique ℓ , F corps p-adique, $p \neq \ell$. Preprint 1987, non publié.

LECTURE NOTES IN MATHEMATICS
Edited by A. Dold and B. Eckmann

Some general remarks on the publication of proceedings of congresses and symposia

Lecture Notes aim to report new developments - quickly, informally and at a high level. The following describes criteria and procedures which apply to proceedings volumes. The editors of a volume are strongly advised to inform contributors about these points at an early stage.

§1. One (or more) expert participant(s) of the meeting should act as the responsible editor(s) of the proceedings. They select the papers which are suitable (cf. §§ 2, 3) for inclusion in the proceedings, and have them individually refereed (as for a journal). It should not be assumed that the published proceedings must reflect conference events faithfully and in their entirety. Contributions to the meeting which are not included in the proceedings can be listed by title. The series editors will normally not interfere with the editing of a particular proceedings volume - except in fairly obvious cases, or on technical matters, such as described in §§ 2, 3. The names of the responsible editors appear on the title page of the volume.

§2. The proceedings should be reasonably homogeneous (concerned with a limited area). For instance, the proceedings of a congress on "Analysis" or "Mathematics in Wonderland" would normally not be sufficiently homogeneous.

One or two longer survey articles on recent developments in the field are often very useful additions to such proceedings - even if they do not correspond to actual lectures at the congress. An extensive introduction on the subject of the congress would be desirable.

§3. The contributions should be of a high mathematical standard and of current interest. Research articles should present new material and not duplicate other papers already published or due to be published. They should contain sufficient information and motivation and they should present proofs, or at least outlines of such, in sufficient detail to enable an expert to complete them. Thus resumes and mere announcements of papers appearing elsewhere cannot be included, although more detailed versions of a contribution may well be published in other places later.

Surveys, if included, should cover a sufficiently broad topic, and should in general not simply review the author's own recent research. In the case of surveys, exceptionally, proofs of results may not be necessary.

"Mathematical Reviews" and "Zentralblatt für Mathematik" require that papers in proceedings volumes carry an explicit statement that they are in final form and that no similar paper has been or is being submitted elsewhere, if these papers are to be considered for a review. Normally, papers that satisfy the criteria of the Lecture Notes in Mathematics series also satisfy this

.../...

requirement, but we would strongly recommend that the contributing authors be asked to give this guarantee explicitly at the beginning or end of their paper. There will occasionally be cases where this does not apply but where, for special reasons, the paper is still acceptable for LNM.

§4. Proceedings should appear soon after the meeeting. The publisher should, therefore, receive the complete manuscript within nine months of the date of the meeting at the latest.

§5. Plans or proposals for proceedings volumes should be sent to one of the editors of the series or to Springer-Verlag Heidelberg. They should give sufficient information on the conference or symposium, and on the proposed proceedings. In particular, they should contain a list of the expected contributions with their prospective length. Abstracts or early versions (drafts) of some of the contributions are very helpful.

§6. Lecture Notes are printed by photo-offset from camera-ready typed copy provided by the editors. For this purpose Springer-Verlag provides editors with technical instructions for the preparation of manuscripts and these should be distributed to all contributing authors. Springer-Verlag can also, on request, supply stationery on which the prescribed typing area is outlined. Some homogeneity in the presentation of the contributions is desirable.

Careful preparation of manuscripts will help keep production time short and ensure a satisfactory appearance of the finished book. The actual production of a Lecture Notes volume normally takes 6 -8 weeks.

Manuscripts should be at least 100 pages long. The final version should include a table of contents and as far as applicable a subject index.

§7. Editors receive a total of 50 free copies of their volume for distribution to the contributing authors, but no royalties. (Unfortunately, no reprints of individual contributions can be supplied.) They are entitled to purchase further copies of their book for their personal use at a discount of 33.3 %, other Springer mathematics books at a discount of 20 % directly from Springer-Verlag. Contributing authors may purchase the volume in which their article appears at a discount of 33.3 %.

Commitment to publish is made by letter of intent rather than by signing a formal contract. Springer-Verlag secures the copyright for each volume.

Vol. 1232: P.C. Schuur, Asymptotic Analysis of Soliton Problems. VIII, 180 pages. 1986.

Vol. 1233: Stability Problems for Stochastic Models. Proceedings, 1985. Edited by V.V. Kalashnikov, B. Penkov and V.M. Zolotarev. VI, 223 pages. 1986.

Vol. 1234: Combinatoire énumérative. Proceedings, 1985. Edité par G. Labelle et P. Leroux. XIV, 387 pages. 1986.

Vol. 1235: Séminaire de Théorie du Potentiel, Paris, No. 8. Directeurs: M. Brelot, G. Choquet et J. Deny. Rédacteurs: F. Hirsch et G. Mokobodzki. III, 209 pages. 1987.

Vol. 1236: Stochastic Partial Differential Equations and Applications. Proceedings, 1985. Edited by G. Da Prato and L. Tubaro. V, 257 pages. 1987.

Vol. 1237: Rational Approximation and its Applications in Mathematics and Physics. Proceedings, 1985. Edited by J. Gilewicz, M. Pindor and W. Siemaszko. XII, 350 pages. 1987.

Vol. 1238: M. Holz, K.-P. Podewski and K. Steffens, Injective Choice Functions. VI, 183 pages. 1987.

Vol. 1239: P. Vojta, Diophantine Approximations and Value Distribution Theory. X, 132 pages. 1987.

Vol. 1240: Number Theory, New York 1984–85. Seminar. Edited by D.V. Chudnovsky, G.V. Chudnovsky, H. Cohn and M.B. Nathanson. V, 324 pages. 1987.

Vol. 1241: L. Gårding, Singularities in Linear Wave Propagation. III, 125 pages. 1987.

Vol. 1242: Functional Analysis II, with Contributions by J. Hoffmann-Jørgensen et al. Edited by S. Kurepa, H. Kraljević and D. Butković. VII, 432 pages. 1987.

Vol. 1243: Non Commutative Harmonic Analysis and Lie Groups. Proceedings, 1985. Edited by J. Carmona, P. Delorme and M. Vergne. V, 309 pages. 1987.

Vol. 1244: W. Müller, Manifolds with Cusps of Rank One. XI, 158 pages. 1987.

Vol. 1245: S. Rallis, L-Functions and the Oscillator Representation. XVI, 239 pages. 1987.

Vol. 1246: Hodge Theory. Proceedings, 1985. Edited by E. Cattani, F. Guillén, A. Kaplan and F. Puerta. VII, 175 pages. 1987.

Vol. 1247: Séminaire de Probabilités XXI. Proceedings. Edité par J. Azéma, P.A. Meyer et M. Yor. IV, 579 pages. 1987.

Vol. 1248: Nonlinear Semigroups, Partial Differential Equations and Attractors. Proceedings, 1985. Edited by T.L. Gill and W.W. Zachary. IX, 185 pages. 1987.

Vol. 1249: I. van den Berg, Nonstandard Asymptotic Analysis. IX, 187 pages. 1987.

Vol. 1250: Stochastic Processes – Mathematics and Physics II. Proceedings 1985. Edited by S. Albeverio, Ph. Blanchard and L. Streit. VI, 359 pages. 1987.

Vol. 1251: Differential Geometric Methods in Mathematical Physics. Proceedings, 1985. Edited by P.L. García and A. Pérez-Rendón. VII, 300 pages. 1987.

Vol. 1252: T. Kaise, Représentations de Weil et GL_2 Algèbres de division et GL_n. VII, 203 pages. 1987.

Vol. 1253: J. Fischer, An Approach to the Selberg Trace Formula via the Selberg Zeta-Function. III, 184 pages. 1987.

Vol. 1254: S. Gelbart, I. Piatetski-Shapiro, S. Rallis. Explicit Constructions of Automorphic L-Functions. VI, 152 pages. 1987.

Vol. 1255: Differential Geometry and Differential Equations. Proceedings, 1985. Edited by C. Gu, M. Berger and R.L. Bryant. XII, 243 pages. 1987.

Vol. 1256: Pseudo-Differential Operators. Proceedings, 1986. Edited by H.O. Cordes, B. Gramsch and H. Widom. X, 479 pages. 1987.

Vol. 1257: X. Wang, On the C*-Algebras of Foliations in the Plane. V, 165 pages. 1987.

Vol. 1258: J. Weidmann, Spectral Theory of Ordinary Differential Operators. VI, 303 pages. 1987.

Vol. 1259: F. Cano Torres, Desingularization Strategies for Three-Dimensional Vector Fields. IX, 189 pages. 1987.

Vol. 1260: N.H. Pavel, Nonlinear Evolution Operators and Semigroups. VI, 285 pages. 1987.

Vol. 1261: H. Abels, Finite Presentability of S-Arithmetic Groups. Compact Presentability of Solvable Groups. VI, 178 pages. 1987.

Vol. 1262: E. Hlawka (Hrsg.), Zahlentheoretische Analysis II. Seminar, 1984–86. V, 158 Seiten. 1987.

Vol. 1263: V.L. Hansen (Ed.), Differential Geometry. Proceedings, 1985. XI, 288 pages. 1987.

Vol. 1264: Wu Wen-tsün, Rational Homotopy Type. VIII, 219 pages. 1987.

Vol. 1265: W. Van Assche, Asymptotics for Orthogonal Polynomials. VI, 201 pages. 1987.

Vol. 1266: F. Ghione, C. Peskine, E. Sernesi (Eds.), Space Curves. Proceedings, 1985. VI, 272 pages. 1987.

Vol. 1267: J. Lindenstrauss, V.D. Milman (Eds.), Geometrical Aspects of Functional Analysis. Seminar. VII, 212 pages. 1987.

Vol. 1268: S.G. Krantz (Ed.), Complex Analysis. Seminar, 1986. VII, 195 pages. 1987.

Vol. 1269: M. Shiota, Nash Manifolds. VI, 223 pages. 1987.

Vol. 1270: C. Carasso, P.-A. Raviart, D. Serre (Eds.), Nonlinear Hyperbolic Problems. Proceedings, 1986. XV, 341 pages. 1987.

Vol. 1271: A.M. Cohen, W.H. Hesselink, W.L.J. van der Kallen, J.R. Strooker (Eds.), Algebraic Groups Utrecht 1986. Proceedings. XII, 284 pages. 1987.

Vol. 1272: M.S. Livšic, L.L. Waksman, Commuting Nonselfadjoint Operators in Hilbert Space. III, 115 pages. 1987.

Vol. 1273: G.-M. Greuel, G. Trautmann (Eds.), Singularities, Representation of Algebras, and Vector Bundles. Proceedings, 1985. XIV, 383 pages. 1987.

Vol. 1274: N. C. Phillips, Equivariant K-Theory and Freeness of Group Actions on C*-Algebras. VIII, 371 pages. 1987.

Vol. 1275: C.A. Berenstein (Ed.), Complex Analysis I. Proceedings, 1985–86. XV, 331 pages. 1987.

Vol. 1276: C.A. Berenstein (Ed.), Complex Analysis II. Proceedings, 1985–86. IX, 320 pages. 1987.

Vol. 1277: C.A. Berenstein (Ed.), Complex Analysis III. Proceedings, 1985–86. X, 350 pages. 1987.

Vol. 1278: S.S. Koh (Ed.), Invariant Theory. Proceedings, 1985. V, 102 pages. 1987.

Vol. 1279: D. Ieşan, Saint-Venant's Problem. VIII, 162 Seiten. 1987.

Vol. 1280: E. Neher, Jordan Triple Systems by the Grid Approach. XII, 193 pages. 1987.

Vol. 1281: O.H. Kegel, F. Menegazzo, G. Zacher (Eds.), Group Theory. Proceedings, 1986. VII, 179 pages. 1987.

Vol. 1282: D.E. Handelman, Positive Polynomials, Convex Integral Polytopes, and a Random Walk Problem. XI, 136 pages. 1987.

Vol. 1283: S. Mardešić, J. Segal (Eds.), Geometric Topology and Shape Theory. Proceedings, 1986. V, 261 pages. 1987.

Vol. 1284: B.H. Matzat, Konstruktive Galoistheorie. X, 286 pages. 1987.

Vol. 1285: I.W. Knowles, Y. Saitō (Eds.), Differential Equations and Mathematical Physics. Proceedings, 1986. XVI, 499 pages. 1987.

Vol. 1286: H.R. Miller, D.C. Ravenel (Eds.), Algebraic Topology. Proceedings, 1986. VII, 341 pages. 1987.

Vol. 1287: E.B. Saff (Ed.), Approximation Theory, Tampa. Proceedings, 1985–1986. V, 228 pages. 1987.

Vol. 1288: Yu. L. Rodin, Generalized Analytic Functions on Riemann Surfaces. V, 128 pages, 1987.

Vol. 1289: Yu. I. Manin (Ed.), K-Theory, Arithmetic and Geometry. Seminar, 1984–1986. V, 399 pages. 1987.

Vol. 1290: G. Wüstholz (Ed.), Diophantine Approximation and Transcendence Theory. Seminar, 1985. V, 243 pages. 1987.

Vol. 1291: C. Mœglin, M.-F. Vignéras, J.-L. Waldspurger, Correspondances de Howe sur un Corps p-adique. VII, 163 pages. 1987

Vol. 1292: J.T. Baldwin (Ed.), Classification Theory. Proceedings, 1985. VI, 500 pages. 1987.

Vol. 1293: W. Ebeling, The Monodromy Groups of Isolated Singularities of Complete Intersections. XIV, 153 pages. 1987.

Vol. 1294: M. Queffélec, Substitution Dynamical Systems – Spectral Analysis. XIII, 240 pages. 1987.

Vol. 1295: P. Lelong, P. Dolbeault, H. Skoda (Réd.), Séminaire d'Analyse P. Lelong – P. Dolbeault – H. Skoda. Seminar, 1985/1986. VII, 283 pages. 1987.

Vol. 1296: M.-P. Malliavin (Ed.), Séminaire d'Algèbre Paul Dubreil et Marie-Paule Malliavin. Proceedings, 1986. IV, 324 pages. 1987.

Vol. 1297: Zhu Y.-l., Guo B.-y. (Eds.), Numerical Methods for Partial Differential Equations. Proceedings. XI, 244 pages. 1987.

Vol. 1298: J. Aguadé, R. Kane (Eds.), Algebraic Topology, Barcelona 1986. Proceedings. X, 255 pages. 1987.

Vol. 1299: S. Watanabe, Yu.V. Prokhorov (Eds.), Probability Theory and Mathematical Statistics. Proceedings, 1986. VIII, 589 pages. 1988.

Vol. 1300: G.B. Seligman, Constructions of Lie Algebras and their Modules. VI, 190 pages. 1988.

Vol. 1301: N. Schappacher, Periods of Hecke Characters. XV, 160 pages. 1988.

Vol. 1302: M. Cwikel, J. Peetre, Y. Sagher, H. Wallin (Eds.), Function Spaces and Applications. Proceedings, 1986. VI, 445 pages. 1988.

Vol. 1303: L. Accardi, W. von Waldenfels (Eds.), Quantum Probability and Applications III. Proceedings, 1987. VI, 373 pages. 1988.

Vol. 1304: F.Q. Gouvêa, Arithmetic of p-adic Modular Forms. VIII, 121 pages. 1988.

Vol. 1305: D.S. Lubinsky, E.B. Saff, Strong Asymptotics for Extremal Polynomials Associated with Weights on ℝ. VII, 153 pages. 1988.

Vol. 1306: S.S. Chern (Ed.), Partial Differential Equations. Proceedings, 1986. VI, 294 pages. 1988.

Vol. 1307: T. Murai, A Real Variable Method for the Cauchy Transform, and Analytic Capacity. VIII, 133 pages. 1988.

Vol. 1308: P. Imkeller, Two-Parameter Martingales and Their Quadratic Variation. IV, 177 pages. 1988.

Vol. 1309: B. Fiedler, Global Bifurcation of Periodic Solutions with Symmetry. VIII, 144 pages. 1988.

Vol. 1310: O.A. Laudal, G. Pfister, Local Moduli and Singularities. V, 117 pages. 1988.

Vol. 1311: A. Holme, R. Speiser (Eds.), Algebraic Geometry, Sundance 1986. Proceedings. VI, 320 pages. 1988.

Vol. 1312: N.A. Shirokov, Analytic Functions Smooth up to the Boundary. III, 213 pages. 1988.

Vol. 1313: F. Colonius, Optimal Periodic Control. VI, 177 pages. 1988.

Vol. 1314: A. Futaki, Kähler-Einstein Metrics and Integral Invariants. IV, 140 pages. 1988.

Vol. 1315: R.A. McCoy, I. Ntantu, Topological Properties of Spaces of Continuous Functions. IV, 124 pages. 1988.

Vol. 1316: H. Korezlioglu, A.S. Ustunel (Eds.), Stochastic Analysis and Related Topics. Proceedings, 1986. V, 371 pages. 1988.

Vol. 1317: J. Lindenstrauss, V.D. Milman (Eds.), Geometric Aspects of Functional Analysis. Seminar, 1986–87. VII, 289 pages. 1988.

Vol. 1318: Y. Felix (Ed.), Algebraic Topology – Rational Homotopy. Proceedings, 1986. VIII, 245 pages. 1988

Vol. 1319: M. Vuorinen, Conformal Geometry and Quasiregular Mappings. XIX, 209 pages. 1988.

Vol. 1320: H. Jürgensen, G. Lallement, H.J. Weinert (Eds.), Semigroups, Theory and Applications. Proceedings, 1986. X, 416 pages. 1988.

Vol. 1321: J. Azéma, P.A. Meyer, M. Yor (Eds.), Séminaire de Probabilités XXII. Proceedings. IV, 600 pages. 1988.

Vol. 1322: M. Métivier, S. Watanabe (Eds.), Stochastic Analysis. Proceedings, 1987. VII, 197 pages. 1988.

Vol. 1323: D.R. Anderson, H.J. Munkholm, Boundedly Controlled Topology. XII, 309 pages. 1988.

Vol. 1324: F. Cardoso, D.G. de Figueiredo, R. Iório, O. Lopes (Eds.), Partial Differential Equations. Proceedings, 1986. VIII, 433 pages. 1988.

Vol. 1325: A. Truman, I.M. Davies (Eds.), Stochastic Mechanics and Stochastic Processes. Proceedings, 1986. V, 220 pages. 1988.

Vol. 1326: P.S. Landweber (Ed.), Elliptic Curves and Modular Forms in Algebraic Topology. Proceedings, 1986. V, 224 pages. 1988.

Vol. 1327: W. Bruns, U. Vetter, Determinantal Rings. VII,236 pages. 1988.

Vol. 1328: J.L. Bueso, P. Jara, B. Torrecillas (Eds.), Ring Theory. Proceedings, 1986. IX, 331 pages. 1988.

Vol. 1329: M. Alfaro, J.S. Dehesa, F.J. Marcellan, J.L. Rubio de Francia, J. Vinuesa (Eds.): Orthogonal Polynomials and their Applications. Proceedings, 1986. XV, 334 pages. 1988.

Vol. 1330: A. Ambrosetti, F. Gori, R. Lucchetti (Eds.), Mathematical Economics. Montecatini Terme 1986. Seminar. VII, 137 pages. 1988.

Vol. 1331: R. Bamón, R. Labarca, J. Palis Jr. (Eds.), Dynamical Systems, Valparaiso 1986. Proceedings. VI, 250 pages. 1988.

Vol. 1332: E. Odell, H. Rosenthal (Eds.), Functional Analysis. Proceedings, 1986–87. V, 202 pages. 1988.

Vol. 1333: A.S. Kechris, D.A. Martin, J.R. Steel (Eds.), Cabal Seminar 81–85. Proceedings, 1981–85. V, 224 pages. 1988.

Vol. 1334: Yu.G. Borisovich, Yu. E. Gliklikh (Eds.), Global Analysis – Studies and Applications III. V, 331 pages. 1988.

Vol. 1335: F. Guillén, V. Navarro Aznar, P. Pascual-Gainza, F. Puerta, Hyperrésolutions cubiques et descente cohomologique. XII, 192 pages. 1988.

Vol. 1336: B. Helffer, Semi-Classical Analysis for the Schrödinger Operator and Applications. V, 107 pages. 1988.

Vol. 1337: E. Sernesi (Ed.), Theory of Moduli. Seminar, 1985. VIII, 232 pages. 1988.

Vol. 1338: A.B. Mingarelli, S.G. Halvorsen, Non-Oscillation Domains of Differential Equations with Two Parameters. XI, 109 pages. 1988.

Vol. 1339: T. Sunada (Ed.), Geometry and Analysis of Manifolds. Procedings, 1987. IX, 277 pages. 1988.

Vol. 1340: S. Hildebrandt, D.S. Kinderlehrer, M. Miranda (Eds.), Calculus of Variations and Partial Differential Equations. Proceedings, 1986. IX, 301 pages. 1988.

Vol. 1341: M. Dauge, Elliptic Boundary Value Problems on Corner Domains. VIII, 259 pages. 1988.

Vol. 1342: J.C. Alexander (Ed.), Dynamical Systems. Proceedings, 1986–87. VIII, 726 pages. 1988.

Vol. 1343: H. Ulrich, Fixed Point Theory of Parametrized Equivariant Maps. VII, 147 pages. 1988.

Vol. 1344: J. Král, J. Lukeš, J. Netuka, J. Veselý (Eds.), Potential Theory – Surveys and Problems. Proceedings, 1987. VIII, 271 pages. 1988.

Vol. 1345: X. Gomez-Mont, J. Seade, A. Verjovski (Eds.), Holomorphic Dynamics. Proceedings, 1986. VII, 321 pages. 1988.

Vol. 1346: O. Ya. Viro (Ed.), Topology and Geometry – Rohlin Seminar. XI, 581 pages. 1988.

Vol. 1347: C. Preston, Iterates of Piecewise Monotone Mappings on an Interval. V, 166 pages. 1988.

Vol. 1348: F. Borceux (Ed.), Categorical Algebra and its Applications. Proceedings, 1987. VIII, 375 pages. 1988.

Vol. 1349: E. Novak, Deterministic and Stochastic Error Bounds in Numerical Analysis. V, 113 pages. 1988.

MIX
Papier aus verantwortungsvollen Quellen
Paper from responsible sources
FSC® C105338

If you have any concerns about our products,
you can contact us on
ProductSafety@springernature.com

In case Publisher is established outside the EU,
the EU authorized representative is:
Springer Nature Customer Service Center GmbH
Europaplatz 3, 69115 Heidelberg, Germany

Printed by Libri Plureos GmbH
in Hamburg, Germany